Introduction to
PRECISION
MACHINE
DESIGN
and Error Assessment

Mechanical Engineering Series

Frank Kreith & Roop Mahajan - Series Editors

MEMS: Applications
Mohamed Gad-el-Hak
MEMS: Design and Fabrication
Mohamed Gad-el-Hak
The MEMS Handbook, Second Edition
Mohamed Gad-el-Hak
MEMS: Introduction and Fundamentals
Mohamed Gad-el-Hak
Multiphase Flow Handbook
Clayton T. Crowe
Nanotechnology: Understanding Small Systems
Ben Rogers, Sumita Pennathur, and Jesse Adams
Optomechatronics: Fusion of Optical and Mechatronic Engineering
Hyungsuck Cho
Practical Inverse Analysis in Engineering
David M. Trujillo & Henry R. Busby
Pressure Vessels: Design and Practice
Somnath Chattopadhyay
Principles of Solid Mechanics
Rowland Richards, Jr.
Thermodynamics for Engineers
Kau-Fui Wong
Vibration Damping, Control, and Design
Clarence W. de Silva
Vibration and Shock Handbook
Clarence W. de Silva
Viscoelastic Solids
Roderic S. Lakes

Introduction to
PRECISION
MACHINE
DESIGN
and Error Assessment

Edited by
Samir Mekid

CRC Press
Taylor & Francis Group
Boca Raton London New York

CRC Press is an imprint of the
Taylor & Francis Group, an **informa** business

CRC Press
Taylor & Francis Group
6000 Broken Sound Parkway NW, Suite 300
Boca Raton, FL 33487-2742

First issued in paperback 2019

ISBN-13: 978-0-8493-7886-7 (hbk)
ISBN-13: 978-0-367-38606-1 (pbk)

Library of Congress Cataloging-in-Publication Data

Introduction to precision machine design and error assessment / editor, S. Mekid.
    p. cm. -- (Mechanical engineering series)
  Includes bibliographical references and index.
  ISBN 978-0-8493-7886-7 (alk. paper)
  1. Machine design. 2. Machine-tools--Monitoring. 3. Measurement. I. Mekid, S. II. Title. III. Series.

TJ230.I65 2008
621.8'15--dc22
                                         2008046284

Contents

Preface

The development of precision engineering has greatly increased our living standards. High-precision manufacturing not only offers quality and reliability for conventional products but also opens windows to entirely new products at the standard, meso-, and microscale with new features such as mechatronics, high density function, and high performance.

Precision engineering involves development at the forefront of current technology. Current advanced technology products are dependent on high-precision manufacturing processes, machines, control technologies, and even nanotechnology. Achieving ultrahigh precision in the manufacture of extremely small devices opens up prospects in several diverse and futuristic fields such as massive computing power, biomedical devices, global personal communication devices, and high-resolution optical devices. Precision engineering is a multidisciplinary field that includes machine tool design, materials, machining processes with novel manufacturing methods, metrology, sensors and actuators, microsystems, biomedical applications, and other relevant fields. Precision engineering requires an in-depth understanding of most physical phenomena within the previous disciplines and their effects at the micro- and nanoscale.

With the current trend toward highest precision micromachining and assembly systems based on continuous miniaturization and functional integration of products, application areas with enormous market growth are extended to automotive goods, optoelectronics, biomedicine, microchemistry, and consumer goods. Microelectronics needs mechanical interfaces, electrical and optical connections with smaller geometries, and structures and tolerances down to nanometers.

This book is a result of several years of teaching the various topics that have been covered. It has been written to meet the growing need of mechanical engineers, and others, to understand design process issues with a particular focus on most errors associated with precision design, machine diagnostics, error modeling, and compensation. This book is necessary as it carries complementary information to existing books on precision machine design. The various chapters have been written by contributors who are international experts in their respective fields. This book is designed to cover key topics for any course in precision machine design. It covers precision machine design principles and related physical aspects, strategies of design for various kinematics concepts and scales, and gives an introduction to most types of errors. This book consists of eight chapters treating specific topics. The content is suitable for students at level 3 and who are pursuing masters in mechanical and aerospace engineering.

The topics presented in this book are as follows:

- Machine design principles for serial kinematic machines at the standard and microscale with an introduction to parallel kinematic machine design
- Precision control required for machines, actuation, and sensing
- Introduction to most available errors in machines and their various aspects with emphasis on thermal errors
- Modeling of errors and global budgeting

Chapter 1 is a brief introduction to precision engineering and applications. Chapter 2 introduces error measurements with fundamental definitions for measurement characterization and error classification. An example of numerical-controlled machine error assessment is discussed in great detail. Chapter 3 is concerned with an in-depth discussion of thermal error sources and transfer, modeling and simulation, compensation, and machine tool diagnostics. Chapter 4 introduces principles and strategies to design standard-size precision machines. Techniques are extended to precision micromachines. A number of second-order phenomena that may affect precision are discussed.

This chapter presents description of several case studies. Chapter 5 considers parallel kinematic machines and techniques of design and modeling of workspace and its corresponding dexterity. Chapter 6 is concerned with the precision control techniques covering linear systems and nonlinear aspects. It includes fundamentals of motion control and control design strategies. Several case studies are discussed at the end of the chapter. Chapter 7 introduces various types of drives, actuators, and sensors required for machines along with several examples. Chapter 8 presents position error compensation modeling, measurements using laser interferometry, and examples and programs for different types of numerical controllers.

Samir Mekid

Acknowledgments

First of all thanks to Allah, the almighty God, the most gracious and the most merciful.

I am thankful to my postgraduate students for their help with suggestions and proofreading of the book: Azfar Kalid (PhD student) with his contribution to Chapter 5, Tunde Ogedengbe (PhD student), Han Sung Ryu (MSc), Mam Aamer Jalil (MSc), and Daniel Ellison (MEng).

I would like to thank Roland Geyl (SAGEM, France), Len Chaloux (Moore Nanotechnology Systems, United States), Gavin Chapman (Focal Dimensions, United Kingdom), Fernando Garramiola (Gurato, Spain), and Professor B. Denkena (IFW, Hannover) for their comments and for some pictures published in this book.

I cannot finish without thanking my father Nadir who molded my mind and advised me early in my life: "In this life you either want to highlight your name in a bank or in history!." My thanks are extended to my children Sheyma and Khawla for their support and precious help in formatting some of the figures.

J. Jedrzejewski would like to express his gratitude to his colleagues: K. Buchman, W. Kwasny, Z. Kowal, W. Modrzycki, and Z. Winiarski (all PhDs), as well as to Doosan Infracore Co. for their cooperation in research activity, contributing to the elaboration of his chapter.

K. K. Tan would like to express his thanks to Singapore Institute of Manufacturing Technology and National University of Singapore for their support for our parts of the work.

C. Wang would like to thank Optodyne for the general support and O. Svoboda at CIMT for performing the measurement and releasing data from his PhD thesis.

Samir Mekid

Editor

Samir Mekid received his MSc in applied mechanics and materials and PhD in precision engineering from Compiegne University of Technology (France) in 1994. He has worked with industry on several projects and in 2001 he joined UMIST (United Kingdom), which, after merging with Victoria University in 2004, came to be known as The University of Manchester. He is currently an associate professor and leads a research team in precision engineering, metrology, and instrumentation. He has gained substantial experience in multidisciplinary research activities. His commitment to industrially collaborative research is supported by public grants including EU funding, e.g., I*PROMS, DYNAMITE, and EPSRC funding in the United Kingdom.

Dr. Mekid has designed and manufactured a number of ultrahigh precision mechanical systems dedicated for long and short strokes such as optical delay line for very large telescopes required by the European Organization for Astronomical Research in the Southern Hemisphere (ESO), and linear slides and micromachines with nanometer positioning and low levels of vibration. He has developed a number of sensors for in-process measurements, mesoscale inspections, and wireless MEMS measurements. Dr. Mekid is a chartered engineer and a member of IMechE and the Institute of Nanotechnology. He also has a great interest in teaching and learning and has been conferred the award of the Postgraduate Certificate in Academic Practice. He has also won several teaching and research awards

Contributors

Sunan Huang
Electrical and Computer Engineering
National University of Singapore
Singapore

Jerzy Jedrzejewski
Institute of Production Engineering
  and Automation
Division of Machine Tools
  and Mechatronic Systems
Wroclaw University of Technology
Wroclaw, Poland

Tan Kok Kiong
Department of Electrical
  and Computer Engineering
National University of Singapore
Singapore

Samir Mekid
School of Mechanical, Aerospace,
  and Civil Engineering
The University of Manchester
Manchester, England,
  United Kingdom

Andi Sudjana Putra
Electrical and Computer Engineering
National University of Singapore
Singapore

Charles Wang
Optodyne, Inc.
Compton, California

Contributor Profiles

Sunan Huang received his PhD from Shanghai Jiao Tong University, Shanghai, China, in 1994. Since 1997, he has been a research fellow in the Department of Electrical and Computer Engineering, National University of Singapore. His research interests include error compensation of high-precision machine, adaptive control, neural network control, and automated vehicle control.

Jerzy Jedrzejewski is a full professor at Wroclaw University of Technology, past vice president. He is a founder and a long-time head of division of Machine Tool Design and Machining Systems. He is also a fellow of the International Academy of Production Engineering CIRP, World Academy of Material Engineering and Manufacturing, and Engineers Academy in Poland. He is an honorary member of SIMP (Polish Association of Mechanical Engineers and Technicians) and SPPO (Polish Association of Machine Tool Manufacturers). He is also the editor-in-chief of *Journal of Machine Engineering* and *Machine Engineering*—in Polish. He has been nominated for the Japan Prize. He is the founder of the International Scientific School of Diagnostics and Supervising of Machining Systems as well as the Scientific School of Designing and Optimization of Machine Tool Thermal Behaviour. He is a member of the editorial committee of a number of national and foreign scientific journals.

Tan Kok Kiong received his PhD in 1995. Prior to joining the National University of Singapore (NUS), he was a research fellow at SIMTech, a national R&D institute spearheading the promotion of R&D in local manufacturing industries, where he was involved in managing industrial projects. He is currently an associate professor with NUS and his current research interests are in precision motion control and instrumentation, advanced process control and autotuning, and general industrial automation. He has produced more than 160 journal papers to date and has written 5 books, all resulting from research in these areas. He has so far attracted research fundings in excess of S\$3 million and has won several teaching and research awards.

Andi Sudjana Putra received his BSc in mechanical engineering from Brawijaya University, Malang, Indonesia, in 2001 and a joint MSc in mechatronic design fromTechnische Universiteit Eindhoven, the Netherlands, and the National University of Singapore, Kent Ridge, Singapore, in 2004, where he is currently pursuing his PhD. His research interests include high-precision actuation, piezoelectricity, self-sensing actuation, mechatronic designs, and semiconductor.

Charles Wang is president of Optodyne, Inc., a manufacturer of laser measurement instruments. He is responsible for the development and application of technology, such as the LDDM technology, the laser vector method, and the noncontact laser/ballbar. He has received eight patents and invention awards. He received his BS in mechanical engineering from Taiwan University and his PhD in aeronautics from California Institute of Technology. He was an adjunct professor at the University of California, San Diego, and taught advanced fluid mechanics, laser development, and applications. He has carried out research work on laser development and application and has more than 100 publications. He was a cochairman of the International Conference on Lasers in Beijing and Shanghai in 1980. He is a fellow of Optical Society of America, an associate fellow of American Institute of Aeronautics and Astronautics, and a member of many professional and honorary societies.

1 Introduction to Precision Engineering

Samir Mekid

CONTENTS

An attitude wherein there is no such thing as randomness, all effects have a deterministic cause.

1.1 BRIEF HISTORY

It is believed that precision engineering had its initial roots in astronomy and sailing when Hipparchus in second century BC and Ptolemy in AD 150 used "graduated" instruments. Later, the Astrolabe was built in AD 984 by Al-Khujandi and Ibrahim Al-Fazari to determine precise directions. Abul Wafa Al-Buzjani (AD 998) was probably the first to show the generality of sine theorem relative to spherical triangles. He gave a new method of constructing sine tables, the value of sine 30, for example, being correct to the eighth decimal place (converted to decimal notation). Al-Nasawi's (AD 1075) arithmetic explained the division of fractions and he may have introduced for the first time the decimal system in place of sexagesimal system. Ghiyahth al-Kashi (early fifteenth century) figured the value of pi to the sixteenth decimal place. He also had an algorithm for computing the nth root. In 1572, the angular diameter of Tycho's star in Cassiopeia was measured to be from 4.5 to 39 arcmin by using the best instruments of the day. The first diffraction grating was made around 1785 by Philadelphia inventor David Rittenhouse, who strung hairs between two finely threaded screws. The first lathes and many other machine tools are rooted in watches and clocks making.

1.2 PRECISION MANUFACTURING

The improvement of precision, products and processes offer substantial benefits to a wide range of applications from ultraprecision to mass production with higher quality and better reliability. The recent development of ultraprecision machines is reaching subnanometer precision under much specified conditions. The ultrahigh precision machines such as the molecular measuring machine, diamond cutting machines, and very large telescope (VLT) interferometers are designed under very tight specifications. Table 1.1 shows a difference between conventional machining and diamond machining requirements.

1

TABLE 1.1

Comparison between Conventional and Diamond Machining

	Conventional Machining	Diamond Machining
Feed	0.1–2 mm	1–20 μm
Depth of cut	0.1–10 mm	1–20 μm
Process forces	>100 N	<1–5 N
Roughness	0.2–5 μm	2–20 nm
Form deviation	>10 μm	0.1–1 μm

The achievement of ultrahigh precision requires extremely advanced technologies and the highest skill level. The development goes beyond known concepts to intelligent systems where a variety of functions are embedded within one machine. Moreover, to achieve multi-machining processes, reconfigurability is added as a new concept for future machines. These aspects are discussed later in this chapter.

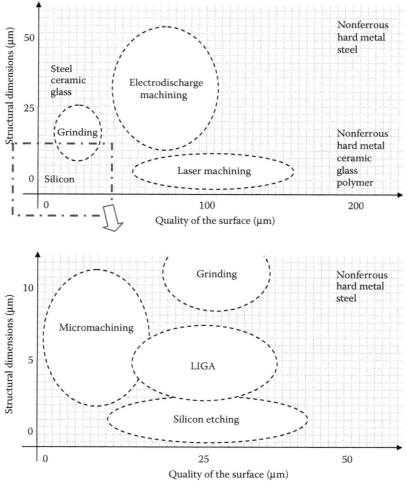

FIGURE 1.1   Overall capabilities of precision manufacturing processes.

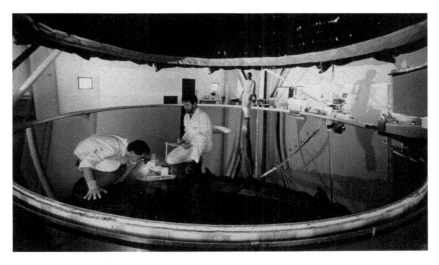

FIGURE 1.2 (See color insert following page 174.) Inspection of the VLT mirror of 8 m. (Courtesy of SAGEM.)

Figure 1.1 shows a mapping of a variety of precision manufacturing processes with respect to the structural dimension of the material versus the expected quality of the surface finish. Various types of materials are also shown at different ratios.

As an example of precision component, a piston precision design has evolved over the last century. The piston cylinder clearance has, for example, improved from 2000 μm to better than 40 μm. Zero clearance can be achieved with special friction coatings. There is currently a large demand for ultraprecision surface for various applications, for example telescopes. The current performance of the surfaces form accuracy is about 20 nm across 200 mm component. REOSC (France) has polished the 8-m wide monolithic mirror for the European VLT (Figure 1.2), installed in Chile, and the 11-m wide mirror for the Gran Telescopio de Canarias with a precision of about 10 nm. The mirrors are composed of several segments (Figure 1.3). SAGEM-REOSC is also applying its skill to extreme UV optics working at a wavelength of 13.5 nm and producing components with root mean square (RMS) figure error of less than 1 nm over 200 mm size.

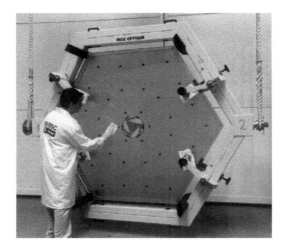

FIGURE 1.3 One segment of a mirror. (Courtesy of SAGEM.)

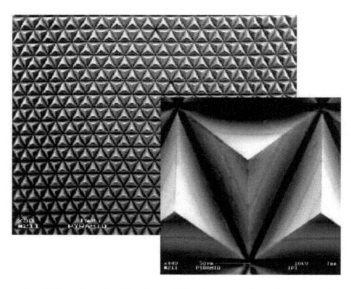

FIGURE 1.4 Microprism 100 μm length with micromilling/raster cutting. (Courtesy of Institute of Production Technology, Germany.)

Ultraprecision machining based on single-point diamond turning and ultraprecision diamond grinding currently enables economic production of optical, mechanical, and electronic components or products with form accuracy and surface roughness in micrometer and nanometer ranges, respectively.

1. Single-point diamond turning:
   • Machining various types of materials, for example nonferrite materials such as Al, Cu, electroless Ni, and plastic materials
   • Spherical and aspherical surfaces
   • Applications: plastic lens, metal mirror, and injection mold
2. Diamond grinding:
   • Machining brittle materials such as glass, quartz, ceramic, or steel
   • Spherical, aspherical, and 3D freeform surfaces
   • Applications: f–θ lens/mold, mold for glass molding, etc.

Both single-point diamond turning and ultraprecision grinding can produce optical surfaces on functional optical materials under very special machining conditions (Figures 1.4 and 1.5). The relation between crystallographic orientation and machining direction is important for producing optical surfaces on nonlinear optical crystals. Freeform surfaces include telescopes, direct contact lenses, lens arrays, and camera lenses. Associated form and surface roughness measurement technology also plays an important role in evaluating the quality of the ultraprecision machined surfaces.

1.3 BEYOND INTELLIGENT MANUFACTURING

Zero defects for complex components with higher performance become a new challenge required for the new generation of machine tools. Increasing the precision and accuracy of machines, products and processes offer substantial benefits to a wide range of applications from ultraprecision to mass production with higher quality and better reliability. The recent development of ultraprecision machines is reaching subnanometer precision under much specified conditions.

FIGURE 1.5 Molds with diamond turning (center for optic manufacturing).

The machine tool industry is responding to a number of current and future requirements, for example just-in-time production and most importantly zero defect component, this is facilitated by integrating new materials, design concepts, and control mechanisms which enable machine tools that operate at high speed with accuracies below 5 μm. But what is widely missing, however, is the integration of human experience in manufacturing toward flexible, self-optimizing machines. This can be achieved by enhancing existing computing technologies and integrating them with human knowledge of design, automation, machining, and servicing into e-manufacturing.

It is also required to build new intelligent RMS, which realizes a dynamic fusion of human and machine intelligence, manufacturing knowledge, and state-of-the-art design techniques. This may lead to low-cost self-optimizing integrated machines with fault-tolerant advanced predictive maintenance facilities for producing high-quality error-free workpieces using conventional and advanced manufacturing processes.

It is intended to address new challenges required by the new generation of intelligent machine tools with built-in in-process inspection and autohealing for better performance and quality. New routes philosophy of possible machine architectures will be presented with great features as well as challenges related to various aspects in the machine.

1.3.1 A Few Expected Objectives

The expected objectives targeted for such machines are

1. Integration: Development of an integrated machine tool being capable of performing both conventional and nonconventional processes in one platform.
2. Bidirectional data flow: definition of a bidirectional process chain for unified data communication exchange between CAD, CAM, CNC, and drive systems.
3. Optimization: Development of a self-configuring, self-optimizing control system for autonomous manufacturing.
4. Predictive maintenance: Specification of a load- and situation-dependent monitoring system as a basis for self-reliant machine operation. This addresses both the better evaluation of existing sensors as well as the implementation of in-process inspection facilities. This will be followed by the formulation of a self-organizing predictive maintenance schedule that is based on self- and remote diagnostics and covers both short- and long-term aspects.

The detailed science and technological targets would therefore be

1. To develop an integrated intelligent machine center dedicated to e-manufacturing
2. To investigate and develop fast, stable, and stiff reconfigurable machines with hybrid machining processes to prepare a new platform for future machine-tools
3. To investigate implementation of total error compensation and in-situ inspection facility
4. To develop and produce new methodologies and concepts of autonomous manufacturing, self-supervision, and self-diagnostic/tuning/healing
5. To develop and incorporate an extendible CAM system capable of recognizing complex features and determining the optimum tools/sets and the tools parts

In order to achieve these objectives, an interdisciplinary approach of machine tool builders, control manufacturers, research institutions, and potential end-users is required.

The breakthroughs could be defined as follows:

1. Delay-free cum zero-downtime production: The proposed e-manufacturing approach will see the use of electronic services based on available data from machined processes, sensor signals, and human experience that is integrated in a zero delay-time system to enable machines with near zero-downtime and production that meets user requirements with zero delay-time.
2. Self-reliant production: Machines will be enabled to operate widely autonomously.
3. Optimal production: Self-configuration and self-optimization will drastically reduce production errors down to the limitations of the in-process measurement devices.

1.4 RECONFIGURABLE SYSTEMS

Reconfigurable systems or reconfigurable science is a new and fairly inexperienced field of research. It deals with the technology of systems that can reconfigure in order to adapt the functionality. This reconfiguration can take many forms, from the adjustment of structure, hardware components, or software. Hence the scope of this research is extremely large as entails it disciplines including mechanics, modelling, electronics, control, economics, design, manufacturing, robotics, and more. A lot of these disciplines are discussed in this chapter, some in more detail than others. Most research done is to improve on, or work on the foundations of known technology. Reconfigurable systems are a brand new development that is intended to reach current technology but from a different starting point. The aim in manufacturing is to have machining tools that can work as fast as modern day production lines but that have been designed from the outset to be reconfigurable, to take currently used production lines but rebuild with small fundamental blocks, or smaller grain size. Some of the difficulties found in the development of reconfigurable systems are reported in the literature and many developments are required to embed precision in the machines.

1.4.1 RECONFIGURABLE MANUFACTURING SYSTEMS

The era of globalization has had many effects on the world. It has popularized a consumer market bringing about global scale businesses and hence changes in economics as well as customer needs and product demand. To be truly competitive a business and indeed industry must be responsive to these changes. Part of this responsiveness needed will be in the field of manufacturing. RMS has been proposed by many as a solution to constant evolution industry. They will allow manufacturing systems to adapt to changes rather than be upgraded after each significant change. RMSs will allow for

- Wider product variety
- Shorter product lifetime
- Large fluctuation in product demand
- Short opportunities for new products to access full production

REFERENCE

1. Mekid, S., Pruschek, P., and Hernandez, P., Beyond intelligent manufacturing: A new generation of flexible intelligent NC machines. *Int. J. Theory Mech. Mach.*, 2008, in press. (doi: 10.1016/j.mech machtheory. 2008. 03. 006)

2 Motion Errors

Samir Mekid

CONTENTS

When you can measure what you are speaking about and express it in numbers, you know something about it.

Lord Kelvin (1883)

It is intended in this chapter to introduce fundamental definitions to characterize a measurement and to describe various aspects of the error. Statistical analysis of the error is discussed at a level sufficient to provide information for engineering measurements related to the content of the book. This chapter introduces most types of errors encountered in mechanical systems.

2.1   INTRODUCTION TO ERROR MEASUREMENTS

The initial step before designing any machine is to read the specifications that have most of the time very tight quantitative values. These are required to be achieved with high accuracy. It is also required to report on the performance of the machine as a final stage to check whether there is any compliance with initial targets otherwise adjustments should be considered in order to meet the specifications.

2.1.1   METROLOGY AND MEASUREMENT

According to the vocabulaire international de metrologie (VIM)c,[*] metrology is a field of knowledge concerned with measurement. It includes all theoretical and practical aspects of measurement,

---

[*] VIM draft version of April 2004.

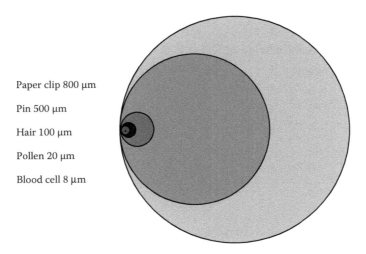

Paper clip 800 μm

Pin 500 μm

Hair 100 μm

Pollen 20 μm

Blood cell 8 μm

FIGURE 2.1 (See color insert following page 174.)   Size comparison of five items.

whichever the measurement uncertainty and field of application. Measurement is the process of experimentally obtaining information about the magnitude of a quantity. The measurement implies a measurement procedure, based on a theoretical model. In practice, measurement presupposes a calibrated measuring system, possibly subsequently verified.

The purpose of any measurement system is to provide the user with a numerical value corresponding to the variable being measured by the system (Figure 2.2). It is well known that this value is generally not the "true value" of the variable. A "true value" would be obtained by a perfect measurement. It is also defined as the value that is approached by averaging an increasing number of measurements with no systematic errors. To characterize these measurements and the accompanying errors, a number of definitions and rules have been set and detailed in the following sections.

One meter is defined as the length traveled by light in 3.335641×10^{-9} s (based on the speed of light in a vacuum). It is also defined as 1,650,763.73 wavelengths in vacuum of the orange red line of the spectrum of Krypton-86. To have an idea about a micrometer, the diameter of a hair is about 45 μm (Figure 2.1), while 1 nm is almost as wide as a DNA molecule and 10 times the diameter of a hydrogen atom. It is about how much your fingernails grow each second. It is also the thickness of a drop of water spread over a square meter. It is one-tenth the thickness of the metal film on your tinted sunglasses or your potato chip bag. The smallest lithographic feature on a Pentium computer chip is about 100 nm. Table 2.1 shows standard scale of sizes.

2.1.2 ERROR

The amount by which a measured (assumed) value deviates from its true value is called error, and it is closely associated with accuracy. Errors belong to two distinct classes: random and systematic errors.

FIGURE 2.2   Measurement process.

TABLE 2.1
Standard Scale of Sizes

Symbol	Corresponding Scale
Femto (f)	10^{-15}
Pico (p)	10^{-12}
Angstrom (Å)	10^{-10}
Nano (n)	10^{-9}
Micro (μ)	10^{-6}
Milli (m)	10^{-3}

Random errors are related to precision of results and may be treated statistically, for example, averaging may be used to improve the precision of a best estimated value. The magnitude of such errors may be judged from the results of a set of repeated measurements and could be reduced with the high number of observations. Systematic errors are those that occur in the same way at every measurement and so cannot be discovered only by examining the results. Bad calibration is a simple example. The search for a reduction of systematic errors is both a major chore and an ultimate source of inspiration for the instrument designer. Systematic errors are also the main reason why it is much more expensive to provide good accuracy than good precision (or repeatability) [1]. Unlike random errors, systematic errors cannot be reduced by increasing the number of observations. Some examples of systematic and random errors are given below.

2.1.2.1 Systematic Errors

Missing factor.　　Writing a model for the behavior to be assessed, the challenge is to account for all possible factors involved in the experiment except the one independent variable that is being analyzed. If you are measuring by using laser interferometry but you do not measure the effects of temperature, pressure, or humidity on the laser beam, correction can be applied sometimes to account for an error that was not detected.

Failure to calibrate or check zero of instrument.　　Instruments used for measurements should be calibrated prior to measurements. Accuracy could be checked with another calibrated instrument if there is no calibration standard for the instrument in use. The zero of the instrument has to be checked whether it is the real zero at initial readings.

Instrument drift.　　It is observed that some electronic instruments have readings that drift over time. The amount of drift is generally not a concern, but sometimes it could be a significant source of error that should be considered.

Hysteresis and lag time.　　Some hysteresis may appear either in the system under inspection or within some add-ons to the instrument. It is important to check whether the instrument has embedded hysteresis via initial measurement tests. Also, some measuring devices have low-response time to take and record a measurement.

2.1.2.2 Random Errors

Resolution of the instrument.　　All instruments have finite precision that limits the ability to resolve small measurement differences.

Physical variations.   Statistical measurements over the entire range are important to detect any variation that may be important to investigate locally.

2.1.2.3 Random or Systematic Errors

Environmental factors.   Many inspections in precision engineering require the protection of the experiment from changes in temperature, pressure, humidity, vibrations, drafts, electronic noise, or other effects from nearby devices.

Parallax factor.   This error can occur whenever there is some distance between the measuring scale and the indicator used to obtain a measurement. The reading may introduce some errors.

User errors.   The user could incorrectly introduce errors for many reasons mainly related to his skills and aptitude to measure properly that day.

There is an increasing need in evaluating and upgrading the performance of standard or high-precision machines. An error assessment or uncertainty analysis is a tool which can be used to determine the performance capability of machine tools and to highlight potential areas of performance and cost improvement. The implementation of this process is of critical interest to machine designers. Aspects related to the characterization of measurements used in this book will be defined later.

2.1.3 EVALUATION OF PRECISION MOTION IN A MACHINE

Usually, a machine tool or a precision mechanical system is designed to satisfy a number of functionalities required with a defined accuracy. These are assessed in the product manufactured by the machine tool, or by the assessment of the tasks performed by the mechanical system according to the specifications (e.g., telescope, robot). From the point of view of precision engineering, the following parameters will be normally assessed for any mechanical system:

Manufactured product:

- Dimensional precision
- Angular precision
- Form precision
- Surface roughness

Machine:

- Kinematic precision
- Surface layer alteration

2.2 FUNDAMENTAL DEFINITIONS FOR MEASUREMENT CHARACTERIZATION

The following definitions are referred to International Vocabulary of basic and general terms in Metrology (VIM) or BS 5725. The measurement is characterized by certain number of criteria defined below. It is important to distinguish between accuracy, repeatability, and resolution. These terms are often used interchangeably, but they have specific meaning according to international standards. It is particularly important not to confuse high resolution with high accuracy. It is known that the accuracy is much more difficult and more expensive to achieve than resolution. Having one does not necessarily guarantee the other!

2.2.1   ACCURACY

The accuracy is defined as the closeness of agreement between a test and the accepted reference value. In precision engineering systems, it is the maximum translational or rotational error between the desired position and the actual position.

Accuracy of Measurement

The accuracy of measurement is the closeness of agreement between a quantity value obtained by measurement and the true value of the measurand (quantity intended to be measured).

Notes:

- Accuracy cannot be expressed as a numerical value.
- Accuracy is inversely related to both systematic error and random error.
- The term "accuracy of measurement" should not be used for trueness of measurement and the term "measurement precision" should not be used for "accuracy of measurement."

Accuracy of a Measuring System

It is the ability of a measuring system to provide a quantity value close to the true value of a measurand.

Notes:

- Accuracy is greater when the quantity value is closer to the true value.
- The term "precision" should not be used for "accuracy.",
- This concept is related to accuracy of measurement.

BS 5725 uses two terms "trueness" and "precision" to describe the accuracy of a measurement method. Trueness refers to the closeness of agreement between the arithmetic mean of a large number of test results and the true or accepted reference value. Precision refers to the closeness of agreement between test results.

Measurement Precision

It is the closeness of agreement between quantity values obtained by replicate measurements of a quantity, under specified conditions.

Notes:
Measurement precision is usually expressed numerically by measures of imprecision, such as standard deviation, variance, or coefficient of variation under the specified conditions of measurement.

The term accuracy, when applied to a set of test results, involves a combination of random components and a common systematic error or bias component. Figure 2.3 gives a good example when assessing measurement quality.

2.2.2   REPEATABILITY, RESOLUTION, ERROR, AND UNCERTAINTY

2.2.2.1   Resolution

The resolution is defined as the error of mobility or the smallest generable movement. In servo control, it is defined as the smallest value (digital, analog) the sensor can indicate (noise level).

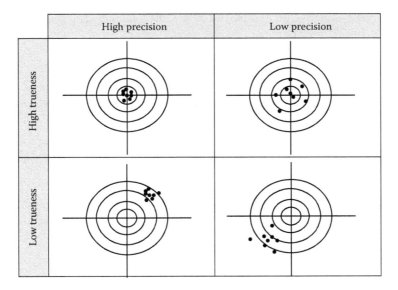

FIGURE 2.3 Example of measurements.

Resolution of a measuring system
It is the smallest change, in the value of a quantity being measured by a measuring system that causes a perceptible change in the corresponding indication.

Note:
The resolution of a measuring system may depend on, for example, noise (internal or external) or friction. It may also depend on the value of the quantity being measured.

Resolution of a displaying device
It is the smallest difference between indications of a displaying device that can be meaningfully distinguished. Figure 2.4 shows the minimum increment between two measurements defining resolution called also noise.

Application.   Using a controller having N bits, the smallest recognized step by the controller over the intended stroke is

$$\text{Step} = \frac{\text{distance}}{2^N} \qquad (2.1)$$

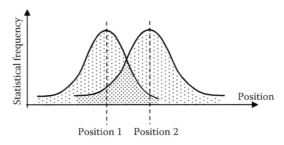

FIGURE 2.4 Resolution.

For a 16 bit controller used to cover a stroke of 100 mm, the minimum displacement resolved by this controller will be $100/2^{16} = 1.5 \mu m$.

2.2.2.2 Measurement Errors

Error of measurement

It is the deviation of quantity value obtained by the measurement and the true value of the measurands.

Note:

It is necessary to distinguish "error of measurement" from "relative error of measurement."

Error sources

Calibration errors. The elements of errors could easily infiltrate the measurement system during the process of calibration. There are two principal sources:

- The bias and precision errors of the reference used in the calibration
- The way the reference is used

Data acquisition errors. Errors due to the actual act of measurement are referred to as data acquisition errors. Power settings, environmental conditions, and sensor locations are some examples of data acquisition errors.

Data reduction errors. The errors due to curve fits and correlations with their associated unknowns are defined as the data reduction errors.

Systematic (bias) and precision errors

Systematic (bias) error. It is the difference of average that would ensue from an infinite number of replicated measurements of the same measurand carried out under repeatability conditions and true value of the measurand.

Notes:

- Systematic error, and its causes, can be known or unknown. Correction should be applied for systematic error, as far as it is known.
- Systematic error equals the difference of error of measurement and random error of measurement.

The systematic (bias) errors cannot be seen with the statistical analysis; therefore, the estimation is made by calibration, inter-laboratory comparisons, or experience.

Precision error (random error). The precision error is affected by

- *Measurement system*: repeatability and resolution
- *Measured system*: temporal and spatial variations
- *Process*: variations in operating and environmental conditions
- *Measurement procedure and technique*: repeatability

2.2.2.3 Uncertainty

It is the parameter that characterizes the dispersion of the quantity values that are being attributed to a measurand, based on the information used. As an example to propagation of uncertainty to a result:

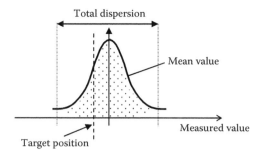

FIGURE 2.5   Repeatability.

- Surface area derived from measured diameter
- Normal stress derived from force and cross-sectional area measurements $\sigma = f(F, A)$

2.2.2.4   Repeatability

The repeatability is defined as the dispersion of readings or the bandwidth of uncertainty (Figure 2.5) obtained when positioning a worktable repeatedly to one chosen position. It is also the error between the successive attempts to move the carriage to the same position.

Repeatability condition of measurement

The condition of measurement in a set of conditions including the same measurement procedure, same operator, same measuring system, same operating conditions and same location, and replicated measurements over a short period.

Measurement repeatability (repeatability)

The measurement repeatability is the measurement precision under repeatability conditions of measurement and it is expressed as standard deviation. The conditions of measurement are as follows:

- With the same method
- On identical test items
- In the same laboratory
- By the same operator
- Using the same equipment
- Within short intervals of time (in particular the equipment *should not* be recalibrated between the measurements)

The repeatability depends on

- Resolution
- Drifts of displacement transducer, thermal, mechanical, and electronics
- Deformation of the machine structure
- Combination of limiting friction of guideways and elastic compliance of actuators

The repeatability does not include hysteresis or drifts and it could be estimated in uni- or bidirectional way depending on the application.

2.2.3 TRACEABILITY

VIM defines traceability as the property of the result of a measurement or the value of a standard whereby it can be related to stated references, usually national or international standards, through an unbroken chain of comparisons all having stated uncertainties. The unbroken chain of comparisons is called a "traceability chain." The traceability chain is composed of a number of instruments linked together to supply measurement. The competence and uncertainty are essential elements in the traceability according to ISO 17025, Section 5.6.

There is always a difference between the true value of a measurand and the output of an instrument. Measurement uncertainty is a quantitative statistical estimate of the limits of that difference. VIM defines measurement uncertainty as a parameter associated with the results of a measurement that characterizes the dispersion of the values that could reasonably be attributed to the measurand.

Essential Components of Traceability

1. Traceable calibration involves comparisons with traceable standards or reference materials.
2. Traceable calibrations can be performed only by laboratories that demonstrate their competence by accreditation to ISO 17025.
3. A traceable calibration certificate must contain an estimate of the uncertainty associated with the calibration.

Example of Traceability of Measurement in a Coordinate Measuring Machine

Traceability is one of the most fundamental and important aspects of the proper use of a coordinate measuring machine (CMM) in a quality management system. The aspects of concern are the actual physical chains by which measurement may be related to the SI unit of length. The physical chain of traceability is described as follows:

1. Laser wavelength ref. to atomic clock frequency ($4/10^{14}$)
2. Laser displacement interferometer ($2.5/10^{11}$)
3. Calibrated line scale or step gage ($2–5/10^{8}$)
4. CMM ($1/10^{6}$)

2.3 MODEL OF MEASUREMENT

Theoretically, the measured value $x_{measured}$ from an instrument must be identical to the true value x_{true}:

$$x_{true} = x_{measured} \tag{2.2}$$

However, in general, these two values are not identical. The deviation from the true value is the *accuracy figure* for an instrument, or more generally, described as "Error."

$$x_{true} = x_{measured} + \text{Error} \tag{2.3}$$

Note that in most textbooks both terms "Error" and "Uncertainty" are used interchangeably but the term Error is chosen here.

The Error can be a combination of many different types of errors, but most of them can be classified as *systematic errors, E_s or random errors, E_r* (Figure 2.6). If estimates for both random and systematic uncertainties are known and because the method of combining random and systematic uncertainties are not completely clear, it is stated that

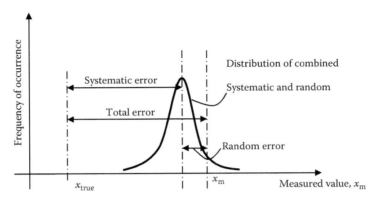

FIGURE 2.6 Systematic error is larger than the typical random error.

$$x_{true} = x_{measured} \pm E_s \pm E_r \qquad (2.4)$$

where
  x_{true} is the true value of measurements
  E_s is the systematic error, or component of laboratory bias
  E_r is the random error under repeatability conditions (Figure 2.7)

Example: The error of measurement could be defined as $E_s + E_r$.

The error depends on the following characteristics:

1. Resolution of the sensor
2. Positioning repeatability
3. Static and dynamic stiffness of machine element
4. Abbé offset error

In case of positioning, Figure 2.8 gives an overview of standard errors. With a displacement of a linear side along a distance D, Figure 2.9 shows the displacement error and systematic error. Since all objects could have six degrees of freedom (DOF) in 3D space, therefore the error could be defined in a planar or volumetric movement as shown in Figure 2.10.

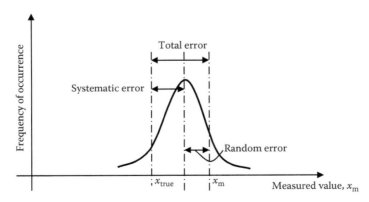

FIGURE 2.7 Typical random error is larger than the systematic error.

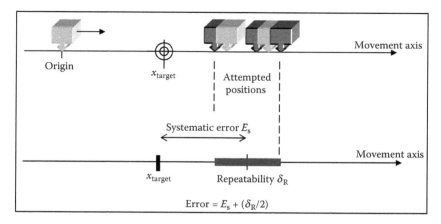

FIGURE 2.8 Example of error on linear movement.

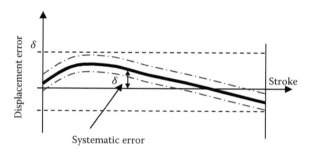

FIGURE 2.9 Total error of a linear slide. One-dimensional uncertainty.

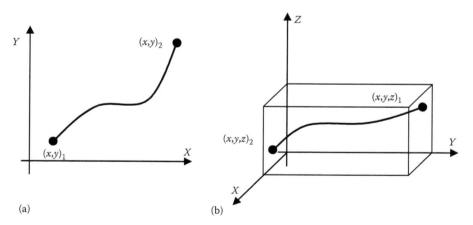

FIGURE 2.10 (a) Planar error and (b) volumetric error.

2.4 STATISTICAL MEASUREMENTS

In the context of machine design within this book, measurements have to be conducted by the efficient application of scientific principles. High accuracy is either achieved by refining existing product, that is machine or system, or designing a new product that meets initial specifications. The engineering method to formulate and to converge to targets would encompass the following steps:

1. To develop a clear and concise description of what to measure
2. To identify key factors that may affect the measurements
3. To identify key instrument(s) required for the measurements
4. To conduct appropriate experiments and acquire data properly
5. To identify the more appropriate model for error assessment
6. To draw conclusions and report quality of the results
7. To set up actions if it is necessary to adjust the system to comply with initial specifications

2.4.1 MEASUREMENTS WITH STATISTICAL ANALYSIS

Statistical analysis is used to understand variability that exists in the system that is under investigation, for example, machine tool. Successive inspection measurements on the latter may not produce exactly the same result. Hence, the existing techniques will be able to construct a clear picture of the existing defect if it exists within the machine.

The variations are usually observed in engineering measurements repeatedly taken under identical conditions. The source of the variation can be identified within:

1. The measurement system in which resolution and repeatability are key parameters
2. The measurement procedure and technique mainly represented by repeatability
3. The measured variables including temporal variation (t) and spatial variation (x)

The statistical analysis provides estimates of

1. Single representative value that best characterizes the data set
2. Some representative value that provides the variation of the data
3. An interval (range) about the representative value in which the true value is expected to be

The statistical analysis helps understanding experimental and interpreting data mainly in

1. Extracting the best value of a quantity from a set of measurements by estimating a parameter from a sample of data. The value of the parameter using all of the possible data, not just the sample data, is called the population parameter or true value of the parameter. An estimate of the true parameter value is made using the sample data.
2. Deciding whether the experiment (i.e., measurements) is consistent (with theory, other experiments, etc.).

Probability

Some aspects of probability are briefly introduced hereafter. The probability for N trials with specified events, occurring e times, is defined as follows:

$$P(E) = \frac{e}{N} \quad N \to \infty \tag{2.5}$$

The probability, P, is a nonnegative number and is $0 \le P \le 1$.

- 0 never occurs
- 1 always occurs

The events A and B are independent if $P(A \cap B) = P(A) \cdot P(B)$. They are mutually exclusive if $P(A \cap B) = 0$ or $P(A \cup B) = P(A) + P(B)$.

The probability could be either discrete or continuous:

a. Discrete probability: P has certain values only

Example: Coin $P(x_i) = \frac{1}{2}$ (x_i head or tail).

b. Continuous: P could have any number between 0 and 1: $0 \leq P \leq 1$
   - Probability density function (PDF); $f(x)$

$$f(x)dx = dP(x \leq a \leq x + dx) \tag{2.6}$$

   - Probability for x to be in the range $a \leq x \leq b$ is defined as

$$p(a \leq x \leq b) = \int_a^b f(x) \cdot dx \tag{2.7}$$

where
   $f(x)$ is normalized to one, $\int_{-\infty}^{+\infty} f(x) \cdot dx = 1$ and $\int_a^a f(x) \cdot dx = 0$
   $P(x)$ is the discrete: binomial or Poisson
   $f(x)$ is the continuous: uniform, Gaussian, exponential, chi-squared

A probability distribution is described by its mean, mode, median, and variance. Mean and variance will be introduced later in this chapter.

2.4.2 SAMPLE AND POPULATION

Before going into further explanation, it is necessary to introduce both terms "sample" and "population," which are generally used in statistical analysis of error. In a manufacture, a batch of ball bearings had been made and an inspector takes a small number of bearings for quality check. So it can be said a sample of size n of bearings had been drawn from a finite population of size p ($n \ll p$), of ball bearings (Figure 2.11).

Note that the following conditions are applied:

1. The sample is used to estimate properties of the population and additional data cannot be added to the population. For example, no more bearings can be added to the bag.
2. The sample size is assumed to be very small compared to the population size: $n \ll p$.
3. A finite number of items, n, is randomly drawn from a population that is assumed to have indefinite size.

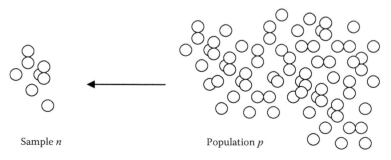

Sample n Population p

FIGURE 2.11   Sample taken from a population.

The most commonly used measure of a measurand is the mean. The true mean is the sum of all the members of the given population divided by the number of members in the population. As it is typically impractical to measure every member of the population, a random sample is drawn from the population. The sample mean is calculated by summing the values in the sample and dividing by the number of values in the sample. This sample mean is then used as the point estimate of the population mean.

2.4.2.1 Formulation of Mean and Variance for Direct Measurements

The population mean is the expected outcome, such that if an infinite number of measurements are made, the average of the infinite measurements is the mean. In real-life situation, it is impractical and impossible to extract statistical information from entire population. Take the bearing manufacturing for example, it is virtually impossible to make measurement on every single bearing if there is more than a million in a batch. Nevertheless this will provide good background information on how to establish the size of error. This represents the *true* value of a measurement. If a large number of measurements are taken then the arithmetic average of these n measurements is

$$\mu \approx \sum_{i=1}^{n} \frac{x_i}{n} \tag{2.8}$$

where
    μ is the population mean
    x_i is the ith value of the quantity being measured
    n is the total number of measurements

The amount of error in a *single* measurement is given by

$$d = x - \mu \tag{2.9}$$

where
    d is the deviation
    x is a particular value of the quantity being measured
    μ is the population mean

The standard deviation of population σ is an estimate of the average error of measurements of a very large sample and can be found from square root of variance (σ^2).

    The variance is the summing of the deviation divided by number of measurements.

$$\sigma \approx \sqrt{\frac{d_1^2 + d_2^2 + d_3^2 + \cdots + d_n^2}{n}} \tag{2.10}$$

The standard deviation is a very important parameter in establishing the likely size of error.

2.4.2.2 Mean and Variance Based on Samples

In real life, it is usual to deal with samples from a population and not the population itself. So sample mean $\bar{x}$ and sample standard deviation S_x would be used as an approximation to population mean μ and population standard deviation σ respectively. The population mean is usually denoted μ, and is the expected value $E(x)$ for a measurement. The sample standard deviation S_x is defined in Equation 2.11.

$$S_x = \sqrt{\frac{1}{n-1}\sum_{i=1}^{n}(x_i - \overline{x})^2}$$ (2.11)

where

    n is the number of measurements
    x_i is the ith value of the quantity being measured
    $\overline{x}$ is the sample mean

The denominator of standard deviation $(n-1)$ is called the number of DOF, which is a measure of how much precision an estimate of variation has. The DOF can be viewed as the number of independent parameters available to fit a model to data. Usually, the more parameters you have, the more accurate your fit will be. However, for each estimate made in a calculation, you remove one DOF. This is because each assumption or approximation you make puts one more restriction on how many parameters are used to generate the model. Put another way, for each estimate you make, your model becomes less accurate.

    The difference between the two definitions of standard deviation for a sample or for a population is almost numerically insignificant. The comparison is between $\sqrt{n}$ and $\sqrt{n-1}$.

2.4.2.3 The Standard Deviation of the Mean

The standard deviation σ_x characterizes the average uncertainty of the separate measurements x_1, x_2, x_3, ..., x_i which average is the best estimate.

    Now, for the whole population, the best estimation is still the population average but with the uncertainty on the average $\sigma_{\overline{x}} = \sigma_x / \sqrt{n}$ also called standard deviation of the mean (SDOM) or standard error of the mean. Hence,

$$x_{\text{value}} = \overline{x} \pm \sigma_{\overline{x}}$$ (2.12)

Example

The inspection of a component diameter has been made automatically 20 times. The values are recorded in Figure 2.12.

    Hence, the mean diameter d is 30.076 mm and the uncertainty in any one measurement is therefore 0.006 mm with 68% confidence that d lies within the range 30.076 ± 0.006 mm.

2.4.3 Formulation of the Standard Uncertainty and Average of Indirect Measurements

The measurand $g(X)$ is not measured directly but determined from the quantity X. The Taylor series expansion gives:

$$g(X) = \underbrace{\frac{(X-a)^0}{0!}\cdot g(a)}_{\text{zero-order}} + \underbrace{\frac{(X-a)^1}{1!}\cdot\frac{\partial g}{\partial X}(a)}_{\text{first-order}} + \underbrace{\frac{(X-a)^2}{2!}\cdot\frac{\partial^2 g}{\partial X^2}(a)}_{\text{second-order}} + \cdots + \underbrace{\frac{(X-a)^n}{n!}\cdot\frac{\partial^n g}{\partial X^n}(a)}_{n\text{th-order}} + R_n \quad (2.13)$$

where a is a constant about which the expansion is carried out. R_n is a reminder term known as the Lagrange remainder, defined by Blumenthal [2] as

$$R_n = \underbrace{\int \cdots \int_{x_0}^{x} g^{(n+1)}(X)(\mathrm{d}X)^{n+1}}_{n+1}$$

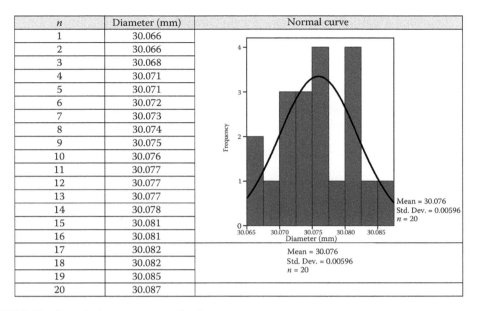

n	Diameter (mm)	Normal curve
1	30.066	
2	30.066	
3	30.068	
4	30.071	
5	30.071	
6	30.072	
7	30.073	
8	30.074	
9	30.075	
10	30.076	
11	30.077	
12	30.077	
13	30.077	Mean = 30.076
14	30.078	Std. Dev. = 0.00596
15	30.081	n = 20
16	30.081	
17	30.082	Mean = 30.076
18	30.082	Std. Dev. = 0.00596
19	30.085	n = 20
20	30.087	

FIGURE 2.12 Statistical measurement of a diameter.

With respect to the current investigation, it seems obvious that the expansion should be carried out about the mean (i.e., central value). The average of measurement is obtained through the expectation value defined by the first statistical moment as

$$x = E(X) = \int_{-\infty}^{\infty} x \cdot f_X(x) \cdot \mathrm{d}x \tag{2.14}$$

where $f_X(x)$ is the PDF.

The standard deviation is known to be the square root of the variance, which in turn is written as the second statistical moment:

$$\mathrm{Var}(X) = E\left(X - E(X)\right)^2 = \int_{\infty}^{-\infty} (x - \mu_x)^2 \cdot f_X(x) \cdot \mathrm{d}x \tag{2.15}$$

where $\mu_x \equiv E(X)$. It also expresses the uncertainty of measurement $u(x)$ defined in *Guide to the Expression of Uncertainity in Measurement* [3], and the standard deviation σ_x, hence:

$$\mathrm{Var}(X) = \sigma_x^2 = u^2(x) \tag{2.16}$$

2.4.4 How to Determine the Measured Value and Random Error?

Random errors, E_r (also referred to as precision errors) are different for each successive measurement but have an average value of zero. If enough readings are taken, the distribution of precision errors will become apparent. A reliable estimate of random errors can be obtained with appropriate statistical methods being applied.

There are different approaches in making measurements, such as

a. Repeated measurements of one single quantity
b. Measurement of two different variables or more which are related

Both approaches will be discussed in detail in the following sections.

2.4.4.1 Repeated Measurements of One Single Quantity

Although this is not the most general situation that will encounter in practice, however, this is the easiest one and it is necessary to fully understand this before moving to a more general and complex one.

2.4.4.1.1 How to Determine Measured Value $x_{measured}$?

As introduced previously, the value of a measurand could be represented as

$$x_{true} = x_{measured} \pm E_s \pm E_r \tag{2.17}$$

For repeated measurements, the previous equation becomes

$$x_{true} = \overline{x} \pm E_s \pm E_r \tag{2.18}$$

where
  x_{true} is the true value
  $\overline{x}$ is the mean value
  E_s is the systematic error, or component of laboratory bias
  E_r is the random error under repeatability conditions

The total uncertainty could be expressed as the quadratic sum of random and systematic errors.

However, there is an alternative method to determine the "best estimate" if the following conditions are met:

1. Total number of measurements n is very large
2. Value in the sample occurs more than once

$$\overline{x} = \frac{\sum_{i=1}^{n}(f_i x_i)}{\sum_{i=1}^{n} f_i} \tag{2.19}$$

where
  f_i is the occurrence of the ith value
  $f_i x_i$ is the product of ith value and its corresponding occurrences

2.4.4.1.2 How to Determine Random Error?

The uncertainty calculated previously for the SDOM is also considered as the random uncertainty component in a measurement. The random uncertainty could also be computed using distribution functions. The latter are used in statistical analysis of error, such as binomial, exponential, hypergeometric, chi-squared, Poisson, etc. Not all distributions have a symmetric bell shape; Poisson distribution is usually not symmetric, for example. It is demonstrated [4] that if a measurement is subject to many small sources of random error and negligible systematic error, the measured value will be distributed in accordance with a bell-shaped curve and this curve will be centered on the true value of x. Two particular distributions will be discussed in this book but others could be found in many other specialized books on statistics and probability for engineers. The distributions considered here are either normal distribution or Student's t distribution.

Normal distribution

When the measurement's sets contain only random errors, their distribution will be a symmetric bell-shaped curve and centered on the mean value with zero error. Such distribution is called *Normal* (also known as *Gaussian*) distribution (Figure 2.13).

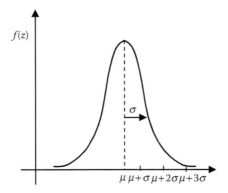

FIGURE 2.13 Normal distribution.

The probability distribution of x values is described by a PDF. For an infinite population, the mathematical expression for the normal PDF is

$$f(x) = \frac{1}{\sigma\sqrt{2\pi}} \exp\left[-\frac{(x-\mu)^2}{2\sigma^2}\right] \tag{2.20}$$

or

$$f(z) = \frac{1}{\sqrt{2\pi}} \exp\left[-\frac{z^2}{2}\right] \tag{2.21}$$

where
   x is a particular value of the quantity being measured

$$z = \frac{x-\mu}{\sigma} \tag{2.22}$$

σ is the standard deviation of population
μ is the population mean

Confidence intervals (CI) (size of random error) for population
As introduced previously, a CI gives an estimated range of values, where the population mean μ is located. The confidence level defines how sure you can be. It is a percentage of certainty that a measurement will lie within the CI. The confidence level is defined in terms of probability to be within the CI as $(1 - \alpha)$, where α is the probability to be outside the interval.

The probability of $(1 - \alpha)$ to select a sample that will produce a range containing the value of x is defined as (Figure 2.14).

$$\begin{cases} P(L \leq x \leq U) = 1 - \alpha \\ \quad 0 < \alpha < 1 \end{cases} \tag{2.23}$$

This range is called a $100(1 - \alpha)\%$ CI for the parameter x. If a sampling distribution, x, is normal with a mean, μ and a variance, σ^2, then if $z = (x - \mu)/\sigma$, the probability could be written as

$$P\left(-z_{\frac{\alpha}{2}} \leq z \leq z_{\frac{\alpha}{2}}\right) \tag{2.24}$$

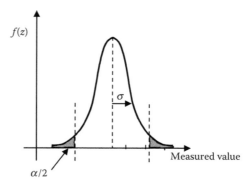

FIGURE 2.14   Confidence interval.

and x could be located as

$$P\left(\mu - z_{\frac{\alpha}{2}} \cdot \sigma \leq x \leq \mu + z_{\frac{\alpha}{2}} \cdot \sigma\right) \qquad (2.25)$$

The mathematical expression for the CI (within which the true population μ is located) is given by the expression

$$\mu - z_{\frac{\alpha}{2}}\sigma < x < \mu + z_{\frac{\alpha}{2}}\sigma \qquad (2.26)$$

where

   σ is the standard deviation of population
   μ is the population mean
   x is a particular value of the quantity being measured
   z is a variable
   $c = 1 - \alpha$ is the confidence level

   The variable z is defined as

$$z_{\frac{c}{2}} = \frac{x - \mu}{\sigma} \qquad (2.27)$$

The width of the CI depends on the confidence level required. As the width of the CI increases, the probability that the population mean μ will fall within the interval increases. Table 2.2 is a summary of various confidence levels for the normal distribution.

For example, if a population has a normal distribution (Figure 2.15 and Table 2.3), then the probability that a single measurement has an error greater than the standard deviation $\pm\sigma$ is 31.7%. On the other hand, it can be said that the probability that the error will be within $\pm\sigma$ is 68%: confidence level, c.

In most precision engineering applications, we would be satisfied with 99.7 confidence level (3σ). With such a confidence level, the odds that measurement x is greater than 3σ is 1 in 370 (Figure 2.15).

Student t distribution

When the sample is very small as is usual in engineering applications, the standard deviation S_x does not provide reliable estimate of the standard deviation σ of the population, therefore the previous equations (normal distribution) should not be used. Hence, when it is required to calculate the size of sample needed to obtain a certain level of confidence in measurements, it requires prior knowledge of the population standard deviation σ but the latter is unknown.

TABLE 2.2
Variable z versus Confidence Level

Variable (z)	Confidence Level (%)
±0.6754	50
±1	68.3
±1.6449	90
±1.96	95
±3	99.7
±4	99.994

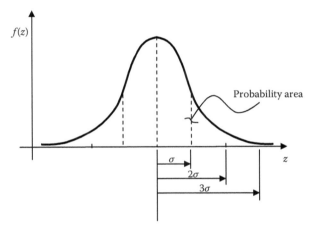

FIGURE 2.15 Areas under the curve between 0 and various values.

Often a preliminary sample will be conducted so that a reasonable estimate of this critical population parameter can be made. If CIs for the population mean are required using an unknown σ, then the distribution known as the Student t distribution[*] can be used.

Some important properties of the Student t distribution are given below.

1. The Student t distribution is different for different sample sizes.
2. The Student t distribution is generally bell-shaped, but with smaller sample sizes shows increased variability (flatter). In other words, the distribution is less peaked than a normal distribution and with thicker tails. As the sample size increases, the distribution approaches a normal distribution. For $n > 30$, the differences are negligible.
3. The mean is zero.
4. The population standard deviation is unknown.

The distribution of the quantity t is defined as

$$t = \frac{\bar{x} - \mu}{S_{\bar{x}}}$$

(2.28)

where
$S_{\bar{x}}$ is the standard deviation of sample mean
μ is the population mean (unknown)
$\bar{x}$ is the sample mean

---

[*] By W. S. Gosset, an amateur statistician, who was writing under the pseudonym "Student" in 1908.

TABLE 2.3
Table of Normal z Distribution

z	0.00	0.01	0.02	0.03	0.04	0.05	0.06	0.07	0.08	0.09
0.0	0.0000	0.0040	0.0080	0.0120	0.0160	0.0199	0.0239	0.0279	0.0319	0.0359
0.1	0.0398	0.0438	0.0478	0.0517	0.0557	0.0596	0.0636	0.0675	0.0714	0.0753
0.2	0.0793	0.0832	0.0871	0.0910	0.0948	0.0987	0.1026	0.1064	0.1103	0.1141
0.3	0.1179	0.1217	0.1255	0.1293	0.1331	0.1368	0.1406	0.1443	0.1480	0.1517
0.4	0.1554	0.1591	0.1628	0.1664	0.1700	0.1736	0.1772	0.1808	0.1844	0.1879
0.5	0.1915	0.1950	0.1985	0.2019	0.2054	0.2088	0.2123	0.2157	0.2190	0.2224
0.6	0.2257	0.2291	0.2324	0.2357	0.2389	0.2422	0.2454	0.2486	0.2517	0.2549
0.7	0.2580	0.2611	0.2642	0.2673	0.2704	0.2734	0.2764	0.2794	0.2823	0.2852
0.8	0.2881	0.2910	0.2939	0.2967	0.2995	0.3023	0.3051	0.3078	0.3106	0.3133
0.9	0.3159	0.3186	0.3212	0.3238	0.3264	0.3289	0.3315	0.3340	0.3365	0.3389
1.0	0.3413	0.3438	0.3461	0.3485	0.3508	0.3531	0.3554	0.3577	0.3599	0.3621
1.1	0.3643	0.3665	0.3686	0.3708	0.3729	0.3749	0.3770	0.3790	0.3810	0.3830
1.2	0.3849	0.3869	0.3888	0.3907	0.3925	0.3944	0.3962	0.3980	0.3997	0.4015
1.3	0.4032	0.4049	0.4066	0.4082	0.4099	0.4115	0.4131	0.4147	0.4162	0.4177
1.4	0.4192	0.4207	0.4222	0.4236	0.4251	0.4265	0.4279	0.4292	0.4306	0.4319
1.5	0.4332	0.4345	0.4357	0.4370	0.4382	0.4394	0.4406	0.4418	0.4429	0.4441
1.6	0.4452	0.4463	0.4474	0.4484	0.4495	0.4505	0.4515	0.4525	0.4535	0.4545
1.7	0.4554	0.4564	0.4573	0.4582	0.4591	0.4599	0.4608	0.4616	0.4625	0.4633
1.8	0.4641	0.4649	0.4656	0.4664	0.4671	0.4678	0.4686	0.4693	0.4699	0.4706
1.9	0.4713	0.4719	0.4726	0.4732	0.4738	0.4744	0.4750	0.4756	0.4761	0.4767
2.0	0.4772	0.4778	0.4783	0.4788	0.4793	0.4798	0.4803	0.4808	0.4812	0.4817
2.1	0.4821	0.4826	0.4830	0.4834	0.4838	0.4842	0.4846	0.4850	0.4854	0.4857
2.2	0.4861	0.4864	0.4868	0.4871	0.4875	0.4878	0.4881	0.4884	0.4887	0.4890
2.3	0.4893	0.4896	0.4898	0.4901	0.4904	0.4906	0.4909	0.4911	0.4913	0.4916
2.4	0.4918	0.4920	0.4922	0.4925	0.4927	0.4929	0.4931	0.4932	0.4934	0.4936
2.5	0.4938	0.4940	0.4941	0.4943	0.4945	0.4946	0.4948	0.4949	0.4951	0.4952
2.6	0.4953	0.4955	0.4956	0.4957	0.4959	0.4960	0.4961	0.4962	0.4963	0.4964
2.7	0.4965	0.4966	0.4967	0.4968	0.4969	0.4970	0.4971	0.4972	0.4973	0.4974
2.8	0.4974	0.4975	0.4976	0.4977	0.4977	0.4978	0.4979	0.4979	0.4980	0.4981
2.9	0.4981	0.4982	0.4982	0.4983	0.4984	0.4984	0.4985	0.4985	0.4986	0.4986
3.0	0.4987	0.4987	0.4987	0.4988	0.4988	0.4989	0.4989	0.4989	0.4990	0.4990
3.1	0.4990	0.4991	0.4991	0.4991	0.4992	0.4992	0.4992	0.4992	0.4993	0.4993
3.2	0.4993	0.4993	0.4994	0.4994	0.4994	0.4994	0.4994	0.4995	0.4995	0.4995
3.3	0.4995	0.4995	0.4995	0.4996	0.4996	0.4996	0.4996	0.4996	0.4996	0.4997
3.4	0.4997	0.4997	0.4997	0.4997	0.4997	0.4997	0.4997	0.4997	0.4997	0.4998

Such distribution depends on the number of samples taken through the DOF v, which is $(n-1)$. As the number of sample increases toward 30, t distribution will be exactly as normal distribution (Figure 2.16).

Confidence intervals for large number of samples ($n > 30$)
It is known that *mean* is the sum of the values of quantity being measured divided by total number of measurements n. However, if additional measurements were made, the new mean value would differ from the first one. Furthermore if the tests (with n number of measurements) were repeated many times, a set of samples for the mean values would be obtained (sampling errors with large population). Such samples of mean would also show dispersion about a central value. A profound theorem of statistics shows that if the number of measurement n for each sample is very large, then the distribution of the mean values is normal and it has a standard deviation.

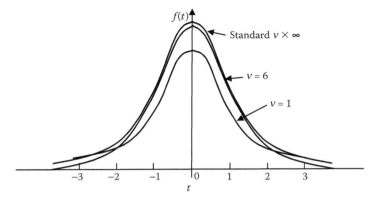

FIGURE 2.16 Probability distribution for the *t*-statistic.

$$\sigma_\mu = \frac{\sigma}{\sqrt{n}} \tag{2.29}$$

where
σ_μ is the standard deviation of population mean
σ is the sample deviation of population
n is the total number of measurements

So the CI now becomes

$$\bar{x} - z_{\frac{c}{2}}\sigma_\mu < \mu < \bar{x} + z_{\frac{c}{2}}\sigma_\mu \tag{2.30}$$

$$\bar{x} \pm \frac{z\sigma}{\sqrt{n}} \tag{2.31}$$

where
σ_μ is the standard deviation of population mean
μ is the population mean
c is the confidence level
$\bar{x}$ is the sample mean

Unfortunately, σ is usually unknown but a reasonable approximation to σ is S_x when n is large. This gives:

$$\bar{x} - z_{\frac{c}{2}}S_{\bar{x}} < \mu < \bar{x} + z_{\frac{c}{2}}S_{\bar{x}} \tag{2.32}$$

with standard deviation of sample mean $S_{\bar{x}}$ equal to

$$S_{\bar{x}} = \frac{S_x}{\sqrt{n}} \tag{2.33}$$

Equations 2.32 and 2.33 provide CI for the sample mean $\bar{x}$ when standard deviation of population σ is known or can be approximated by standard deviation of sample mean $S_{\bar{x}}$. However, these equations are only valid when the sample is "large," that is $n > 30$. In most engineering experiments, n is usually less than that and it would be more suitable to employ Student's t distribution (Table 2.4).

Confidence intervals for small number of samples on the unknown mean ($n < 30$)
The CI is given by

$$\bar{x} - t_{\frac{\alpha}{2}, v}S_{\bar{x}} < \mu < \bar{x} + t_{\frac{\alpha}{2}, v}S_{\bar{x}} \tag{2.34}$$

TABLE 2.4
Table of Student _t_ Distribution

ν	$t_{0.10}$	$t_{0.05}$	$t_{0.025}$	$t_{0.01}$	$t_{0.005}$	$t_{0.001}$
1	3.078	6.314	12.706	31.821	63.657	318.313
2	1.886	2.920	4.303	6.965	9.925	22.327
3	1.638	2.353	3.182	4.541	5.841	10.215
4	1.533	2.132	2.776	3.747	4.604	7.173
5	1.476	2.015	2.571	3.365	4.032	5.893
6	1.440	1.943	2.447	3.143	3.707	5.208
7	1.415	1.895	2.365	2.998	3.499	4.782
8	1.397	1.860	2.306	2.896	3.355	4.499
9	1.383	1.833	2.262	2.821	3.250	4.296
10	1.372	1.812	2.228	2.764	3.169	4.143
11	1.363	1.796	2.201	2.718	3.106	4.024
12	1.356	1.782	2.179	2.681	3.055	3.929
13	1.350	1.771	2.160	2.650	3.012	3.852
14	1.345	1.761	2.145	2.624	2.977	3.787
15	1.341	1.753	2.131	2.602	2.947	3.733
16	1.337	1.746	2.120	2.583	2.921	3.686
17	1.333	1.740	2.110	2.567	2.898	3.646
18	1.330	1.734	2.101	2.552	2.878	3.610
19	1.328	1.729	2.093	2.539	2.861	3.579
20	1.325	1.725	2.086	2.528	2.845	3.552
21	1.323	1.721	2.080	2.518	2.831	3.527
22	1.321	1.717	2.074	2.508	2.819	3.505
23	1.319	1.714	2.069	2.500	2.807	3.485
24	1.318	1.711	2.064	2.492	2.797	3.467
25	1.316	1.708	2.060	2.485	2.787	3.450
26	1.315	1.706	2.056	2.479	2.779	3.435
27	1.314	1.703	2.052	2.473	2.771	3.421
28	1.313	1.701	2.048	2.467	2.763	3.408
29	1.311	1.699	2.045	2.462	2.756	3.396
30	1.310	1.697	2.042	2.457	2.750	3.385

$$\overline{x} \pm \frac{ts}{\sqrt{n}} \tag{2.35}$$

where
$S_{\overline{x}}$ is the standard deviation of sample mean
c is the confidence level usually provided as an input
$\alpha = 1 - c$ is the level of significance
n is the sample size
$v = n - 1$ is the DOF

Example

If 12 values in a sample have a sample mean $\overline{x} = 1.009$ and a standard deviation $S_x = 0.04178$, what is the CI for the true mean value m with confidence level = 95%?

- DOF $v = 12 - 1 = 11$
- Level of significance $\alpha = 1 - 0.95 = 0.05$

Following is part of Table 2.4:

ν	$t_{0.10}$	$t_{0.05}$	$t_{0.025}$	$t_{0.01}$	$t_{0.005}$	ν
10	1.372	1.812	2.228	2.764	3.169	10
11	1.363	1.796	2.201	2.718	3.106	11
12	1.356	1.782	2.179	2.681	3.055	12

From the Table, value for $t_{0.025,11} = 2.201$, therefore the CI is

$$\bar{x} - 2.201\frac{S_x}{\sqrt{12}} < \mu < \bar{x} + 2.201\frac{S_x}{\sqrt{12}}$$

$$0.98245 < \mu < 1.03555$$

2.5   PROPAGATION OF ERRORS

2.5.1   First Order Average and Uncertainty of Measurement

With the first order truncation of Taylor's expansion (Equation 2.13), and application of the definition of the average (Equation 2.14), the first order average is given in Equation 2.15. Evaluating the expectation on either side of Equation 2.13 truncated to the first order will give

$$y = g(x) \tag{2.36}$$

where x is the best estimate of measurement (i.e., average of measurement).

The combined variance and standard uncertainty of $g(X)$ are calculated by subtracting $E(g(X))$ from either sides of Equation 2.36 truncated to the first order, squaring both sides, and writing the expectation from either side to finally obtain in terms of uncertainty:

$$u_c^2(y) = \left(\frac{\partial g}{\partial x}\right)^2 \cdot u^2(x); \text{ hence, } u_c(y) = \left(\frac{\partial g}{\partial x}\right) \cdot u(x) \text{ or } \delta Q = \left|\frac{dQ}{dx}\right| \cdot \delta x \tag{2.37}$$

Example 1

If $Q = A \cdot x$ therefore, $dQ = |A| \cdot dx$

Example 2

Find the uncertainty of $\sin(\alpha)$ in radian if the measured angle is $\alpha = 15 \pm 1°$

$$\sin(\alpha) = 0.258819$$

The uncertainty of $\sin(\alpha)$ is

$$\delta(\sin(\alpha)) = \left|\frac{d}{d\alpha}\sin(\alpha)\right| \times \delta\alpha = \left|\cos(\alpha)\right| \times \delta\alpha$$

$$\delta\alpha = 1° = 0.0174532 \text{ rad}$$

$$\delta(\sin(\alpha)) = \left|\cos(15)\right| \times 0.0174532 = 0.9659258 \times 0.0174532 = 0.0168584$$

Hence, $\sin(\alpha) = 0.258819 \pm 0.0168584$

Function with two variables or more

Given a functional relationship between several measured variables (x,y), $Q = f(x,y)$, the uncertainty in Q if the uncertainties in x,y are known is as follows:

The variance in Q:

$$\sigma_Q^2 = \sigma_x^2 \left(\frac{\partial Q}{\partial x} \right)^2 + \sigma_y^2 \left(\frac{\partial Q}{\partial y} \right)^2 + 2\sigma_{xy} \left(\frac{\partial Q}{\partial x} \right) \left(\frac{\partial Q}{\partial y} \right) \tag{2.38}$$

If the variables x and y are uncorrelated, $\sigma_{xy} = 0$, therefore:

$$\sigma_Q^2 = \sigma_x^2 \left(\frac{\partial Q}{\partial x} \right)^2 + \sigma_y^2 \left(\frac{\partial Q}{\partial y} \right)^2 \quad \text{where} \quad \sigma_Q = \sqrt{\sigma_x^2 \left(\frac{\partial Q}{\partial x} \right)^2 + \sigma_y^2 \left(\frac{\partial Q}{\partial y} \right)^2} \tag{2.39}$$

For more variables $(x, y, \ldots, z)$ if independent (uncorrelated)

$$\delta Q = \sqrt{\left(\frac{\partial Q}{\partial x} \right)^2 \delta x^2 + \left(\frac{\partial Q}{\partial y} \right)^2 \delta y^2 + \cdots + \left(\frac{\partial Q}{\partial z} \right)^2 \delta z^2} \tag{2.40}$$

Q at the average values

If x and y have several measurements, $x(x_1, x_2, x_3, \ldots, x_n)$ and $y(y_1, y_2, y_3, \ldots, y_n)$, the average of x and y will be

$$\mu_x = \frac{1}{n} \sum_1^n x_i \quad \text{and} \quad \mu_y = \frac{1}{n} \sum_1^n y_i \tag{2.41}$$

The evaluation of Q at the average values: $Q \equiv f(\mu_x, \mu_y)$

The expansion of Q_i about the average using Taylor expansion (see Equation 2.13):

$$Q_i = f(\mu_x, \mu_y) + (x_i - \mu_x) \left(\frac{\partial Q}{\partial x} \right) \bigg|_{\mu_x} + (y_i - \mu_y) \left(\frac{\partial Q}{\partial y} \right) \bigg|_{\mu_y} + \text{higher order terms} \tag{2.42}$$

If measurements are close to the average values

$$Q_i - Q = (x_i - \mu_x) \left(\frac{\partial Q}{\partial x} \right) \bigg|_{\mu_x} + (y_i - \mu_y) \left(\frac{\partial Q}{\partial y} \right) \bigg|_{\mu_y}$$

and its variance is

$$\sigma_Q^2 = \frac{1}{n} \sum_{i=1}^n (Q_i - Q)^2 \tag{2.43}$$

or

$$\sigma_Q^2 = \sigma_x^2 \left(\frac{\partial Q}{\partial x} \right)_{\mu_x}^2 + \sigma_y^2 \left(\frac{\partial Q}{\partial y} \right)_{\mu_y}^2 \tag{2.44}$$

for *uncorrelated variables*.

Higher order expressions for average and uncertainty are developed for two variables in Ref. [5].

Example 1

Calculate the variance in the power using error propagation: $P = I^2R$ with $I = 1.0 \pm 0.1$ A and $R = 10.0 \pm 1.0\,\Omega$.

The power is calculated; $P = 10.0\,W$

$$\sigma_P^2 = \sigma_I^2 \left(\frac{\partial P}{\partial I}\right)_{I=1}^2 + \sigma_R^2 \left(\frac{\partial P}{\partial R}\right)_{R=10}^2$$
$$= \sigma_I^2 (2IR)^2 + \sigma_R^2 (I^2)^2$$
$$= (0.1)^2 (2\times 1\times 10)^2 + (1)^2 (1^2)^2 = 5\,W^2$$

Hence, $P = 10.0 \pm 2.2\,W$

If the true value of P is 10.0 W measured several times with an uncertainty of ± 2.2 W, from the previous statistics (e.g., normal) 68% of the measurements would lie in the range [7.8–12.2] W.

Example 2

The variables x and y were measured for the quantity $Q = x^2y - xy^2$ as follows: $x = 3.0 \pm 0.1$ and $y = 2.0 \pm 0.1$. What is the value of Q and its uncertainty?

Ans. $Q = 6.0 \pm 0.9$

Example 3

An object is placed at a distance p from a lens and an image is formed at a distance q from the lens, the lens's focal length is expressed as follows:

$$f = \frac{pq}{p+q}$$

Using the general rule for error propagation, derive a formula for the uncertainty df in terms of p, q and their uncertainties. Calculate the quantity f and its uncertainty if $p = 10.1 \pm 0.3$ and $q = 4.0 \pm 0.5$.

Ans. $f = 2.9 \pm 0.3$

2.6 MOTION ERRORS PRINCIPLE

2.6.1 TRANSLATIONAL BODY

A body has six DOF in 3D space (Figure 2.17); three translations along x, y, and z, and, three rotations that are pitch, yaw, and roll, respectively about x, y, and z. If this body is assigned to move in a linear movement along only y-axis, for example, the three angles become angular errors

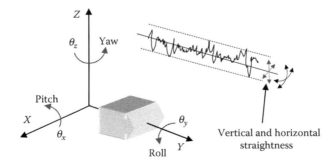

FIGURE 2.17 Possible motion errors of a rigid body translating along Y axis.

(roll, pitch, and yaw) around x, y, and z, respectively, and, any movement along x- or z-axis will be horizontal and vertical straightness errors respectively. The remaining error along y-axis will be position error. There are six geometric errors for one axis. If a system has three axes such as a machine tool, the squareness error between two axes is added. This will bring all the errors present in this machine to 21 parametric errors. As an example, conventional computer numerical controlled (CNC) milling machines are required to have the positional accuracy within $10\,\mu m$, while modern high-precision CNC milling machines are required to have the positional accuracy within $2\,\mu m$ for all axes with very tight specifications on the other remaining errors, for example, 18.

Associated errors with translation

A nonnegligible number of errors are associated to the translation of a body in space. These errors are generated by either the geometric errors of the components or the chosen design.

A number of these errors encountered in mechanical systems are described in this chapter using standard definitions [6,7].

Positioning errors.   This error is concerned with the uncertainty of the linear movement. Irregularities of the leadscrew can cause such errors which are normally static or quasistatic when the temperature effect is taken into consideration.

Straightness errors of each axis in ITS perpendicular axes.   Straightness is a deviation from the true line of travel perpendicular to the direction of travel in the horizontal plane. A straightness deviation in the travel of the X-axis stage will cause a positioning error in the Y direction. A straightness deviation in the travel of the Y-axis stage will cause a positioning error in the X direction.

Pitch angular error.   Pitch is a rotation around an axis in the horizontal plane perpendicular to the direction of travel. If the position of interest being measured is not located at the center of rotation, then the pitch rotation will cause an Abbé error in two dimensions. For the X-axis, a pitch rotation will cause an Abbé error in both the X and Z directions. For the Y-axis, a pitch rotation will cause an Abbé error in both the Y and Z directions. The magnitude of these errors can be determined by multiplying the length of the offset distance by the sine and 1-cosine of the rotational angle.

Roll angular error.   Roll is a rotation around an axis in the horizontal plane parallel to the direction of travel. If the position of interest being measured is not located at the center of rotation, then the roll rotation will cause an Abbé error in two dimensions. For the X-axis, a roll rotation will cause an Abbé error in both the Y and Z directions. For the Y-axis, a roll rotation will cause an Abbé error in both the X and Z directions. The magnitude of these errors can be calculated by multiplying the length of the offset distance by the sine and cosine of the rotational angle.

Yaw angular error.   Yaw is a rotation around an axis in the vertical plane perpendicular to the direction of travel. If the position of interest being measured is not located at the center of rotation, then the yaw rotation will cause an Abbé error in two dimensions. For X- or Y-axis stages, yaw rotation will cause an Abbé error in both the X and Y directions. The magnitude of these positioning errors can be calculated by multiplying the length of the offset distance by the sine and cosine of the rotational angle.

Abbé error.   Displacement error caused by angular errors in slide ways and an offset distance between the point of interest and the drive mechanism (ballscrew) or feedback mechanism (linear encoder).

Reversal errors.   Caused by mechanical lash (see Section 2.6.1.1.8).

Backlash errors.   This is a component of positioning caused by the reversal of travel direction. Backlash is caused by clearance between elements in the drive train. As the clearance increases, the amount of input required to produce motion is greater. This increase in clearance results in increased backlash error. Backlash also affects bidirectional repeatability. For example, linear motor-based stages that are direct-driven can have zero backlash.

Flatness error. This is a deviation from the true line of travel perpendicular to the direction of travel in the vertical plane. A flatness deviation in the travel of the X-axis or Y-axis stage will cause a positioning error in the Z direction.

Squareness error (or orthogonality error) between two axes. This is a condition of a surface or axis which is perpendicular to a second surface or axis. Orthogonality specification refers to the error from 90° from which two surfaces of axes are aligned.

Contouring errors. Circular hysteresis, circular deviation, radial deviation, periodic deviations, scale deviation, reversal error, acceleration of axes, and mismatch of position loop gain.

Hysteresis errors. Hysteresis error is a deviation between the actual and commanded position at the point of interest caused by elastic forces in the motion system. Hysteresis also affects bidirectional repeatability. Elastic forces in the machine base, load, and load coupling hardware are not accounted for and must also be examined and minimized for optimal performance.

Nonperpendicularity of Axes. For the two stages to travel precisely along the X and Y axes, the line of travel for the Y-axis must be orthogonal to the line of travel of the X-axis. If the two travel lines are not orthogonal, Y-axis travel creates a position error in the X direction. The maximum value of this error can be determined by multiplying the travel length of the stage by the sine of the angular error.

Friction and stick slip motion errors. Friction encountered when accelerating an object from a stationary position. Static friction is always greater than moving friction, and limits the smallest possible increment of movement.

Inertia force errors while braking/accelerating. Inertia is the physical property of an object to resist changes in velocity when acted upon by an outside force. Inertia is dependent upon the mass and shape of an object. It can cause hysteresis or reversal errors.

Machine assembly errors. Errors induced by the amount of torque applied to assembly elements, for example screws resulting in slight elastic deformations of the components.

Parasitic movements. These are unwanted movement errors generated by mechanical systems due to internal or external constraints or improper design.

2.6.2 ROTATIONAL BODY

In the case of a spindle rotating about the z-axis, the datum point is the center of the spindle. Angular deviation may happen around the x- and y-axes, radial displacement along x- and y-axes (Figure 2.18), and axial error along the z-axis. Hence, errors of radial throw, run-out as well as out-of-round, are errors of form (Figure 2.19).

2.6.2.1 Associated Error with Rotation

Out-of-round. Out-of-round is the error of a circular form of a component in a plane perpendicular to its axis at a given point of the latter. For a shaft, the value of the out-of-round is given by the difference between the diameter of the circumscribed circle and the smallest measurable diameter of the shaft. For a hole, it is given by the difference between the diameter of the inscribed circle and the largest measurable diameter of the hole, each of them measured in a plane perpendicular to the axis. With ordinary methods of measurement, this definition cannot be strictly applied in practice.

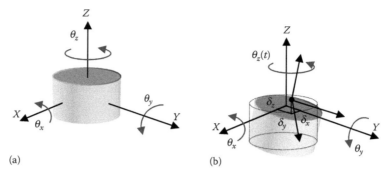

(a)   (b)

FIGURE 2.18   Possible motion errors of a rigid body rotating around Z axis.

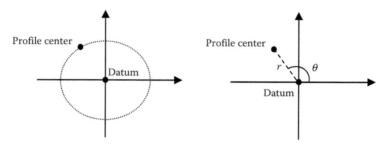

FIGURE 2.19   Radial throw.

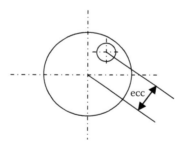

FIGURE 2.20   Eccentricity is a distance subject to tolerances.

However, when the out-of-round of a component is measured, this definition should be kept in mind and the method used should be chosen so that the results are in as close accordance as possible with the definition.

Eccentricity.   This is not an error but a distance between two parallel axes when one is rotating around the other (Figure 2.20).

Radial throw of an axis at a given point.   It is the deviation due to a geometric axis that does not coincide with the rotary axis.

Spindle radial deviation.   Positioning error of the rotary stage in the horizontal direction when the tabletop is oriented in the horizontal plane.

Spindle axial deviation.   Error of the rotary stage axis of rotation in the vertical direction when the stage is oriented in the horizontal plane.

Acceleration of axes.   If the feed rate for a path is increased, the acceleration of the axes increases accordingly. The drive of an axis may behave in such a way that the amplitude of the movement decreases at a higher frequency at higher-feed rates.

Spindle radial play (run-out error).   The deviation from the desired form of a surface during full rotation (360°) about an axis. For a rotary stage, axis run-out refers to the deviation of the axis of rotation from the theoretical axis of rotation.

Spindle inclination (tilt or wobble error).   Wobble is defined as the angular error between the actual axis of rotation and the theoretical axis of rotation.

2.6.3 GEOMETRIC ERRORS AND KINEMATIC ERRORS

Usually these errors constitute the largest source of inaccuracy and are dominant under machine cold-start conditions. Their usual sources are

1. Within the machines due to its design
2. Inaccuracies built-in during assembly
3. Results from tolerance of components used on the machine
4. Concerned with quasistatic accuracy of motion surfaces relative to each other

And their characteristics are

1. Smooth and continuous
2. Can exhibit hysteresis
3. Random or systematic behavior

They depend upon

1. Straightness
2. Surface roughness
3. Misalignment of machine components
4. Bearing preloads

2.7 OTHER TYPES OF ERRORS IN MACHINES

2.7.1 THERMAL ERRORS

The thermal error account for 40%–70% of the total dimensional and shape errors in precision machines. It is more cost effective to compensate for thermal errors rather than using expensive and high-precision components for the machine construction. Thermal error could have either quasistatic or dynamic behavior. The possible origins are known here:

- Environmental temperature changes
- Heat from cutting action and swarf
- Heat from bearings
- Gears and hydraulic oil
- Drives and clutches
- Pumps and motors
- Guideways
- External heat sources

- Heating or cooling provided by cooling systems
- Thermal memory from previous conditions

Thermal errors review

1. *Spindle axial growth*:   Heat induced into the spindle will cause a thermal axial expansion thus lifting the tool position in the *Z*-axis.
2. *Spindle radial drift*:   Heat will cause a radial expansion of the spindle combined with a radial drift of the spindle axis in *X*- or *Y*-axis. This is a result of the complex effect of heat to the spindle bearings and structure.
3. *Spindle displacements holder deformation*:   The thermal distortion of the spindle will cause two inclination angles around the *X*- and *Y*-axis. Thus two error components of the tool tip will come up in *X*- and *Y*-axis.
4. *Expansion of the leadscrew drive*:   The heat produced on the leadscrew as a result of the friction on bearings and the screw nut will induce thermal positional errors, additional to the geometric positional error mentioned before.
5. *Expansion and bending of machine column*:   The machine column can have several distortion modes depending upon the pattern of the heat distribution in it. In general, this will induce a volumetric error vector.
6. *Expansion and bending of machine bed*:   This can have a serious effect over positional, angular, squareness, parallelism, etc., accuracy of the machine bed. For a 3-axis milling machine, these errors can cause large Abbé error components and must be searched along the *X* and *Y* axis drives.
7. *Workpiece thermal deflection*:   The heat generated from the cutting action can significantly deform the workpiece in complex 3D modes, especially for thin-walled components or hard-cut conditions. FE analysis can give a rough prediction of these errors, depending on the degree of the model detail.
8. *Thermal parasitic errors*:   The heat distribution in a machine cannot be precisely predicted thus there can be some special condition which will affect drive components, thus causing unexpected parasitic errors. Thermal cameras can be engaged to locate such errors sources.

2.7.2   Cutting Force-Induced Errors

Cutting force-induced errors become more and more severe nowadays while hard cutting conditions are engaged for productivity reasons. In such conditions the part is machined from its raw form directly to the finished shape involving deep depths of cut and single operations. Therefore, cutting forces as well as temperature become an important source of part distortion and subsequently dynamic or quasistatic errors. Also increased forces will have an effect on the machine causing distortion along the line of compliance thus degrading the accuracy.

Cutting force-induced errors review

1. *Vibration*:   The transmission of the vibrations generated during the cutting action can excite some of the eigen-frequencies of machine components. Isolation measures or component redesign can significantly reduce these parasitic shifts.
2. *Material instability errors*:   Most of the nonceramic materials have the tendency to leap from an unstable state to a stable one, inducing variations in geometry. Hence, it is important to use heat-treated components to secure increased stability.
3 *Instrumentation errors*:   Unwanted sensor errors can occur when the measurement loop is not separated from the force loop. Abbé errors would be amplified for larger Abbé offsets.

4. *Cantilevered loading errors*: When a cantilevered load is placed on a translation stage, nonsymmetrical moment loads are created. Shear and bending forces induce deflection in the stage structural elements. In an *X–Y* assembly, the cantilevered load, acting on the lower axis, increases as the load traverses to the extremes of the upper axis. A position error in the *Z* direction occurs due to a combination of *Y*-axis deflection and *X*-axis roll.

5. *Tool deflection*: One of the major causes of machining error is cutting deflection of the tool or the tool shaft due to cutting force [8]. This error is observed especially when small diameter mill cutters are being used. The tool deflection can consist of cantilever loading bending and axial compression. That error can lead to contouring inaccuracies that are not negligible when net-shape NC machining is attempted. Adaptive control combined with tool path modification systems can correct this problem.

2.7.3   ENVIRONMENTAL ERRORS

The wavelength of light emitted by He–Ne lasers is equal to 632.9 nm in vacuum. Interferometer accuracy in vacuum is ±0.1 ppm. However, most applications operate in atmospheric conditions, so accuracy degrades. The refraction index of air effectively changes the frequency of the laser light which appears as a path length difference. Fortunately, the effects of temperature, pressure, and humidity as they affect the wavelength of light are well elaborated by Edlen's equation, so wavelength correction factors can be produced. An environmentally corrected system will have ±1.5 ppm accuracy or better. The final accuracy is a function of the stability of the environmental conditions. The most effective but also most expensive means of compensating for changes in the refractive index of air is by utilization of a wavelength tracker. Also known as a refractometer, a wavelength tracker measures the relative change in the refractive index of air. Because it is a relative measure only, initial environmental conditions must be accurately measured and computed to establish an initial wavelength scale factor. The wavelength tracker is a pure optical device that is highly accurate, mostly used in cutting-edge applications due to its high cost.

Mechanical vibrations and air turbulence errors

Mechanical vibration or air turbulence can cause perturbations in the positioning feedback system that will limit overall system performance. Mechanical vibration errors can be minimized through proper design of the machine base vibration isolation system. Thermal gradients across long beam paths are created due to turbulence in the air so careful design of the machine microenvironment is critical to subnanometer performance. A simple and effective means of minimizing these effects is to "shield" the beam by placing a tube around the system or simply by minimizing the flow of air.

2.7.4   COMMON GEOMETRIC ERRORS

2.7.4.1   Cosine Error

A measurement error that results when there is an angular misalignment between the desired line of measurement and the actual line of measurement. The following example (Figure 2.21) shows a

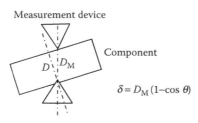

FIGURE 2.21   Cosine error.

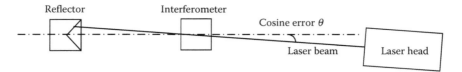

FIGURE 2.22 Cosine error illustration in laser interferometry.

component to be inspected but could be inclined (random position) with respect to the measurement device with an angle θ.

Assuming that the mechanical subsystem is sound, and environmental correction is properly implemented, the final pieces to the puzzle are the optics themselves and their alignment. All optics have inherent inaccuracies in the form of optical nonlinearity. This error cannot be controlled by the user, and is a function of the quality of the optics. All interferometer optics will have some amount of nonlinearity, so this error cannot be completely eliminated but is minimized by the use of high-quality optics. An optical error that can be controlled by the user is a misalignment that is commonly known as cosine error (Figure 2.22).

Cosine error occurs when the laser beam path and the axis of stage motion are not completely parallel. The relationship is best modeled as a triangle where the laser beam represents one leg of the triangle, and the actual motion is the hypotenuse. This error can be minimized through careful alignment of the optics to the stage.

2.7.4.2 Abbé Errors

Ernst Abbé has defined this error as follows: "If errors in parallax are to be avoided, the measuring system must be placed coaxially with the axis along which displacement is to be measured on the work-piece." Angular errors are also amplified by the distance from the source. The actions to be taken to minimize such effects are

1. Design with bearings and actuators near the movement or the scene where the machine is performing.
2. Measure near the source.
3. Make sure you compensate for it if 1 and 2 cannot be done.

This error could appear in various ways. Figure 2.23 shows an example of a caliper where any force on the jaw will induce an error proportional to the offset. Similar error occurs when the trajectory

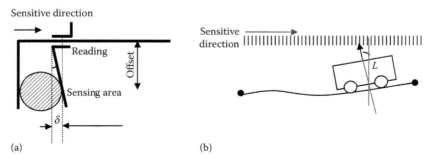

(a)   (b)

FIGURE 2.23 Abbé error illustrations: (a) Caliper Vernier error reading, (b) positioning error when rolling on wavy surface.

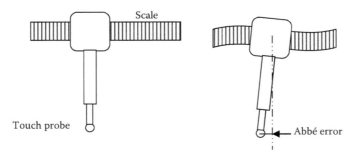

FIGURE 2.24   Abbé error in CMM machine.

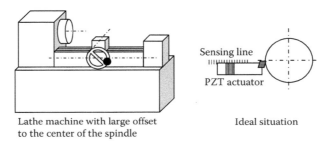

FIGURE 2.25   Current situation with lathes versus ideal situation.

is not parallel to the measuring line. Thermal effect could also amplify the Abbé error depending on the design (Figure 2.24).

Bryan has extended the Abbé error to the following:

> The displacement measuring system should be in line with the functional point whose displacement is to be measured. If this is not possible, either the slideways that transfer the displacement must be free of angular motion, or angular motion data must be used to calculate the consequences of the offset. [9].

Figure 2.25 shows the current situation with standard Abbé error in machine tools, for example, lathe. The ideal situation is to design a tool cutter actuated through its axis and displacement read from a linear encoder mounted along the tool axis. Another situation could arise with a milling machine. The corresponding amount of Abbé error is shown also in Figure 2.26.

2.7.4.3   Dead-Path Error

This error shown in Figure 2.27a is caused by portions of the beam that are effectively uncompensated between the interferometer and the reflector. This is a less obvious source of error that occurs as a result of the placement of the optics away from the measurement length. While the moveable reflector translates throughout the measurement path, the compensation and correction for the change in the index of refraction of air are made. The dead path is a distance that the laser beam travels where it undergoes no relative motion. Since the compensation is made only for relative motion, this distance remains uncorrected. The compensation of index of air is required for current environmental conditions, for example pressure, temperature, and humidity. Figure 2.27b shows a minimized dead path. The error due to dead path is negligible when the optics are placed within 50 mm of each other.

The compensation for the dead-path error requires an additional calculation to be performed that not only accounts for temperature, pressure and humidity, but for the dead-path distance as well.

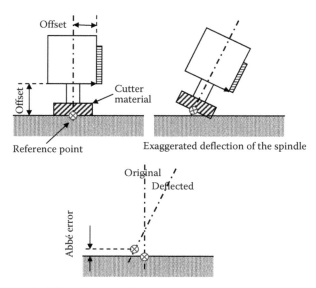

FIGURE 2.26 Abbé error in NC milling machine.

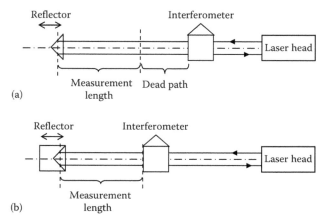

FIGURE 2.27 Illustration of the dead path in configuration (a) but not in (b).

The dead-path error is the product of the dead-path distance by the change in wavelength compensation number during the measurement period. The correction error is the product of dead-path distance by wavelength compensation number repeatability.

2.8 CLASSIFICATION OF ERRORS

The errors that are known so far could fall into general categories such as geometric errors, thermal errors, axis straightness and orthogonality, Abbé errors, controller errors, and vibration errors. Some are beyond the control of the machine builder. It is important to make an accurate list of errors and to measure or estimate their magnitude (Table 2.5). Once the errors are identified and weighted, decisions must be made as to how the errors will be combined, based upon the category of errors (Table 2.7).

Some of the more advanced error considerations are machine compliance, bearing spindle and table errors, thermal transients, acoustical vibrations, natural frequencies, material dampening, motor variations, hydraulic pressure pulses, tool tip profile, and feedback algorithms. By building upon previous experience, extensive testing, and different operating conditions, it is possible to analyze the uncertainty of any machine. The error budget is compared to the actual machine

TABLE 2.5
Error Types

Geometric and Kinematic Errors	Thermal Errors
Positioning errors	Spindle axial growth
Straightness errors of each axis in its perpendicular axes	Spindle radial drift
Pitch angular error	Spindle thermal deflection
Roll angular error	Distortion of the spindle region
Yaw angular error	Expansion of the leadscrew drive
Abbé error	Expansion and bending of machine column
Reversal errors	Expansion and bending of machine base
Backlash errors	Workpiece and bed thermal deflection
Straightness error of each axis in its perpendicular axes	Thermal parasitic errors
Squareness error between two axes	*Cutting force-induced errors*
Contouring error of each axis	Material instability errors
Hysteresis errors	Vibration
Nonperpendicularity of axes	Workpiece and bed elastic deflection
Spindle radial drift	Instrumentation errors
Spindle axial deviation	Tool wear
Acceleration of axes	Spindle elastic deflection
Spindle inclination (Tilt or wobble error)	Cutting force parasitic errors
Spindle radial play (run-out error)	*Fixturing errors*
Friction and stick slip motion errors	Axes offset error
Inertia force errors while braking/accelerating	Workpiece slippage
Machine assembly errors	Workpiece fixturing deformation
Parasitic movements	*Controller errors*
	Servo errors
	Interpolation algorithmic errors
	Mismatch of position loop gain
	Instrumentation errors
	Noise

performance, and errors are varied to assess the sensitivity of each component and determine which of them could be improved. Changes are implemented on an experimental basis, compared to the predictions, and the uncertainty analysis is updated with the validated data [10].

The simulation with these aspects will deliver a model complying with reality which will help in improving the design. The goals of this sort of design tool are to

- Locate and understand the error sources
- Combine them in a proper way
- Weigh the impact of each source of error in a cost-effective manner

2.8.1 Classification in Systematic and Random Errors

Systematic and random errors have been introduced previously. The following section will consider most of the errors known in machines.

A systematic error is defined as the difference of the average that would ensue from an infinite number of replicated measurements of the same measurand carried out under repeatability conditions and true value of the measurand. For a systematic error:

- Causes can be known or unknown. As far as systematic error is known, correction can be applied.
- Systematic error is the difference: $(x_{true} - x_{measurement}) - E_{random}$.

Systematic error cannot be seen by statistical analysis. Thus, estimation is made by

- Calibration
- Interlaboratory comparisons
- Experience

Random errors are affected by

- Measurement system (repeatability and resolution)
- Measured system (temporal and spatial variations)
- Process (variations in operating and environmental conditions)
- Measurement procedure and technique (repeatability)

Criteria of classification

The classification (Table 2.6) is made for the two generic classes of errors for which each class has its own criteria that are shown hereafter.

TABLE 2.6

Systematic and Random Error Classification

Systematic Errors	Random Errors
Geometric and kinematic errors	*Geometric and kinematic errors*
Positioning errors	Backlash errors
Straightness errors of each axis in its perpendicular axes	Contouring error of each axis
Pitch angular error	Hysteresis errors
Roll angular error	Spindle axial play
Yaw angular error	Spindle radial play
Straightness error of each axis in its perpendicular axes	Friction and stick slip motion errors
Squareness error between two axes	Inertia force errors while braking/accelerating
Nonperpendicularity of axes	Parasitic movements
Reversal errors	Machine assembly errors
Abbé error	*Thermal errors*
Acceleration of axes	Spindle radial growth
Spindle radial deviation	Spindle thermal deflection
Spindle axial deviation	Workpiece and Bed thermal deflection
Thermal errors	Distortion of the spindle region
Spindle axial growth	Thermal parasitic errors
Expansion of the leadscrew drive	*Cutting force-induced errors*
Expansion and bending of machine column	Vibration
Expansion and bending of machine base	Material instability errors
Cutting force-induced errors	Instrumentation errors
Workpiece and bed elastic deflection	Cutting force parasitic errors
Tool wear	*Fixturing errors*
Spindle elastic deflection	Workpiece slippage
Fixturing errors	Workpiece fixturing deformation
Axes offset error	*Controller errors*
Controller errors	Servo errors
Mismatch of position loop gain (different following errors)	Interpolation algorithmic errors
Instrumentation systematic errors	Instrumentation random errors
	Noise

- Systematic errors: progressive, cyclic, and reversal errors
- Random errors: repeatability

Purpose of classification in systematic and random errors

The reason for this categorization of errors in systematic and random is that it would help the error identification and compensation strategy. As long as we have specified the systematic errors, we can easily and permanently compensate for them by performing correction actions, concerning calibration. Most of the systematic errors are not costly to compensate for. Unfortunately, random errors cannot be permanently compensated for, because of their random nature. They can exhibit random characteristics such as amplitude, frequency, and static and dynamic response, etc. Thus, it is difficult to model and predict them. The only way to compensate for these errors is to monitor them during the process. Thus, eliminating random errors can be elaborate and costly.

2.8.2   SYNCHRONOUS AND ASYNCHRONOUS ERRORS

Synchronous errors are components of the total error motion occurring at integer multiple frequencies. It is the average data recorded for each angular position. Figure 2.28 shows a synchronous error rotative motion.

Asynchronous errors in the context of machine tools mean that deviations are not repetitive against time or periodic at subharmonics or other spindle or ball screw drivers frequency. Asynchronous errors are not necessarily statistically random. They are defined as components of the total error motion occurring at noninteger multiple frequencies. In practice, it is the largest pic-to-valley band for each angular position that constitutes the asynchronous error.

Purpose of classification in synchronous and asynchronous errors

It is often useful to consider the total error motion as the sum of the asynchronous and the repetitive error motion. The repetitive or synchronous error is composed of only errors with frequencies that are integer multiples of the axis of rotation frequencies. Asynchronous errors are often due to nonrandom

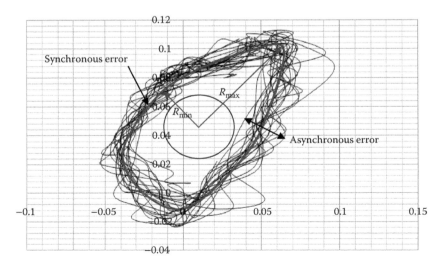

FIGURE 2.28   (See color insert following page 174.)   Graphical error representation of synchronous and asynchronous errors.

sources (e.g., motors, bearings, and gears) with frequencies which are not integer multiples of the axis of rotation frequencies.

The asynchronous error motion values can be used to help prediction of the deviation from theoretical surface finish that will be produced on the machine when a wear-free tool is used, thus assessing the performance of different types of motion, such as rotary and linear motion. Errors could be classified depending on their effect and measurability (Table 2.7).

TABLE 2.7
Severity Classification of Errors

Geometric and Kinematic Errors	Severity	Measurability
Positioning	High	Easy
Straightness	High	Easy
Pitch	Medium	Easy
Yaw	Medium	Easy
Roll	Medium	Easy
Abbé	High	Difficult
Reversal	Medium	Medium
Backlash	Medium	Medium
Straightness	High	Easy
Squareness	High	Easy
Contouring	Medium	Easy
Hysteresis	Low	Difficult
Spindle radial drift	Low	Difficult
Spindle axial deviation	Medium	Difficult
Acceleration of axes	Low	Easy
Spindle axial play	Low	Difficult
Spindle inclination	Medium	Difficult
Friction and stick slip motion errors	Low	Difficult
Inertia force errors	Low	Difficult
Machine assembly	High	Easy
Parasitic movements	?	?
Thermal Errors	**Severity**	**Measurability**
Spindle axial growth	High	High
Spindle radial drift	Low	High
Spindle thermal deflection	Medium	High
Distortion of the spindle region	High	High
Expansion of the leadscrew drive	High	High
Expansion and bending of machine column	High	High
Expansion and bending of machine base	High	High
Workpiece and bed thermal deflection	High	High
Thermal parasitic errors	?	?
Cutting Force-Induced Errors	**Severity**	**Measurability**
Material instability errors	Medium	Difficult
Vibration	Medium	Difficult
Workpiece and bed elastic deflection	High	Medium
Instrumentation errors	Low	Difficult
Tool wear	Low	Medium
Spindle elastic deflection	High	Difficult
Cutting force parasitic errors	?	?

TABLE 2.7 (continued)
Severity Classification of Errors

Fixturing Errors	Severity	Measurability
Axes offset error	High	Easy
Workpiece slippage	Medium	Easy
Workpiece fixturing deformation	Medium	Difficult
Controller Errors	**Severity**	**Measurability**
Servo errors	Medium	Difficult
Interpolation algorithmic errors	Medium	Difficult
Mismatch of position loop gain (different following errors)	Medium	Medium
Instrumentation errors	Low	Difficult
Noise	?	?

By measuring error signals and applying fast Fourier transform routines to them, it gives a detailed frequency spectrum enabling identification of vibrations and pinpointing their source. Most possible errors that can be identified with the method described above are

1. Straightness errors
2. High-frequency straightness errors of linear drives (roughness errors)
3. Friction errors
4. Spindle errors
5. Errors due to faulty bearings
6. Preload and hysteresis errors of ballscrew and nut mechanisms
7. Shaft misalignment and coupling errors
8. Backlash error and noise due to gear wear errors
9. Noise induced by motor and servo errors
10. Vibrations caused by hydraulic and pneumatic systems
11. Position error affected by control stability and encoder performance
12. Feedrate accuracy, stability, and interpolation accuracy
13. Control-loop following errors

2.8.3 Classification in PITE and PDTE

Other possible classifications [11] are possible using position-independent thermal errors (PITE) where the errors are function of the temperature and these have the following characteristics:

- These errors change the machine offsets
- The effects of PITE are dependent upon rate of change of PITE

The other class is position-dependent thermal errors (PDTE) where the errors depend on the temperature and position.

2.9 METHODOLOGIES OF ERROR ELIMINATION

The errors could be dealt with in two aspects:

1. Source of error generation due to physical phenomena
2. Transformation and amplification of errors through coordinate transformation

The budget of error is one of the requirements to define the global uncertainty of positioning through uncertainty propagation according to standards. As an example, precision machining requires compensation of cutting force-induced errors that are more significant because of the increase of hard machining, for example. This method of compensation is very efficient to increase accuracy in machine tools. There are various ways to minimize or eliminate errors described hereafter starting by measuring properly key specifications in a machine.

2.9.1   SUITABLE METROLOGY: DIRECT MEASUREMENT, SUITABLE ACCURACY, AND REPEATABILITY OF SENSORS

Better measurement is the one taken directly from the source with the right instrument without the machine distortions and free of Abbé errors and its options. Hence, an extreme importance must be given to choose the right sensor for the right location. Three general conditions of measurements were identified to be taken into account when performing a measurement.

1. Sensor location
   - Avoid applying any force and avoid creating any Abbé offset.
2. Condition of measurement
   - Temperature must be known at the time of measurement including pressure and humidity if lasers interferometry is in use.
3. Condition on the instrument
   - The precision of the measuring system must be higher than the expected precision of the measured object.
   - The user should know perfectly the instrument of measurement and its behavior.

The sensor should never be located in area where the material is subject to forces or constraint as this may give erroneous measurement. It is also extremely important to perform a dynamic analysis of the modes of vibrations to check whether the location of the sensor is not affected by the mode of vibration. Abbé error is one of the most fundamental mistakes usually found in mechanical systems that should be addressed.

2.9.2   CONTROL STRATEGY AND ACTUATION

Issues related to this topic are discussed in Chapters 6 and 7.

2.9.3   ERROR AVOIDANCE TECHNIQUES

These techniques are applied during designing and manufacturing in order to keep the sources of inaccuracy to minimum. The machine cost rises exponentially with the level of accuracy required. In such case, a high degree of investment is needed but usually leads to over-designed machines which may still exhibit a significant degree of errors. The whole attempt is done to avoid errors by eliminating their source at first place.

2.9.4   ERROR EVALUATIONS AND ERROR COMPENSATION TECHNIQUES

Comprehensive techniques for error compensation were reviewed; some of them are detailed in Ref. [12].

2.9.4.1   Error Evaluations

Grid calibration method

Calibrating volumetric errors in discrete points are defined by a 3D grid and interpolates for the intermediate tool positions. The distribution of the error vectors in space is calculated and defined, thus correction of the data can be done with a predefined error vector table. This technique may provide an improvement of the accuracy in mass production where the same part is manufactured constantly but it is not suitable for flexible manufacturing environments. Appropriately geometric errors are measured and generally static errors which are inherited to the system.

Error synthesis model
Individual error components are obtained using appropriate measurement systems and methods and then the total error is calculated by the means of a synthesis model, such as homogenous transformation matrix (HTM).

Designed artifact method
This method measures dimensions of specially designed artifacts instead of direct error measurements. The artifacts are strategically designed in order to facilitate the individual error or total error identification, usually by the means of a CMM. Suitable for geometric and thermal errors.

Metrology method frame
It uses optical or other systems mounted on the machine to monitor errors online. Thus the errors can be compensated for online. The drawback of this method is that not all error components can be measured simultaneously and an increased bandwidth is required for the data acquisition. Also measurements are subject to limitations (such as geometric orientation or coolant flow) and noise. It measures partial geometric and thermal errors.

Finite element method
The machine tool is modeled up to favorable detail using finite elements. The thermal and mechanical boundary conditions are estimated or measured. This stage implements a great deal of inaccuracy to the model due to the fact that thermal sources and their distributions are difficult to be identified precisely. Errors are estimated through thermoelastic deformation and heat transfer analysis on machine structure. Error maps for different situations are generated and loaded to the machine as correction values. It involves no experiments.

2.9.4.2   Error Compensation

Compensation techniques can deliver accurate machines at lower cost. No attempts are done to avoid errors in that case. The errors are measured by the means of an accurate device and then compensated for.

1. Precalibrated error compensation
   - Errors are measured before or after the machining process and they are used to calibrate subsequent operations. This assumes that the entire process of machining and measurement is highly repeatable which is not true.
2. Active error compensation
   - Errors are monitored during the machining process and alternations are made online. Higher accuracy in lower-machine grade can be achieved. This technique is highly desired from industry because it is more cost effective and can extend equipment life.
3. Static error compensation
   - Deals with identification and correction of the basic and internal machine errors.
4. Dynamic error compensation (real-time error compensation)
   - Generally used to correct kinematic, thermal, and cutting force-induced errors. This is conducted continuously, while machine is in operation.
5. Master part tracking approach
6. Parametric error measurement approach

- Identification or modeling of the error in question
- Measurement and mapping error components
- Development of control system or network for compensation

The process of error compensation

1. Identification of the machine linkage structure and derivation of an error synthesis model using HTM
2. Determination of the optimal location for mounting of temperature measuring devices
3. Measurement of each error component, slide position, and temperature under different machine conditions
4. Creation of the error component model and synthesis of all individual models into the error synthesis model
5. Installation of the error compensation system
    - Only the error synthesis model offers an overall scheme to correct the overall quasistatic errors.
    - Empirical modeling (like multiple regression analysis and neural networks) can approximate machine tool error maps.

Problems

- Error compensation systems should take into consideration the interaction between errors rather than considering each error in isolation.
- Modern machine tools are equipped with pitch and backlash error compensation under static conditions. Real-time systems needed to account for errors in continuous basis considering error interaction.
- Multivariant regression analysis (MRA) and multiple feedforward networks (MFN) are sensitive to sensor locations.
- MRA and MFN require a lot of experimental data for modeling (long calibration time).

2.9.5 Example of Techniques to Eliminate Thermal Errors

1. Control of the heat flow into the machine tool environment
    - High volume of coolant used
    - Temperature-controlled boxes that enclose the machine
2. Redesign machine tool to reduce sensitivity to heat flow
    - Use materials with low expansion coefficient
    - Optimize design before compensation
3. Compensation through controlled movement
    - Real-time thermal compensation

Prediction models can engage the following tools:

1. Finite element analysis
2. Coordinate transformation methods
3. Neural networks
4. Regression analysis
5. MRA
6. Power losses in kinematic system components
7. HTM and kinematic chain
8. Small error assumptions to simplify by neglecting higher-order errors [13]
9. Rigid body kinematics

10. Inverse kinematic algorithms to obtain error components [13]
11. Mathematical expression models
12. Error map method (time consuming)
13. General delta rule model
14. Real-time estimation correlating measured thermal errors with temperature at critical points on machine
15. Optimized positioning of the temperature reading devices in order to achieve linearity

Measurements

1. Thermocouples to directly measure the temperature of a point on or in the machine. A few critical points are described in Ref.[12] to locate thermocouples for temperature monitoring. These could be
   a. Along guideways
   b. Ball screw bearings and nuts
   c. Spindle bearings
   d. Spindle housing
   e. Column front face
   f. Along the spindle guideway
   g. Region of the spindle motor
   h. Region of the guideway motors
   i. Environment (for reference)
2. Laser interferometry to obtain the development of error components for various machine thermal states [14]
3. Noncapacitance sensors to measure spindle errors in specific directions

Problems

It is not possible to simulate exactly the cutting conditions when simulating the thermal distortion [12]. A solution for this problem would be to use considerably higher machining parameters than normal so that the generated heat could be comparable to that generated in an actual production environment.

Real-time error compensation (RTEC) approaches are suitable for dealing with cutting force-induced errors. These approaches can associate the force readings from the sensors with distortion modes of the machine and probably the machined part.

Force sensors suitable for monitoring the cutting forces are

1. Piezoelectric sensors
2. Strain gauges
3. Proximity sensors (to measure vibrations)
4. Sensors or gauges usually mounted on/inside the spindle assembly, under tool turret, and inside the part carriage

Compensation action

Measurements are taken from sensors and loaded real time onto a PC. The PC is used in conjunction with the CNC or it is embedded to open architecture software of the machine to implement the error compensation real time. Mathematical models calculate error components from position, force measurements, and slide positions.

Fixturing errors

- Static or dynamic errors
- Clamping errors can be measured off-line
- Fixturing drift due to process forces should be measured online

Displacements of part during the clamping can be measured by using:

- Camera
- Eddy current displacement sensors or
- Laser interferometry

Compensation techniques:

- Tool path compensation
- Real-time error compensation as mentioned above

Machine base mounting error

The machine base plays an important role in the performance of the linear translation stage. Mounting the stage to a machine base with flatness deviations greater than the specification can deflect the stage. Distortion in a translation stage can cause pitch, roll, yaw, flatness, and straightness deviations greater than the specifications listed.

2.10   OTHER ERRORS

CONTOURING ERROR PROBLEMS

- Contouring accuracy deals with errors derived by coupling stages. Even combining two or more perfectly calibrated stages, the resulted accuracy while moving in complex contours is not perfect [10].
- Motion inconsistency will produce overshoots of the profile in direction change points.

SOLUTION

- Cross-coupled control for multiaxis motion (will reduce backlash errors)
- Static and dynamic compensation components (with optimized parameters)
- Optimize motion parameters (e.g., adaptive PID controllers, PIC controller)
- Use of backlash-free gears and ballscrew drives where applicable

Encoder errors

Imperfections in the operation of the encoder such as absolute scale length, nonuniform division of the grating scale, imperfections in the photodetector signal, interpolator errors, hysteresis, friction, and noise can affect the positioning capabilities of the linear translation stage. The accuracy and repeatability information in the specification tables takes all of these errors into account except absolute scale length. Absolute scale length is affected by thermal expansion of the encoder scale. Temperature considerations must be accounted for during system design and specification.

PROBLEMS RELATED TO ERROR ACQUIRING

- Large amounts of data are gathered in machine calibration. Methods and software for reducing or manipulating the data are needed.
- Accurate structural deformation of the machine is difficult to be modeled or measured.
- Standard measurement set for off-line/online monitoring of multiple error components and quick installation must be created (e.g., x–y stage).
- Some error components cannot be predicted without serious assumptions.

- Most of compensation systems utilize external computer to perform data acquisition, error modeling and compensation values are fed back into the CNC system. PC-based CNC systems must be applied.
- Optimum number and position of thermal and force sensors must be studied.

REALIZATION OF COMPENSATION

- By interrupting the encoder feedback and adding/subtracting quadrature pulses
- Shifting the origin of the axes during cutting
- Using leadscrew compensation tables provided with the CNC controller
- Correction of the NC program by postprocessor or by extra processor
- Using microcomputer board and inject signal to CNC
- Using open architecture control loops

2.11   FUTURE VISION IN MACHINE ERROR INSPECTION

There is a need to rapidly converge toward new paradigms of production and new concepts of product-services. The intention is to move from resource-based toward knowledge-based approaches, to remotely manufacture on-demand, multiuse, and, upgradeable product-services. Intelligent manufacturing systems (IMS) is a research area with worldwide interest encompassing all industrialized nations. Progress has already been made in countries such as the United States, Japan, Canada, and Australia; European countries have also initiated some work in this area. In order to make a significant impact, the next generation of IMS should embody not only the processes but also associated "intelligence," so that they can be used to manufacture high-value components with near-zero faults.

The scientific and technological objectives identified will be

1. To develop an integrated intelligent machine center dedicated to e-manufacturing
2. To investigate and develop fast, stable, and stiff reconfigurable machines with hybrid machining processes to prepare a new platform for future machine tools
3. To investigate implementation of total error compensation and in situ inspection facility
4. To develop and produce new methodologies and concepts for autonomous manufacturing, self-supervision, and self-diagnostic/tuning/healing
5. To develop and incorporate an intelligent computer aided modeling (CAM) system comprising automatic feature recognition of complex parts, determination of optimum tool sets, and tool paths

Figure 2.29 shows an overall architecture embedding several processes including inspection.

Nowadays, machine tool builders are trying to meet the market requirements by selling two types of machines:

- Standard machines usually focus on a specific process, that is a single production step. In order to complete the production of a workpiece, the machine has to be connected to a production line. Programming the production off-line is done automatically partly and feasible for only large and medium batch sizes.
- Machining centers on the other hand are able to carry out more than one process on one workpiece in an integrated fashion. Existing approaches to build machining centers are, however, very complex and much targeted to a specific range of goods that need a certain sequence of processes to be applied. Modularization, configurability, and adaptability of machine centers and their components are missing.

All modern machines are usually equipped with a numerical controller, which allows off-line programming. This is not only a great help in reproducing one piece several times but also in starting

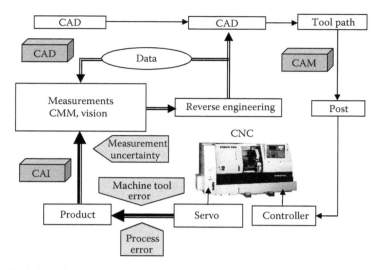

FIGURE 2.29 Global architecture embedding in-process inspection.

the programming of the processes for the next workpiece that will be manufactured on the target machine. Nevertheless, the previous sentence has already emphasized one of the major drawbacks of standard numerical controllers. What are really programmed are the movements of the tool and process-related commands, not the contours of a workpiece. Since all of this is written in a language that has been standardized about 40 years ago, the procedural code used to drive a machine tool is very poor regarding encapsulation of functions, modularity, and reusability compared to modern computer programming languages. For complex parts, in addition, the generation of the code is performed automatically by a postprocessor of the computer aided design (CAD) system. As the code can easily consist of some thousand lines, tracking of suboptimal tool paths by comparing the code to the planned part contours becomes almost impossible. STEP-NC already plans to compensate these problems by defining feature-based parts and processes. The integration in numerical controls and the exploitation of new possibilities related to the resulting workpiece and process models, however, are still missing.

EXAMPLES OF ERROR INSPECTION IN MACHINE TOOLS

Laser interferometry is one of the key instruments used to assess a number of geometric errors in machine tools. The following figure (Figure 2.30) shows a configuration to measure the straightness

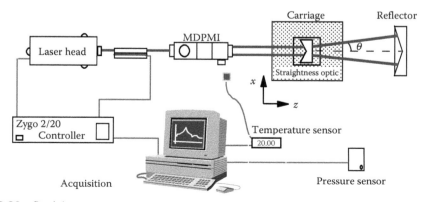

FIGURE 2.30 Straightness measurement.

over a stroke of a carriage. The laser compensation is taken into account with the acquisition of pressure, temperature, and humidity.

2.12 CNC MACHINE* ERROR ASSESSMENT

2.12.1 POSITIONING ACCURACY OF A CNC MACHINING CENTER USING LASER INTERFEROMETER WITH WAVELENGTH COMPENSATION

The purpose of the test is to inspect a CNC machine performance by characterizing its axes in terms of positioning, repeatability, systematic errors, accuracy, lost motion, and the affects of Abbé-offset during calibration and compensation using laser interferometry. Only one axis will be treated as an example.

The CNC machine used for this test is not a precision machine and it is intended with this case study to present a procedure of error assessment to a general audience.

The performance of machine tool improves product quality. To attain desired level of accuracy good calibration of equipment is required. Before taking the measurement, the following are considered:

1. The behavior of the instrument must be considered.
2. The precision of measuring instrument must be 5–10 times more than expected precision of the measured object.
3. When taking the measurement of the object, one must know atmospheric conditions like temperature, coefficient of expansion, radiation from light and sunlight, etc.

The objectives of the test are

1. To determine the positioning accuracy performance of an axis of a CNC machine tool
2. To identify possible causes of accuracy degradation

Experimental procedure

1. The HP laser interferometer was set to take readings along x-axis of the CNC machining center with MDSI open CNC controller at a height to the table surface. Figure 2.31 shows the arrangement of the inspection equipment.
2. Target positions: 30 mm equally spaced target positions were selected over a stroke of 480 mm. Five runs in each direction along the X-axis.
3. Positions at previous targets over the stroke were recorded for several runs without any compensation values in the CNC controller.
4. Wavelength compensation: Since the wavelength of light is dependent upon reflective index of air, it is necessary to compensate for environmental conditions. Air temperature, pressure, and humidity were recorded to obtain wavelength compensation using Edlen's equation. Compensation for machine temperature deviation from 20°C reference temperature was considered.
5. Results of the systematic and random errors along the X-axis in terms of BS3800 (replaced by ISO 230–1 to 230–4) specifications were plotted and print outs of the error data obtained.
6. Using the Talyvel electronic level, the pitch motion of X-axis at 30 mm intervals was measured to evaluate possible Abbé error corresponding to that axis.

---

* The CNC machine used for this test is not a precision machine and it is intended with this case study to present a procedure of error assessment.

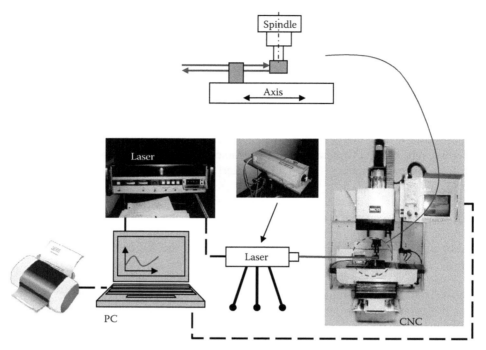

FIGURE 2.31   Experimental setup.

Figure 2.31 Shows the setup and the required equipment for the test:

- CNC machining center with Fanuc 6 CNC controller
- HP laser interferometer head
- Laser display
- Personal computer with inspection software
- Printer

Nomenclature of the variables used in the measurement analysis

P_i	target position
i	the particular position among other selected target positions along the axis
P_{ij}	actual position
j	number of approaches
x_{ij}	deviation of actual position from target position $(P_{ij} - P_i)$
$\uparrow$	refers to data collected from a measurement after a forward (positive) approach on the target
$\downarrow$	refers to data collected from a measurement after a reverse (negative) approach on the target
$\bar{x}_i\downarrow$	mean unidirectional positional deviation, either reverse or forward depending on specified symbol, $\uparrow$ or $\downarrow$
$\bar{x}_i$	mean bidirectional positional deviation, with no direction symbol specified
B_i	reversal value at a position
B	reversal value of an axis
$\underline{B}$	mean reversal value of an axis
$S_i\downarrow$	the standard deviation of positioning at a position, i. In this case it is in the reverse direction indicated by the symbol, $\downarrow$, but it could also be in the forward direction indicated by, $\uparrow$
P	air pressure

T	air temperature
H	relative humidity
N	refraction index
C	correct wavelength of light compensation factor for the measurement conditions
M	range of the mean bidirectional positional deviation of an axis
$R_i\downarrow$	unidirectional repeatability of positioning at a position, i. The approach direction is indicated by either, $\uparrow$ or $\downarrow$
R_i	bidirectional repeatability of positioning at a position, i
$R\downarrow$	unidirectional repeatability of positioning in the reverse or forward direction depending on the symbol, $\uparrow$ or $\downarrow$
R	bidirectional repeatability of positioning of an axis
$E\downarrow$	unidirectional systematic positional deviation of an axis. Approach direction is indicated by the arrow; forward $\uparrow$ or reverse $\downarrow$
E	bidirectional systematic positional deviation of an axis
$A\downarrow$	unidirectional accuracy of positioning of an axis. Approach direction is indicated by the arrow; forward, $\uparrow$ or reverse, $\downarrow$
A	bidirectional accuracy positioning of an axis

It is important to state that the test is run according to standards (BS3800 replaced by ISO 230–1/4). The test cycle is set out to do five runs stopping at 17 different positions and each run will consist of a forward and reverse element as can be seen clearly in Figure 2.32—test cycle executed.

Performance criteria according to ISO standards

Mean unidirectional positional deviation at a specified position

$$\overline{x}_i \uparrow = \frac{1}{n}\sum_{j=1}^{n} x_{ij} \uparrow \tag{2.45}$$

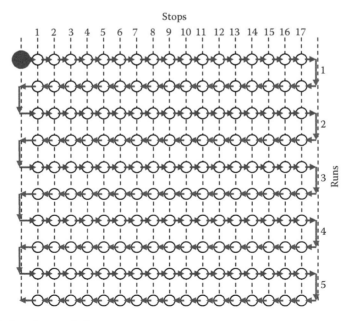

FIGURE 2.32   Test cycle executed.

Equation 2.45 has been used to calculate the mean unidirectional positional deviation at a position. The values used to determine the mean unidirectional positional deviation at a specified position are the raw values recorded from measurements. An example of this calculation is detailed for run 1: $\bar{x}_i \uparrow = \frac{1}{n}\left(x_{11} \uparrow + x_{12} \uparrow + x_{13} \uparrow + x_{14} \uparrow + x_{15} \uparrow\right)$

$$\bar{x}_i \downarrow = \frac{1}{n}\sum_{j=1}^{n} x_{ij} \downarrow \tag{2.46}$$

Mean bidirectional positional deviation at a specified position

$$\bar{x}_i = \frac{\bar{x}_i \uparrow + \bar{x}_i \downarrow}{2} \tag{2.47}$$

Equation 2.46 has been used to find the mean of the values found from Equations 2.46 and 2.47 to provide a bidirectional mean. Below is an example of this calculation for run 1.

Reversal value at a specified position

$$B_i = \bar{x}_i \uparrow - \bar{x}_i \downarrow \tag{2.48}$$

Equation 2.48 has been used to calculate the difference between the mean unidirectional positional deviations obtained from both the forward and reverse approach for each run.

Reversal value of an axis

$$B = \max \cdot \left[|B_i|\right] \tag{2.49}$$

Equation 2.49 has been used to determine the maximum of the absolute reversal values determined by Equation 2.48 at all target positions along the x-axis.

Mean reversal value of an axis

$$\bar{B} = \frac{1}{m}\sum_{i=1}^{m} B_i \tag{2.50}$$

Equation 2.50 simply determines the mean of the values calculated by Equation 2.49 to give the arithmetic mean of the reversal values B_i at all the target positions on the x-axis. For the current exercise, the average is written as follows:

$$\bar{B} = \frac{1}{17}\left(\begin{array}{c} B_1 + B_2 + B_3 + B_4 + B_5 + B_6 + B_7 + B_8 + B_9 \\ + B_{10} + B_{11} + B_{12} + B_{13} + B_{14} + B_{15} + B_{16} + B_{17} \end{array} \right)$$

Estimator of the unidirectional standard uncertainty of position at a specified position

$$S_i \uparrow = \sqrt{\frac{1}{n-1}\sum_{j=1}^{n}\left(x_{ij} \uparrow - \bar{x}_i \uparrow\right)^2} \tag{2.51}$$

Equation 2.51 determines the standard uncertainty of the positional deviation. This equation has been used as demonstrated below for run 1:

$$S_i \uparrow = \sqrt{\frac{1}{5-1}\left[\begin{array}{c} \left(x_{11} \uparrow - \bar{x}_i \uparrow\right)^2 + \left(x_{12} \uparrow - \bar{x}_i \uparrow\right)^2 + \left(x_{13} \uparrow - \bar{x}_i \uparrow\right)^2 \\ + \left(x_{14} \uparrow - \bar{x}_i \uparrow\right)^2 + \left(x_{15} \uparrow - \bar{x}_i \uparrow\right)^2 \end{array} \right]}$$

Equation 2.52 is used in a very similar way to Equation 2.51:

$$S_i \downarrow = \sqrt{\frac{1}{n-1}\sum_{j=1}^{n}\left(x_{ij}\downarrow - \bar{x}_i \downarrow\right)^2} \tag{2.52}$$

Unidirectional repeatability of positioning at a specified position

$$R_i \uparrow = 6S_i \uparrow \tag{2.53}$$

Equation 2.53 is used to determine the unidirectional repeatability of positioning at a specified position. A coverage factor of 2 has been applied. The opposite approach direction is presented by the following equation $R_i \downarrow = 6S_i \downarrow$.

Bidirectional repeatability of positioning at a specified position

$$R_i = \max \cdot \left[3S_i \uparrow + 3S_i \downarrow + |B_i|; R_i \uparrow; R_i \downarrow \right] \tag{2.54}$$

Unidirectional repeatability of positioning

$$R \uparrow = \max \cdot \left[R_i \uparrow \right] \tag{2.55}$$

The reverse direction is written as $R \downarrow = \max \cdot [R_i \downarrow]$

Bidirectional repeatability of positioning of an axis

$$R = \max \cdot \left[R_i \right] \tag{2.56}$$

Unidirectional systematic positional deviation of an axis

$$E \uparrow = \max \cdot \left[\bar{x}_i \uparrow \right] - \min \cdot \left[\bar{x}_i \uparrow \right] \tag{2.57}$$

Equation 2.57 determines the difference between the algebraic maximum and minimum of the mean unidirectional positional deviations for one approach direction at any specified position.
The following expression is used for reverse motion $E \downarrow = \max \cdot [\bar{x}_i \downarrow] - \min \cdot [\bar{x}_i \downarrow]$

Bidirectional systematic positional deviation of an axis

$$E = \max \cdot \left[\bar{x}_i \uparrow; \bar{x}_i \downarrow \right] - \min \cdot \left[\bar{x}_i \uparrow; \bar{x}_i \downarrow \right] \tag{2.58}$$

Equation 2.58 is the difference between the algebraic maximum and minimum of the mean unidirectional positional deviations for both the forward and reverse approaches.

Mean bidirectional positional deviation of an axis

$$M = \max \cdot \left[\bar{x}_i \right] - \min \cdot \left[\bar{x}_i \right] \tag{2.59}$$

Equation 2.59 is used to determine the difference between the algebraic maximum and minimum of the mean bidirectional positional deviations at any specified position along the x axis.

Unidirectional accuracy of positioning of an axis

$$A \uparrow = \max \cdot \left[\bar{x}_i \uparrow + 3S_i \uparrow \right] - \min \cdot \left[\bar{x}_i \uparrow - 3S_i \uparrow \right] \tag{2.60}$$

The opposite direction is covered by the following expression

$$A\downarrow = \max\cdot\left[\bar{x}_i\downarrow +3S_i\downarrow\right]-\min\cdot\left[\bar{x}_i\downarrow -3S_i\downarrow\right]$$

Bidirectional accuracy of positioning of an axis

$$A = \max\cdot\left[\bar{x}_i\uparrow +3S_i\uparrow;\bar{x}_i\downarrow +3S_i\downarrow\right]-\min\cdot\left[\bar{x}_i\uparrow -3S_i\uparrow;\bar{x}_i\downarrow -3S_i\downarrow\right] \tag{2.61}$$

Equation 2.61 is the most important equation used in this exercise as it describes the range derived from combining the bidirectional systematic deviations and the estimator of the standard uncertainty of bidirectional positioning using a coverage factor of 2.

Edlen's Equation for Wavelength of Light Compensation

The refractive index of the air can be calculated given the environmental conditions estimated and measured at the start of the exercise (although not monitored throughout the test).

$$N = 0.3836391P\left[\frac{1+10^{-6}P\left(0.817-0.0133T\right)}{1+0.0036610T}\right]-3.033\times10^{-3}\times H\times e^{0.057627T}$$

For the values of T, P, and H shown in Table 2.8, $N = 245.14154$.
From the refractive index above the wavelength of light compensation factor can be calculated and used in the corrective calibration as follows:

$$C = \frac{10^6}{N+10^6}, \text{ hence } C = 0.999975486$$

Tests and data acquisition

Environmental parameters were measured (Table 2.8). Position values were recorded and shown in Table 2.9. As 5 runs were performed for 17 stops, only one stop was considered to show examples of machine performance in one axis.

A Talyvel inclinometer was used to measure the axis inclination (deflection) for each position and results are shown in Table 2.10. The final assessment of the axis is found in Tables 2.11 and 2.12.

Figure 2.33a shows as anticipated that x-axis positional error increases as the machine table moves further away. Figure 2.33b shows the same behavior as in Figure 2.33a with reversal error. The two graphs are very similar to one another. Figure 2.33c contains the forward error, reverse error, and systematic (bidirectional) error in the x-axis of the CNC machining center. From Figure 2.33c a direct comparison can be seen between the forward and reverse error.

TABLE 2.8
Initial Environmental Conditions

Air temperature	20.2°C
Air pressure	732 mm/Hg
Humidity	44% relative
Machine temperature	20.0°C
Machine expansion coefficient	11.5 μm/m/°C
Wavelength compensation	755.1 ppm
Total compensation	707.0 ppm

TABLE 2.9
Raw Data Collected by Laser Interferometer Measurements

Target Position	Run Number 1		Run Number 2		Run Number 3		Run Number 4		Run Number 5	
	Forward	Reverse	Forward	Reverse	Forward	Reverse	Forward	Reverse	Forward	Reverse
	x_{ij}	x_{ij}	x_{ij}	x_{ij}	x_{ij}	x_{ij}	x_{ij}	x_{ij}	x_{ij}	x_{ij}
0	0.11	3.06	0.25	3.02	0.40	2.04	−0.27	2.02	−0.47	1.74
30	−0.33	1.59	−0.30	0.76	−0.71	1.42	−0.23	0.45	−0.91	0.74
60	−1.03	−0.18	−1.03	−1.05	−2.15	−1.20	−2.31	−0.68	−2.09	−1.38
90	−1.63	−0.59	−1.26	−0.83	−1.47	−0.96	−2.27	−1.41	−2.46	−1.44
120	−2.32	−1.53	−2.51	−1.35	−2.89	−1.39	−2.42	−1.88	−3.08	−1.98
150	−4.57	−2.99	−4.81	−3.81	−4.99	−3.78	−4.86	−3.71	−4.84	−3.93
180	−5.89	−3.89	−5.81	−4.06	−6.21	−5.47	−6.52	−5.37	−6.44	−4.93
210	−7.32	−6.37	−7.24	−6.29	−7.43	−6.49	−7.24	−6.70	−8.12	−6.94
240	−8.88	−7.94	−9.40	−7.64	−9.49	−8.29	−9.07	−7.87	−9.18	−8.08
270	−10.76	−9.19	−10.27	−8.66	−10.83	−9.46	−10.84	−9.33	−11.11	−9.43
300	−13.07	−11.71	−13.27	−11.46	−12.69	−10.77	−13.26	−11.75	−13.09	−11.39
330	−14.77	−12.97	−15.06	−13.38	−15.58	−13.27	−15.04	−12.87	−15.85	−13.41
360	−16.55	−13.94	−17.13	−13.98	−17.08	−14.31	−16.87	−14.27	−16.98	−13.72
390	−18.48	−14.90	−18.46	−14.84	−18.32	−14.96	−18.63	−15.16	−18.30	−15.05
420	−20.09	−15.97	−20.54	−16.48	−20.45	−15.80	−20.39	−15.52	−20.39	−16.05
450	−23.26	−17.46	−23.31	−18.13	−23.59	−18.30	−23.72	−18.04	−23.50	−17.51
480	−26.24	−18.91	−26.03	−18.69	−25.81	−18.79	−26.40	−19.24	−26.34	−19.25

TABLE 2.10
Pitch Motion Error (Raw Data)

Target Position (mm)	Angle (Secarc)	Angle (°)	Error in Position (µm)
0	0	0	0
30	1.0	0.000277778	0.077160492
60	2.5	0.000694444	0.482253009
90	4.0	0.001111111	1.234567393
120	5.5	0.001527778	2.334103122
150	7.0	0.001944444	3.780859433
180	8.5	0.002361111	5.574835319
210	10.0	0.002777778	7.716029537
240	11.5	0.003194444	10.2044406
270	13.0	0.003611111	13.04006678
300	14.5	0.004027778	16.2229061
330	16.0	0.004444444	19.75295636
360	17.0	0.004722222	22.29921696
390	18.0	0.005000000	24.99979167
420	19.5	0.005416667	29.33999083
450	21.0	0.005833333	34.02739182
480	22.0	0.006111111	37.34521411

TABLE 2.11
Further Calculated Values from the Calibration

Reversal value of an axis, B	7.19 μm
Mean reversal value of an axis	−2.36 μm
Bidirectional repeatability of positioning at a position, R_i	8.69 μm
Unidirectional repeatability of positioning $R\uparrow$	2.53 μm
Unidirectional repeatability of positioning $R\downarrow$	2.93 μm
Bidirectional repeatability of positioning R of an axis	8.69 μm
Unidirectional systematic positional deviation of an axis, $E\uparrow$	26.17 μm
Unidirectional systematic positional deviation of an axis, $E\downarrow$	21.35 μm
Bidirectional systematic positional deviation of an axis, E	28.54 μm
Mean bidirectional positional deviation of an axis, M	23.76 μm
Unidirectional accuracy of positioning of an axis, $A\uparrow$	28.46 μm
Unidirectional accuracy of positioning of an axis, $A\uparrow$	23.16 μm
Bidirectional accuracy of positioning of an axis, A	31.14 μm

TABLE 2.12
Forward and Reverse Error for One Target Position

Target Position 120 mm	Run Number 1	Run Number 2	Run Number 3	Run Number 4	Run Number 5
Forward error (μm)	−5.62	−6.70	−7.01	−6.81	−7.54
Reverse error (μm)	−4.51	−5.11	−5.28	−5.19	−5.32

Example of calculations

1. X-axis positional results for target position @ 210.00 mm
2. Forward average error

$$\bar{X}_f = \frac{(-5.62)+(-6.7)+(-7.01)+(-6.81)+(-7.54)}{5} = -6.736 \ \mu m$$

3. Standard deviation 3σ

$$\text{Forward} = 3\sqrt{\frac{\Sigma(\bar{X}_f - X_i)^2}{n-1}} = 3\sqrt{\frac{1.97282}{4}} = 2.0168 \ \mu m$$

4. Reverse average error

$$\bar{X}_f = \frac{(-4.51)+(-5.11)+(-5.28)+(-5.19)+(-5.32)}{5} = -5.028 \ \mu m$$

5. Standard deviation 3σ

$$\text{Reverse} = 3\sqrt{\frac{\Sigma(\bar{X}_f - X_i)^2}{n-1}} = 3\sqrt{\frac{0.47436}{4}} = 1.0331 \ \mu m$$

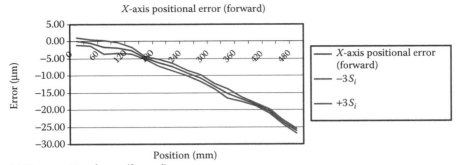

(a) *X*-axis positional error (forward)

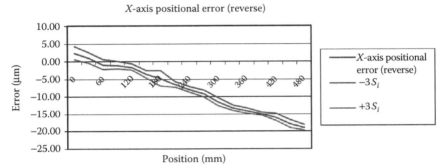

(b) *X*-axis positional error (reverse)

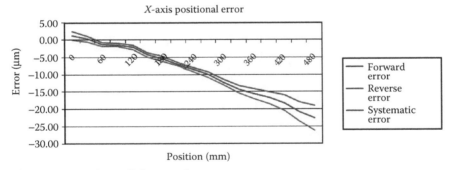

(c) *X*-axis positional error (bidirectional)

FIGURE 2.33 (See color insert following page 174.) Positional error for (a) forward, (b) reverse, and (c) bidirectional.

6. System error $=\dfrac{\text{forward average error} \pm \text{reverse average error}}{2}$

Therefore

$$\text{System error} = \frac{(-6.736)\pm(-5.082)}{2} = -5.909 \,\mu\text{m}$$

Also the reverse error = reverse average error − forward average error

Therefore

Reverse error = $(-5.082) - (-6.736) = -1.654\,\mu\text{m}$ (Tables 2.13 through 2.15)

TABLE 2.13
Forward Error

	X_i	$\overline{x}_f$	$(\overline{x}_f - X_i)$	$(\overline{x}_f - X_i)^2$
Run 1	−5.62	−6.736	−1.116	1.244556
Run 2	−6.70	−6.736	−0.036	0.001296
Run 3	−7.01	−6.736	0.274	0.075076
Run 4	−6.81	−6.736	0.074	0.005476
Run 5	−7.54	−6.736	0.804	0.646416
				Total = 1.97282

TABLE 2.14
Reverse Error

	X_i	$\overline{x}_r$	$(\overline{x}_r - X_i)$	$(\overline{x}_r - X_i)^2$
Run 1	−4.51	−5.082	−0.572	0.327184
Run 2	−5.11	−5.082	0.028	0.000784
Run 3	−5.28	−5.082	0.252	0.063504
Run 4	−5.19	−5.082	0.162	0.026244
Run 5	−5.32	−5.082	0.238	0.056644
				Total = 0.47436

TABLE 2.15
Experimental Contour Tests on the CNC Machine

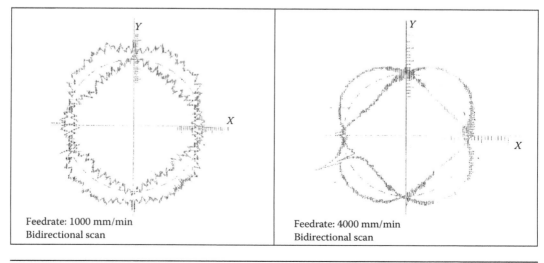

Feedrate: 1000 mm/min
Bidirectional scan

Feedrate: 4000 mm/min
Bidirectional scan

Comments
On-axis accuracy
Accuracy, in layman's language, is the ability to tell the truth. Whereas on-axis accuracy is the uncertainty of position after all sources of linear error are eliminated. Linear (or monotonically increasing) errors include displacement inaccuracy, planar inaccuracy, volumetric inaccuracy, and inaccuracy of the leadscrew pitch, the angular deviation effect at the measuring point (Abbé error) and thermal expansion effects. Graphically these errors are represented by the slope of a best-fit line on a plot of position versus error (Figure 2.33a). Knowing the slope of this line (error/travel), we can approximate *absolute accuracy* as

$$\text{Absolute accuracy} = \text{on-axis accuracy} \pm \text{Abbé error}$$

Repeatability
Repeatability is the ability of a motion system to reliably achieve a commanded position over many attempts. Manufacturers often specify unidirectional repeatability, which is the ability to repeat a motion increment in one direction only. This specification side-steps issues of backlash and hysteresis, and therefore is less meaningful for many real-world applications where reversal of motion direction is common.

As given by the computer analyses the unidirectional repeatability is 4.22 µm at target position of 210,000 mm in forward direction. A more significant specification is bidirectional repeatability, or the ability to achieve a commanded position over many attempts regardless of the direction from which the position is approached. In this case the bidirectional repeatability achieved at target position of 480,000 mm is 10.16 µm.

Systematic errors
Systematic errors do not lend themselves to averaging out over the time of measurement and it may be constant. Accordingly, a systematic error is particularly dangerous in that no matter how many times a measurement is repeated that error can remain undetected. Techniques for identifying systematic errors include measuring the quantity repeatedly over such a period that some of the systematic errors may have had time to become random.

In this test the systematic errors (average) can be measured by taking the average of the forward error average and the reverse error average at each target position.

Correlation of the pitch motion with the Abbe' offset
and the measured positioning performance
To achieve a high-precision length measuring system, the measured length and the measurement scale must lie on a single line. Additional linear off-axis error introduced through amplification of tilt and wobble with a long moment arm (offset). The distance from the source amplifies angular errors. Figure 2.34 between the pitch motion error with Abbé offset as constant and changing-pitch angle with the change in target positions and the target positions is linear; that is with the increase in the displacement there is increase in the pitch motion error.

Influence of selection of target position, weight of work piece,
and thermal errors
Target positions should not be selected randomly. Thirty millimeter equally spaced target positions were selected so that any cyclic error present in the positioning system would be detected along with the progressive component of systematic error.

If the machine fails to return to the same target position, within certain tolerances, then any compensation based upon prediction will of course fail to operate with any reliability. The standard deviation results for each test were therefore of the utmost interest. The removal of systematic error

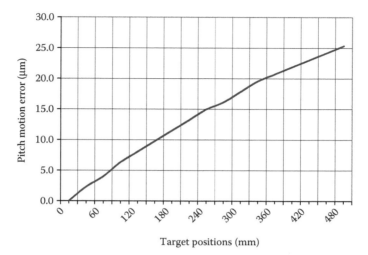

FIGURE 2.34 Pitch motion error.

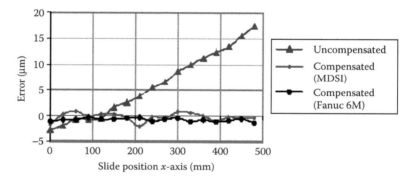

FIGURE 2.35 Compensated positioning.

is now quite commonplace, but the more demanding task of maintaining this accuracy still depends largely upon the initial design of the machine tool structure, guide ways, etc. If backlash, hysteresis, systematic errors, etc., are compensated for the Machine center, the structure of the machine has to be stiff enough to withstand any deformation induced by the machine or components own weight.

Errors induced due to the change in temperature during machining will have adverse effect on the positional accuracy. It is therefore mandatory to keep the temperature constant during the machining operations.

Closed loop compensation

On the following figure, error compensation routine has been activated to compensate for Abbé error mainly[*] (Figure 2.35).

Abbé error could be a significant source of error. The angular error could be caused by any of the following issues:

1. Curvature of ways
2. Entry and exit of balls or rollers in recirculating ways

---

[*] The results used here were done few days later following the test results presented previously.

3. Variation in preload along a way
4. Insufficient preload or backlash in a way
5. Contaminants between rollers and the way surface

The CNC machining center has a hysteresis error aspect. This is because once a position is reached the machine can never return exactly back to that position again once it has moved away. Hysteresis is a problem when present in a CNC machine as it means the program carried out when machining may not achieve an acceptable accuracy and suggests that an optimal way to program a CNC machine for dimensional accuracy as each point should be visited only once. For example if a circular pocket is made to cut the workpiece then later in the same program a hole is to be drilled at the same point the error would have increased and the machine would not be able to achieve the same position again.

The machine has also uncertainties in the calibration and initial setup of the machine. The data inputted to the software to determine the wavelength compensation may not have been accurate enough and is certainly a possible source of error. This is because some of the values inputted were just the teams "best estimates" of environmental conditions such as humidity, pressure, and temperature.

2.12.2 Contouring Assessment of CNC Machine Using Kinematic Ball Bar System

The purpose of using a ball bar link is to characterize circular motion provided by either a milling machine with the combination of two axes, or, in lathe machine. It is expected to verify the eccentricity and geometric errors such as squareness and deformation errors. These may have various sources in a CNC machine.

A number of features will be discussed later to have a clear idea of what to expect from such a test.

The objectives of the test are

1. To investigate the contouring characteristics of a CNC Machining Center using a kinematic ball link system.
2. To investigate the salient features of the contouring/polar results in respect of error sources of the machine.

Experimental procedure

1. Software* compensation from CNC machining center with MDSI open controller was removed.
2. A 100 mm kinematic ball bar system complete with calibrated setting bar was used with the quick setting sleeve from the hardware. The center of the two reference spheres coincide with each other in the center of the x–y reference plane of the machine. Figure 2.36 shows the experimental setup.
3. The machine was programmed to produce a 360° clockwise circle with a 100 mm radius and a feed of 1000 mm/min with a tangential approach to the start point and a tangential exit. The start and exit point was at 202° in the XY plane.
4. Repeated (c) for anticlockwise movement.
5. The above test procedure was repeated for a feed rate of 4000 mm/min.
6. Measured results were displaced (polar plot) on PC and observed the lost motion (backlash) in the x- and y-axes using a software and PC for online data acquisition and analysis of the results.

The results show clearly that there is a cyclic error in the machine. The plots also show a negative backlash. Servo mismatch characteristic in the XY plane and Radial deviation.

There are two cases recorded in Table 2.16 in which the feed-rate is changed from 1000 to 4000. The diagrams show the influence of feed-rate on parameters such as circularity and backlash.

---

* The software CONTISURE was initially developed in house (UMIST) by Dr.M Burdekin as well as the very First ball bar link.

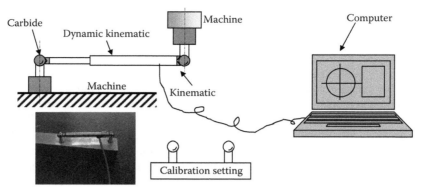

FIGURE 2.36 Experimental setup.

TABLE 2.16
Diagrams Resulting from Contouring Assessment

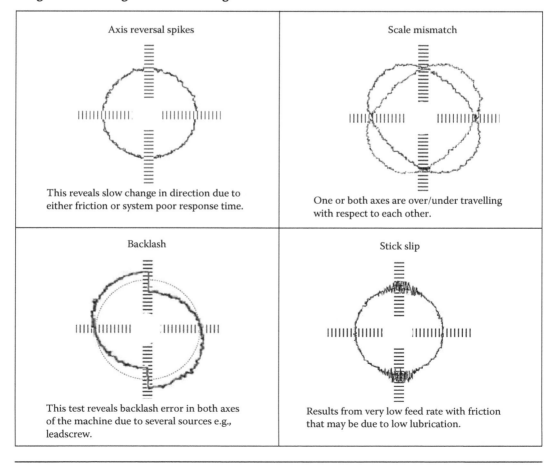

TABLE 2.16 (continued)

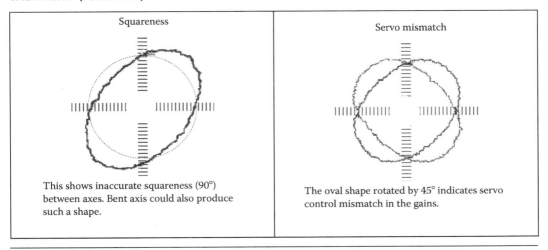

Squareness	Servo mismatch
This shows inaccurate squareness (90°) between axes. Bent axis could also produce such a shape.	The oval shape rotated by 45° indicates servo control mismatch in the gains.

Looseness in guideways appears at low speed while some servo mismatch and some axes reversal spikes appear at high speed.

The Ball-link tests show how the two axes work together to move the machine in a circular path. Two axes would make perfect circles in a perfect machine. The Ball-link measures any deviations the machine makes from a perfect circle and displays the data in a polar format.

A few definitions are introduced here after to recognize a number of known errors. More details could be found in the international standard ISO 230. Most of the known errors are summarized in Table 2.17.

Backlash
The backlash error appears in two different ways, as positive and negative backlash. When moving to the defined test direction, positive backlash can be seen on the plot as a step outwards starting on the axis (deformation or play in the machine structure). Negative backlash appears as a step inwards starting on the axis (backlash overcompensation at the controller, that is the compensation value exceeds the value of the actual backlash error).

Cyclic error
Cyclic error appears as a cyclically changing motion error on the circular plot. It is important to know the machine tool and its driving mechanisms to understand source of errors. The cyclic error may come from ball screws, encoders, transmission, etc.

Radial play
Lateral play usually occurs as a tangential deviation caused by looseness in the machines guide ways. The graphs are usually symmetrical about both axes.

Servo mismatch
The magnitude of this error informs about the response time of the machines position control loop to the given commands. An absolute difference between the nominal and actual values can be obtained. Otherwise, an unbalance will be noticed as a relative value.

For an overshoot or lag of the response time, an oval trace pattern will be seen at ±45° with the machine axes. At low feed-rates, the gain of the axis which is lagging could be turned up, or the gain of the axis that is leading could be turned down.

TABLE 2.17 (See color insert following page 174.)
Diagrams Resulting from Contouring Assessment

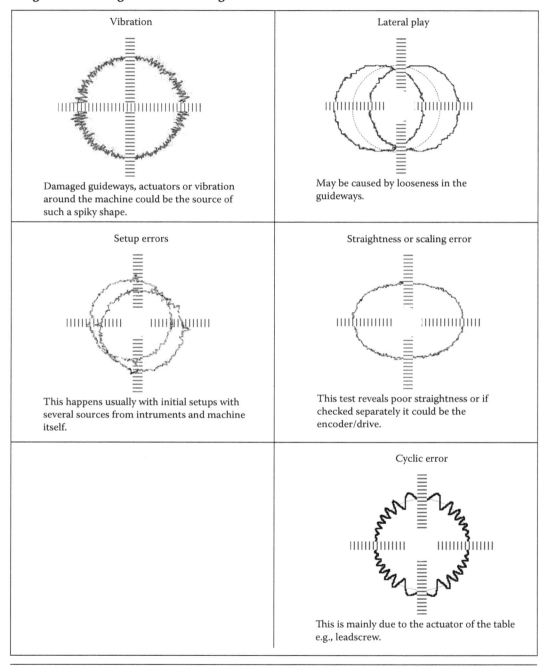

Vibration	Lateral play
Damaged guideways, actuators or vibration around the machine could be the source of such a spiky shape.	May be caused by looseness in the guideways.
Setup errors	Straightness or scaling error
This happens usually with initial setups with several sources from intruments and machine itself.	This test reveals poor straightness or if checked separately it could be the encoder/drive.
	Cyclic error
	This is mainly due to the actuator of the table e.g., leadscrew.

Squareness error

The squareness error is the deviation in the perpendicularity of the axes at the test location. The test reveals whether the axes are bent locally or the axis is misaligned along its whole length with repeated tests.

Directional vibration

If vibration is caused by a particular reason, the location and influencing direction of the error source will cause a directional bias in the vibration pattern.

REFERENCES

1. Smith, S.T. and Chetwynd, D.G., *Development in Nanotechnology*, Volume 2: Foundations of Ultraprecision Mechanism Design, ISBN: 2881248403, 1992.
2. Blumenthal, L.M., Concerning the remainder term in Taylor's formula. *Amer. Math. Monthly,* 33, 424–426, 1926.
3. *Guide to the Expression of Uncertainty in Measurement (GUM).* International Organisation of Standardization (ISO), Geneva, 1995.
4. Taylor, J.R., *An Introduction to Error Analysis—The Study of Uncertainties in Physical Measurements*, University Science Books, ISBN 0–935702–75-X, 1997.
5. Mekid, S. and Vaja, D., Propagation of uncertainty: expressions of second and third order uncertainty with third and fourth moments. *Measurement,* 41(6), 600–609, 2008.
6. BS ISO 230-2: Test code for machine tools. Determination of accuracy and repeatability of positioning of numerically controlled machine tool, *International Organisation for Standardisation* (ISO), 2005.
7. *International Vocabulary of Metrology—Basic and General Concepts and Associated Terms* (VIM), ISO/IEC Guide 99-12: 2007.
8. Raksiri, C. and Parnichkun, M., Geometric and force errors compensation in a 3-axis CNC milling machine. *Int. J. Mach. Tool Manu.,* 44, 1283–1291, 2004.
9. Bryan, J., The Abbé principle revisited: An updated interpretation. *Prec. Eng.,* 1(3), 129–132, 1989.
10. www.precitech.com, 2007
11. Ramesh, R., Mannan M.A., and Poo A.N., Error compensation in machine tools—a review. Part I: geometric, cutting-force induced and fixture-dependent errors. *Int. J. Mach. Tool Manu.,* 40, 1235–1256, 2000.
12. Ramesh, R., Mannan, M.A., and Poo A.N., Error compensation in machine tools—a review. Part II: thermal errors. *Int. J. Mach. Tool Manu.,* 40, 1257–1284, 2000.
13. Yuan, J. and Ni, J., The real-time error compensation technique for CNC machining systems. *Mechatronics,* 8, 359–380, 1998.
14. Donmez, M.A. et al., A general methodology for machine tool accuracy enhancement by error compensation. *Prec. Eng.,* 8(4), 187–196, 1986.

3 Thermal Problems in Machine Tools Design and Operation

Jerzy Jedrzejewski

CONTENTS

When agreed, that a machine equipped with self-acting movement of a machined part and tool is called a machine tool—then it must be assumed, that England is machine tool's homeland, where in year 1794 Maudsley patented the first device for mechanical guidance of a tool—a compound slide—and where basic types of machine tools have been created, that is screw-cutting lathes, drilling, boring and planing machines and others.

Prof. Geisler
Conference on Machine Tools, Warsaw, 1939

3.1 PURPOSE, IMPORTANCE, AND COMPLEXITY OF SHAPING ADVANTAGEOUS THERMAL PROPERTIES OF MACHINE TOOLS

Thermal phenomena in machine tools are the source of machining errors the minimization of which has an important bearing on the effectiveness of machining processes. The higher the required machining accuracy, the more important the reduction of thermal errors. The steady increase in machining speed contributes to the intensification of the thermal phenomena. The growing requirements for machining accuracy necessitate a reduction in thermal errors. Since today thermal errors are the main determinants of machining accuracy, research aimed at reducing them through design and compensation is of vital importance.

As a result of the heating up and thermal deformation of components, units, and whole machine tools, the geometry imparted to them through machining and assembly changes. Consequently, the work and feed motion paths warp whereby the cutting tool's position relative to the workpiece changes directly affecting the latter. Thermal deformations may also adversely affect the resistance to motion of sliding machine tool assemblies and generate additional power losses, increase wear, and shorten the lifetime of machine tool accuracy.

Thermal errors change with the behavior of the machine tool during its operation. They are determined by the intensity of the heat sources, the heat transfer within the machine tool, and between the latter and the environment, and by factors associated with the machine tool design and the performed machining processes.

Formerly, when machining processes were relatively slow and spindle and feed motion speeds low, the thermal processes taking place in machine tools could be regarded as slowly variable. In today's high-speed machining, the dynamic processes do not lend themselves to precise analysis, modeling, simulation, or control. Therefore, it is very difficult to investigate the effect of thermal phenomena on the precision of machine tools and workpieces.

The influence of thermal phenomena on the precision of machine tools and workpieces is reduced through their optimum design which assumes minimization of heat sources and thermal displacements and possibly most precise compensation of spatial thermal errors during the operation of the machine tool. If such measures are insufficient to ensure the required precision, forced cooling of some of the machine tool assemblies is introduced. Design is aided by the numerical modeling, simulation, and optimization of the thermal behavior of the machine tool alone or together with the machining process. Also virtualization—a highly effective but costly tool—can be used in designing. The virtualization of the precision machine tool's thermal behavior in real time, however, is still very difficult to achieve in practice.

When trying to reduce the thermal errors of machine tools one should take into account many design-technological factors and also, because of market competition, the cost factor (of the machine tool and its operation). A complex multilayer model of knowledge processing in the creation of high-performance precision machine tools is shown in Figure 3.1.

The complexity of machine tool design and optimal operating properties' (including high precision dependent on the optimum thermal behavior) assurance arises not only from the complexity of the (static and dynamic) thermal phenomena and the hybrid nature of the machining processes, but also from the fact that the designer acquires many of the modules from the

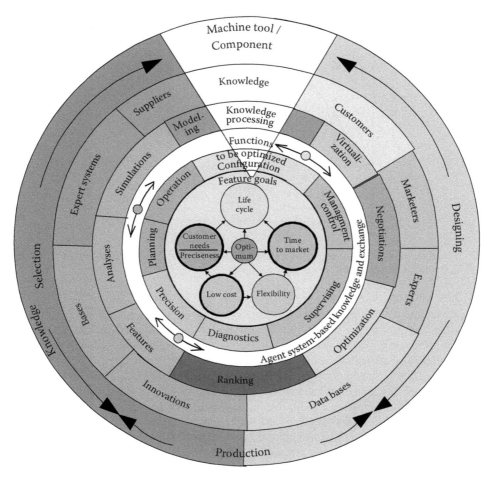

FIGURE 3.1 (See color insert following page 174.) Levels of knowledge and data processing by machine tools creation and optimization.

global market whereby he/she cannot improve their properties and does not know their models needed in the complex process of modeling and optimization. The machine tool's quality is also greatly affected by the manufacturing process and the cost reduction demanded by marketing on one hand and the assurance of high technological flexibility on the other hand. Therefore, the machine tool designer must optimize the machine tool's thermal properties taking into account the combined effect of technical and economic factors.

3.2 HEATING UP AND THERMAL DISPLACEMENT PATTERNS TYPICAL FOR MACHINE TOOL ASSEMBLIES

The machine tool's heat behavior is determined by the output of the heat sources (mostly in the form of power losses in the friction nodes and in the electric motors) located within its structure, the increase in the temperature of the machine tool components at certain points, the temperature distribution, the thermal deformations of the load-bearing system, and the mutual displacements of the assemblies. The mutual positions of the tables, slides, and spindles in or on which cutting tools and workpieces are mounted change over time with ambient temperature, operating conditions, and speed of motion of the work units. The intensity of thermal phenomena is recognized from the

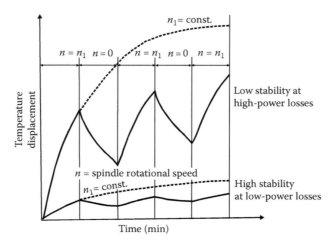

FIGURE 3.2 Example of sharp changes in spindle bearing temperature and displacement at low-machine tool thermal stability and gentle changes at high stability.

temperature diagrams of the machine tool's main components and assemblies and from the thermal displacements diagram of the machine tool's main assemblies, particularly of the cutting tool relative to the workpiece. Recognition is conducted during idle running and machining, at a fixed or naturally variable ambient shop floor temperature.

The machine tool's sensitivity to temperature variation and thermal displacements is defined through thermal stability. The higher the sensitivity, the lower the thermal stability. Not only the heat sources and the heat transmission conditions, but also structural factors, such as the materials, the structure's symmetry, the joints between drive and body parts, the linkage and basing of the parts, the heat capacity of the elements which heat up and other, have a great influence on thermal stability.

Typical patterns of the heating up of spindle bearings, their cooling after spindle stoppage, and the corresponding spindle (cutting tool/workpiece) displacements for a machine tool with, respectively, low and high thermal stability are shown in Figure 3.2.

When the machine tool is started and idles at a spindle speed $n = n_1$, initially temperature sharply rises and then the rate of change gradually decreases until a thermal equilibrium is reached at which temperature and displacement stabilize at a certain level. When the spindle is stopped ($n = 0$), a temperature drop occurs (in a reverse way to heating up). At a variable speed of the spindle (which usually is the main source of heat), the machine tool while it operates may be permanently in a thermally unsteady state, whereby it can be quite difficult to ensure the required machining accuracy, particularly if the machine tool's thermal stability is low. Typical thermal displacements of the lathe spindle axis, measured for 2h since the lathe was started, and the vertical and horizontal thermal displacements of the bed during idle running are shown in Figure 3.3.

The displacements of the tool relative to the workpiece, caused by changes in the machine tool's geometry, directly affect the dimensional-geometric accuracy of the workpiece. The displacements along each controllable axis of the machine tool are the sum of individual positive and negative time-variable displacements.

Figure 3.3 shows that heat sources located in the lathe's headstock have a strong effect on the displacement of the spindle axis and the deformation of the guides. Excessive deformations of the guides may result in seizure of the slide in the guides and damage to the machine tool.

It is apparent that such increases in thermal displacements along axes x and y in the initial stage of operation of the machine tool make it very difficult to achieve the required dimensional-shape accuracy of the turned objects. If the machine tool is not equipped with an efficient thermal error displacement compensation system, machining begins after a period of particularly intensive displacements. As a result, the time of effective machine tool operation is reduced.

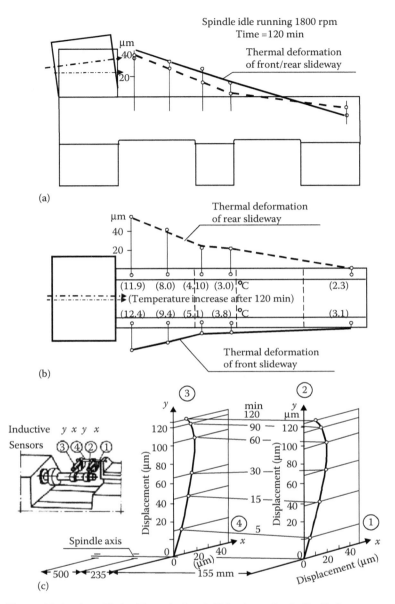

FIGURE 3.3 Thermal behavior of lathe (a) vertical and (b) horizontal spindle displacements and (c) spindle axis displacements along axes x and y during 120 min of machine tool operation.

3.3 HEAT SOURCES IN MACHINE TOOLS

3.3.1 GENERAL DESCRIPTION OF HEAT SOURCES

The sources of heat in machine tools are moving drive system components located inside their bodies or in their immediate vicinity. These include electric motors, toothed and belt transmissions, couplings, brakes, pumps and hydraulic servomotors, electric, hydraulic, and pneumatic control system elements, and bearings. Their effect on the heating up and thermal deformation of machine tools depends on their position relative to the body and to the elements supporting the cutting tool and the workpiece and on their intensity. The distribution of power losses in the spindle drive of a lathe with a conventional headstock and a gearbox, during idle running, is shown in Figure 3.4.

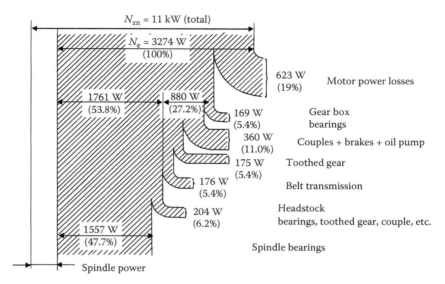

FIGURE 3.4   Distribution of machine tool main driver power losses for spindle idle running at $n = 1800\,$rpm.

One can see a large idle running power loss share N_e in the drive motor's rated power N_{zn}. This share significantly increases with the rotational speed of the spindle and comprises all the drive components. When the drive is loaded with machining power, power losses in the motor, in the toothed gears, and in the belt transmissions increase greatly, especially in the case of V-belt transmissions.

Considering the heating up and thermal deformation of headstocks it is highly important to reduce the heat sources (bearings, gears, and couplings) located inside them. Power losses in a toothed gear transmission depend on the meshing efficiency and the power applied to the driving wheel. The relation for meshing efficiency comprises between 10 and 20 parameters having a bearing on power losses, among them are (besides parameters describing gear transmissions and loads) the relative viscosity of the lubricating oil, the transmission operating temperature, the gear fabrication accuracy class numbers, and the coefficients of friction dependent on the kind of gear wheels and the lubricating oil. It is very difficult and costly to increase meshing efficiency by increasing transmission gear fabrication accuracy and there are limits to it. The effect of the headstock's toothed gear transmissions on the power losses arising inside it versus machining power is illustrated in Figure 3.5.

It is apparent that the toothed gear transmission is the dominant source of heat. The losses can be reduced through optimum (minimal) lubrication, but only slightly. At high spindle speeds toothed gear transmissions are inadequate because of the high power losses, difficulties in ensuring proper lubrication, and the generation of excessive vibration and noisiness. In such cases, the drive is limited to a motor, a cogbelt transmission, and a spindle mounted on bearings in the headstock. Then the spindle bearings have a dominant influence on the heating up of the headstock and the displacement of the spindle. Couplings have been practically eliminated from machine tool drives since they constituted very large sources of heat generated by the friction of the plates during coupling and uncoupling. The required intensive oil cooling of the couplings during power loss generation periods was costly and reduced the spindle drive's energy efficiency. Also power losses in the windings of the electromagnets, couplings, and brakes contributed to the heating up of the headstocks.

3.3.1.1   Power Losses in Belt Transmissions

Belt and ball screw transmissions, the electric motors which drive them, hydraulic system components, and installations transporting the coolant and the lubricating-cooling medium indirectly, but significantly, contribute (depending on their position relative to the housings or to the motion

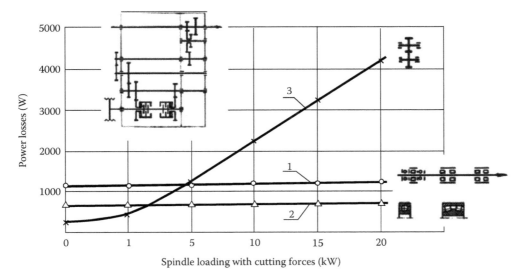

FIGURE 3.5 Influence of cutting force on power losses in lathe headstock: 1—in headstock bearings, 2—in other bearings, and 3—in gear wheels.

transmitting elements) to the heating up of machine tools. Today belt transmissions with V-belts are very seldom used in machine tools since they are a source of high power losses arising from belt slip (which is unavoidable due to the different length of the belts and their improper tension). Belt transmissions with cogbelts are highly efficient and stiff which is of major importance when they drive the machine tool's steerable axes. For transmitted torques $M < 25\,\text{N} \cdot \text{m}$, the efficiency is $\eta = e^{-\frac{0.55}{M}}$ and for $M > 25\,\text{N} \cdot \text{m}$ the efficiency is $\eta = 0.978$. For known input power N_{we}, the power losses in such a transmission are given by the relation

$$\Delta N = N_{we} \cdot (1 - \eta), \text{W} \tag{3.1}$$

3.3.1.2 Power Losses in a Ball Screw

Power losses in a ball screw only to a small degree contribute to the heating up of the machine tool's main bodies, but should be taken into account in precise analyses. They affect the deformations and positioning errors of sliding tables, tool slides, and headstocks. The efficiency of a preloaded ball screw–nut set depends on its design (the ball/race contact and guidance and the ball screw support) and the tension. Therefore, in order to calculate power losses it is best to use the relations given by the manufacturers of ball screws in the catalogs for designing servodrives where one can also find information on when in the course of designing transmissions with high rates of travel of the driven units one should consult the transmission manufacturer. According to SFK company specifications, the limiting friction moment, which takes into account the return motion of the rolling elements and depends on the tension of the ball screw's components, can be calculated from the relation

$$M_B = \frac{F \cdot P_n \cdot \eta}{2000 \Pi}, \text{N} \cdot \text{m} \tag{3.2}$$

where
  F is the rolling elements' tension force, N
  P_n is the guiding length, mm
  η is the intermediate ideal efficiency of the system

motor–transmission–screw $= 2 - (1/\eta)$, where η is an ideal efficiency expressed by the formula

$$\eta = \frac{1}{1 + (k \cdot d_o)/P_n} \tag{3.3}$$

where
$k = 0.018$—for most transmission designs
d_o is the nominal diameter of the screw

For the operating conditions of ball screws, practical efficiency $\eta_p = 0.9\,\eta$ is used. It should be noted that in precision machine tools the transmissions are cooled in order to eliminate the influence of power losses arising in them on machining accuracy. Usually the nut alone is cooled but in some cases also the ball screw from the inside.

3.3.1.3 Power Losses in Slide and Rolling Guide Joints

The power losses can be determined from the relation

$$\Delta N = \mu_S \cdot P \cdot v, \text{W} \tag{3.4}$$

where
μ_s is the guide's friction coefficient
P is the load, N
v is the speed of motion, m/s

The value of motion coefficient μ_s is

- For a steel/steel sliding pair: 0.1 and under lubrication conditions −0.07, while the resting coefficient value is, respectively, 0.18 and 0.15
- For a steel/cast iron pair motion coefficient μ_s is, respectively, 0.18 and 0.05, while resting coefficient μ_s amounts to 0.19 and 0.15

Under heavy loading of the machine tool's sliding parts and high speeds, such losses should be taken into account in the overall balance of machine tool power losses and in their contribution to temperature and deformation increases. This applies particularly to precision machine tools for high-speed cutting (HSC). The losses are usually calculated using specialized calculation programs.

3.3.1.4 Power Losses of Rolling Guides

Power losses arising within the field of action of rolling elements in linear guide carriages depend on the tension of the rolling elements, the lubrication conditions, the speed of motion of the carriage, and the design of the rolling guide joint. The load of the pretensioned rolling joint between the carriage and the guide can be determined according to THK company specifications from the following:

$$P_w = P_0 + k \cdot f_w \cdot P_n \tag{3.5}$$

when

$$\frac{f_w P_n}{P_0} \le 2.83$$

or

$$P_w = f_w \cdot P_n \qquad (3.6)$$

when

$$\frac{f_w P_n}{P_0} > 2.83$$

where

P_{wl} is the resultant loads, N

P_0 is the preload, N

f_w is a load factor

P_n is the changing load, N

k is the external load factor assuming values from 0.5 to 0.65 from the linear dependence on the product $z_{fw} \cdot (P_n/P_0)$ variable in the interval 0–2.8.

Preloading is needed to ensure sufficient stiffness so that carriage displacements under external loads are small.

3.3.1.5 Power Losses in Motors Driving CNC Machine Tool Spindles and Electrospindles

The electrospindle (the shortest form of spindle drive) has been increasingly used in machine tools, except for heavy machine tools for which no acceptable, effective electrospindles have been developed yet. The motors of electrospindles are large sources of heat which directly contribute to the deformation and displacement of electrospindles. Therefore, this heat must be carried away by cooling systems. Three-phase AC motors controlled (through the field vector) by processors are used in the intermediate drives of spindles and in electrospindles. Power losses in such motors depend on the latter's design. In order to estimate the power losses it is best to use the relations given by the manufacturers.

Today spindles with a numerically controllable axis are driven by field vector controlled three-phase AC induction motors. The rate of rotation is controlled through current frequency. The motors have one or two speed ranges (the second range covers high speeds, e.g., above 25,000 rpm). Total input motor power Nel_{in} is given by the relation

$$\text{Nel}_{in} = \sqrt{3} \cdot U \cdot I \cdot \cos \varphi \qquad (3.7)$$

where φ is the angle between voltage U and current I.

Motor power losses comprise mechanical losses (arising in bearings), which will be discussed in detail in Section 3.3.2, losses generated by friction in the air gap between the rotor and the stator, and electrical losses.

Electrical losses arise in the copper and the iron. Losses in the copper stator windings are a function of the current which flows in them. The output motor power comprises machining power and acceleration power. The heat generated by the motor can be calculated from the following relation:

$$Q_{mot} = 2\Pi \cdot f_{mot} \cdot M_{mot} \frac{1 - \eta_{mot}}{\eta_{mot}} \qquad (3.8)$$

where

f_{mot} is the motor's frequency

M_{mot} is the motor's torque

η_{mot} is the motor's efficiency given by the relation:

$$\eta_{mot} = \eta_{mot} \max \cdot \eta_{speed} \cdot \eta_{load} \qquad (3.9)$$

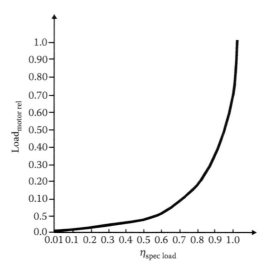

FIGURE 3.6 Dimensionless load efficiency factor.

where

η_{mot} max is the motor's maximum efficiency defined as a ratio of the sum of bearing losses and air gap losses to the (rated) input motor power. The input power comprises all the power components

η_{speed} is the efficiency dependent on the rotational speed

η_{load} is the efficiency dependent on the load

The speed-dependent efficiency coefficient reflects the reduction in power losses which occurs when shifting from a lower to higher speed range. The load acting on the motor affects the latter's efficiency. For example, for a motor with a maximum speed of 25,000 rpm (spindle frequency 426 Hz) under a relative load of zero (idle running and no machining or acceleration power load), the load-dependent efficiency is 0.1. It reaches the maximum value of 1 under a relative load of 0.8 and then decreases (Figure 3.6).

For a strategy of motor and electrospindle heat load reduction through forced cooling, it is helpful to know the rotor's share in the total motor heat load, which directly affects the behavior of the spindle, and the stator's share which affects the temperature and thermal deformation of the headstock.

According to Bernd Bosmanns and Jay F. Tu

$$Q_{rot} = Q_{mot} \cdot \frac{f_{slip}}{f_{sync}} \qquad (3.10)$$

and

$$Q_{stat} = Q_{mot} - Q_{rot} \qquad (3.11)$$

where

Q_{rot} is the heat transferred to the rotor

Q_{stat} is the heat transferred to the stator

f_{slip} is the slip frequency

f_{sync} is the motor's synchronous frequency

Figure 3.7 shows exemplary electrospindle motor power losses as the sum of stator and rotor losses versus machining power load in a spindle speed range of 10,000–40,000 rpm for steady states. The included relation allows one to approximately calculate rotor power losses.

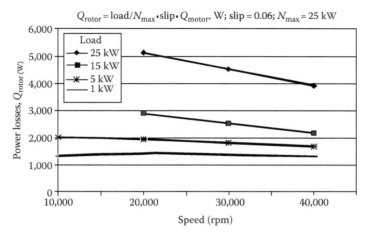

FIGURE 3.7 Power losses generated in rotor depending on load and rotational speed.

Air friction losses arising in the gap between the stator and the rotor have a relatively small but notable effect on electrospindle motor power losses. The air friction power in the gap is a function of the gap's parameters, the air friction coefficient, and the rotor's frequency and its diameter. The power of the losses can be calculated from the relation

$$N_{air} = \frac{\Pi^3 \cdot d^3_{rot} \cdot L_{rot} \cdot V_{air} \cdot f^2_{rot}}{h_{gap}} \tag{3.12}$$

where
d_{rot} is the rotor's diameter
L_{rot} is the rotor's length
V_{air} is a dynamic air viscosity coefficient equal to 18.5×10^{-6} N s/m^2
f_{rot} is the rotor's frequency
h_{gap} is the gap's height

Test results reported for a typical electrospindle motor at 25,000 rpm show that the losses amount to 25 W.

The heat generated in the air gap can be calculated directly from the relation

$$Q_{air} = \omega_{rot} \cdot M_{air} \tag{3.13}$$

where
M_{air} is the moment of friction in the air gap
ω_{rot} is the angular velocity of the rotor

In comparison with the other losses, little power is lost in the air gap.

3.3.2 Power Losses in Rolling Bearings of Machine Tool Spindles

In particular, spindle bearings strongly affect the accuracy of machine tools in their operating conditions. The thermal phenomena and deformations which occur in them under the action of forces and moments determine the axial and radial changes in the position of the spindle in the course of machining processes. Bearing loads are complex and variable. Internal bearing loads depend on the bearing's preload, the cutting forces, and the deformations inside the bearings. The deformations result from the highly complex processes of heat generation and transfer within the bearing and to/from the mating parts and the environment. The spatial distribution of rolling bearing loads also depends on the geometrical and dimensional accuracy of the spindle and body surfaces on which the bearing rings are mounted and on the changes in this accuracy due to thermal deformations as well as on the forces loading the spindle assembly. Thus momentary states of loading and deformation of the individual bearings, variable in time with their operating conditions contribute to the power losses arising in the bearings. Another aspect of bearing load which needs to be considered is the effect of the state of loading on the durability (lifetime) of the bearings.

It follows from the above that knowledge of the phenomena which occur in bearings is essential for machine tool designers since it allows them to rationally design spindle assemblies in order to prevent their overloads and increase their lifetime. It also facilitates the design of the diagnostics and monitoring the behavior of spindle units in operating conditions.

In machine tool spindle bearings, power losses are generated due to friction moment M_0 in the lubricating film and to moment M_1 produced by the bearing load. Spindle bearings are loaded by the forces which accompany spindle tensioning (a negative clearance is generated) and by the forces originating from the machining process which are carried by the bearings.

As a result of Palmgren's studies published in 1957, the total friction moment can be calculated from the following Snare–Palmgren relation:

$$M = M_0 + M_1 \tag{3.14}$$

The original relation for the moment of elastohydrodynamic friction derived for the abundant lubrication of spindle bearings is as follows:

$$M_0 = 10^{-7} f_0 \cdot (\nu n)^{2/3} \cdot d_m^3, \text{N} \cdot \text{mm} \tag{3.15}$$

where
   f_0 is a coefficient dependent on the bearing design and the lubrication mode
   ν is the lubricating medium's kinematic viscosity (mm³/s); in the case of grease it is the
      viscosity of the base oil; the viscosity changes with temperature
   n is the rotational speed, rpm
   d_m is the average diameter of the bearing, mm

The research on the modeling and reduction of power losses in machine tool bearings carried out by J. Jedrzejewski, W. Kwasny, and J. Potrykus in the 1960s and 1970s showed that from the power loss reduction point of view it is rational to substantially reduce the amount of oil supplied to them, select a proper oil, and use grease as the lubricating medium. Consequently, the Palmgren relation for the moment of elastohydrodynamic friction was modified. It was found that in order for the Palmgren formula to remain valid it was enough to write it in the following general form

$$M_0 = 10^{-7} f_0 \cdot (\nu \cdot n)^P d_m^3, \text{N} \cdot \text{mm} \tag{3.16}$$

and put in a proper (for a low oil flow rate) value of index exponent p.

For lubrication ensuring minimum power losses (by feeding possibly smallest oil volumes in a unit time to maintain the continuity of the lubricating film), 0.3 or even as low value as 0.25, depending on the type of bearing and the kind and mode of lubrication, should be substituted for index exponent p. The following relation describes well the power losses in a lubricating medium, such as a properly refined oil fed in minimal quantities to the bearings and a properly matched grease in a wide range of rotational speeds.

$$M_0 = 10^{-7} f_0 \cdot (\nu \cdot n)^{1/3} d_m^3, \text{N} \cdot \text{mm} \tag{3.17}$$

The difference between grease and oil is that the former has the features of an intelligent lubricating medium, i.e., it releases such a volume of base oil for lubrication which is essential to ensure a proper oil wedge and minimal power losses. Feeding large amount (volumes) of oil to the bearing

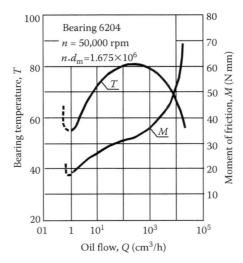

FIGURE 3.8   Moment of friction and bearing temperature versus oil flow.

results in a large increase in the moment of friction. Once a certain volume of oil being fed in a unit time is exceeded the increase in power losses is hazardous for the bearing's durability (lifetime) despite the cooling of the bearing (reflected in the falling temperature of the latter). This is confirmed by the empirical graph drawn by Palmgren (Figure 3.8) for bearing 6204 and the extensive investigations carried out for many bearings by J. Jedrzejewski, W. Kwasny, and J. Potrykus.

If the optimum rate of flow of oil fed to the bearing is reduced below the rate which ensures minimum power losses, the oil film becomes discontinuous, power losses sharply increase, and the bearing quickly seizes up.

Very high pressures and elastic deformations of the mating parts' materials are produced in the convergent lubricating film between the curvatures of rolling elements and races. In addition, absolute oil viscosity (defined by a pressure coefficient) arises. The thinnest critical oil film between the curvatures of mating surfaces amounts to 0.75–0.87 of the rest of the thickness. The elastohydrodynamic film's minimal thickness increases with rotational speed and with the deformation of the rolling elements and the bearing rings while the load slightly decreases, which is described by relevant parameters in electrohydrodynamics (EHD) theory. The oil's viscosity has a significant effect on the film's thickness. The load has little effect on the latter, since as the load increases so does the viscosity of oil in the loaded zone of the Hertz contacts. It is essential that the roughness of the mating surfaces of the bearing components be low.

The friction moment due to loading with forces is described by the relation

$$M_1 = f_1 \cdot \Gamma \cdot d_{\mathrm{m}}, \mathrm{N} \cdot \mathrm{mm} \tag{3.18}$$

where
   f_1 is a coefficient dependent on the type of bearing and its relative load
   Γ is an equivalent load taking into account the magnitude and direction of the forces loading the bearing and of the internal forces

If transverse (radial) force F_r or longitudinal (axial) force F_a acts on the bearing, then the value of this force is substituted for Γ. If both the transverse and the longitudinal forces act on the bearing, then equivalent load $F_e(F_r, F_a)$ (one can find it in catalogs of rolling bearings) should be substituted for Γ.

The bearings which support the spindle (particularly in the front bearing node) operate with negative clearance. Then in approximate power loss calculations, equivalent load F_e is calculated depending on the type of the bearing:

- For the roller bearing:

$$F_e = 5z \cdot L_r^{1.1} \cdot L_w^{0.9}, \text{N} \tag{3.19a}$$

- For the ball bearing:

$$F_e = 0.8z \cdot L_r^{1.5} \cdot d_k^{0.5}, \text{N} \tag{3.19b}$$

where
L_r is the absolute value of negative clearance, μm
L_w is the length of the roller, mm
z is the number of rolling elements
d_k is the ball's diameter, mm

In the case of a complex load generated by a negative clearance and an external load, load Γ is a function of one of the loads, as a rule the one which has a higher value:

$$\Gamma = \begin{cases} F_r \\ f(F_r, F_a) \\ F_e \end{cases} \tag{3.20}$$

In the last three decades there has been a rapid development of designs and technologies of rolling bearings for spindles, resulting in longer bearing lifetimes, higher accuracy, and power loss reduction. This has been accompanied by the development of lubricants for reducing the coefficient of friction and increasing load and temperature variation resistance. The above developments together with economical lubrication techniques are discussed in Section 3.3.3.

The total moment of friction in bearing can be calculated from the values of coefficient f_0 and equivalent load Γ given in Table 3.1

TABLE 3.1
Coefficient f_0 and Equivalent Loads Γ for Spindle Bearings

Type of Bearing	f_0		Γ
	$P = 2/3$	$P = 1/3$	
Cylindrical roller bearings, double row, series number			
NN30 ... K	1	10	F_r
NNU49 ... K	1.5	12	F_r
Angular contact ball bearings			
$\alpha = 15°$	0.5	8	$3.2F_a - 0.1F_r$
$\alpha = 25°$	0.7	10	$1.8F_a - 0.1F_r$
$\alpha = 40°$	1	13	$F_a - 0.1F_r$
Angular contact thrust ball bearings, series number			
234 ..., 2347 ...	1.5	14	$0.4F_a - 0.1F_r$
Thrust ball bearings	1	12	F_a
Taper roller bearings	1.8	18	$9F_a$

P is exponent in formula for M_0 value calculation.

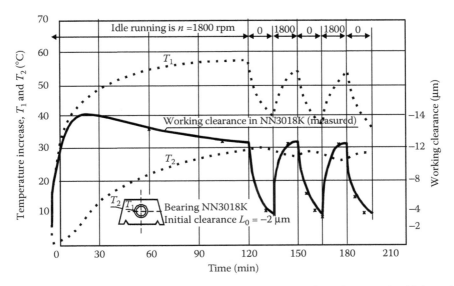

FIGURE 3.9 Relationship between bearing temperature and working clearance by high variation in rotational speed.

Running clearance in the spindle bearing, i.e., the latter's tension, has an effect on both the power losses and the bearing's operating stiffness and so indirectly on the dimensional-shape accuracy of the workpiece.

Running clearance changes with the rotational speed of the bearing, even if the speed is low and changes little. Immediately after an increase in rotational speed, the running clearance increases with time. This can be explained by the quick deformation of the rolling elements, the internal ring and the spindle, and the slow deformation of the housing, which usually has a large mass and a high heat capacity.

Running clearance reaction to large changes in rotational speed, e.g., caused by spindle stopping, and to changes in headstock temperature close to the bearing and at a large distance from it over time is illustrated in Figure 3.9 which is based on measurements. After assembly the tested cylindrical bearing was preloaded with a clearance of $-2\,\mu m$. Such a running clearance reaction to changes in rotational speed is evidence of the low-thermal stability of the headstock unit.

If the working clearance in the bearing changes during its operation, the bearing's stiffness changes at the same time, most dramatically immediately after switching on or when the rate of rotation is increased. After a certain time of operation at a constant rotational speed, both clearance and stiffness stabilize at a lower level. Changes in bearing running clearance and stiffness in the initial period of bearing operation are shown in Figure 3.10. The bearing housing design has a significant effect on changes in bearing stiffness and bearing power losses due to internal and external loads.

If, for example, a double-row cylindrical bearing is housed in the headstock's wall which has a large mass and high stiffness, then such a housing undergoes thermal deformation (as a result of power losses arising in the bearing) at a much slower rate than the spindle, the internal ring, or the rolling elements. Consequently, the tension of the bearing rapidly increases as do the power losses and the temperature of the bearing. The latter may quickly seize up as a result. Such a case of adverse changes in running clearance, power losses, and temperature versus bearing rotational speed is shown in Figure 3.11. This thermal behavior of the spindle bearing/bearings determines the maximum permissible rotational speed for a given spindle bearing assembly structure together with the housing.

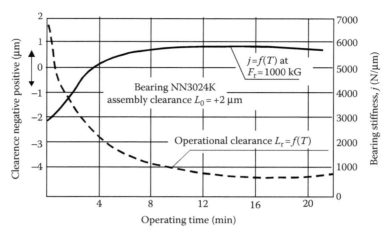

FIGURE 3.10   Thermal changes in spindle bearing working clearance and stiffness over operating time.

In order to increase the critical rotational speed of the bearings in high-speed spindles, the former are embedded in the sleeve which is less rigid than the wall and by radially deforming due to the radial thermal deformation of the bearing prevents too much negative clearance from arising in the latter. Thanks to this solution, high-rotational speeds can be used.

The element which rapidly reacts to an increase in power losses and bearing temperature is the spindle. Undergoing thermal deformation it increases its diameter and thereby increases bearing tension and in consequence, power losses. The high thermal deformability of the spindle is due to its low-heat capacity and hindered heat transfer to the environment. This means that the machine tool designer must know the precise value of the coefficient of thermal expansion of the spindle material and take it into account in a computational analysis of the machine tool's thermal behavior.

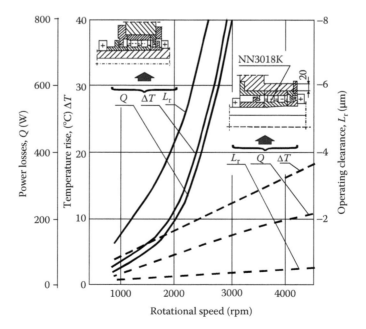

FIGURE 3.11   Influence of housing stiffness on double-row bearing power losses, temperature rise, and operating clearance with increasing rotational speed. (From Jedrzejewski, J., *ASME*, 30, 171, 1998. With permission.)

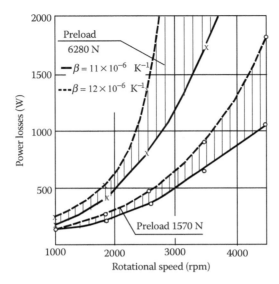

FIGURE 3.12 Dependence of bearing power losses in preloaded angular bearing on rotational speed and thermal expansion coefficient of spindle material. (From Jedrzejewski, J., *ASME*, 30, 171, 1998. With permission.)

How much this coefficient can contribute to the power losses at different preload values is shown in Figure 3.12. As one can see, the use of a spindle material with a high thermal expansion coefficient results in high tension of the bearings and may considerably reduce their life.

In two-row bearings, also the uniformity of the tension in the two rows has an influence on the stiffness of the spindle assembly and on power losses. In the case of a misfit between the spindle journal cone angle and the cone angle of the mounting hole in the internal bearing ring of type NN…K, a negative clearance may arise in one row of the bearing rollers and a positive clearance in the other row when the internal bearing ring is pressed onto the spindle journal. Then the stiffness of the bearing will be appropriately lower than when the two rows are equally tensioned (Figure 3.13). Also power losses will then be smaller. When as a result of the thermal deformation of the bearing negative clearance arises in both rows, stiffness will increase, but so will power losses.

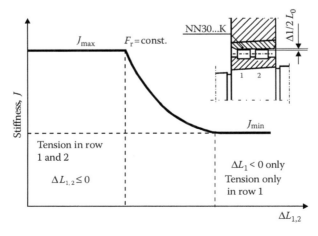

FIGURE 3.13 Influence of clearance difference $L_{1,2}$ on stiffness of bearing type NN…K stiffness.

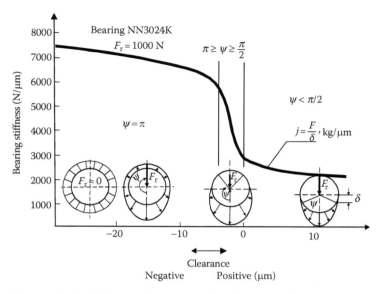

FIGURE 3.14   Influence of spindle bearing clearance and load zone on stiffness.

The thermal behavior of such a bearing node will be variable and unpredictable. Such cases can be prevented by lowering the cone angle fabrication accuracy class. This does not apply to ball bearings used in high-speed bearing nodes with cylindrical internal rings.

One should note that generally when the bearing load is reduced from uniform on the whole circumference to only partially uniform at a negative clearance (Figure 3.14), the bearing's stiffness decreases significantly. If the bearing load zone is between Π and $\Pi/2$, the stiffness varies widely. Therefore, an appropriate preload should be applied to ensure tension in zone $\Pi - (\Pi/2)$ even under maximum external radial load F_r.

A highly complex bearing load distribution occurs in high-speed spindles and in electrospindles. The load distribution can be determined only by means of a precise model. Then the centrifugal forces acting on the rolling elements have a strong influence on tension. The forces relieve the bearing's load.

The forces can reach 200 N and they cause not only an increase in the friction moment produced by the load but also deformations of bearing rings and housings. Bearings for very high-spindle speeds have a special design to ensure their effective lubrication, i.e., despite the centrifugal forces the lubricant should reach the balls/races contact zone.

3.3.3   OPTIMUM CHOICE OF LUBRICANTS AND MINIMAL FRICTION NODE LUBRICATION METHODS

Special lubricating media with suitable operating properties, viscosity and its variation with temperature, load resistance, and lubricating qualities (strength and corrosion resistance), are used to lubricate the critical friction nodes of today's machine tools, such as spindle rolling bearings, linear guide carriages, and ball screws. The viscosity of a lubricant (usually oil) increases with the load carried by the lubricating film. As the viscosity increases so do power losses and the load-carrying capacity of the lubricant. Owing to the constant improvement of lubricating media the latter increasingly perform their lubricating functions better in friction nodes to which they are dedicated. Therefore, when selecting lubricating media one should follow the guidelines provided by the leading manufacturers.

Improved oils for minimal lubrication contain additives which form compounds with the materials of the mating parts, producing wear and seizure resistant coatings on their surfaces. They also contain lifetime and corrosion resistance increasing additives.

TABLE 3.2

Characteristics of Selected Recommended Greases Based on Mineral Oil

	Mineral Oil Grease		Synthetic Oil Grease	
Type/Item	Composite Aluminum Grease	Lithium Grease	Composite Lithium Grease	Base Oil
Lubricant replacement	Awkward	Awkward	Awkward	Awkward
Performance	Good	Good	Good	Good
Maintenance	Easy	Easy	Easy	Easy
Common name	—	Lithium grease	—	Urea grease
Appearance	Butter-like	Butter-like	Butter-like	Butter-like
Dropping point (°C)	>200	170–205	>200	>200
Maximum temperature (°C) for normal use	150	100–130	150	150
Water resistance	Very good	Good	Good–very good	Good
Mechanical stability	Good–very good	Good–very good	Good–very good	Good–very good

One can explore the variety of lubricating media taking as an example the range of greases recommended by THK for use in linear drive systems (Table 3.2). These include greases based on mineral and synthetic oils.

When selecting oil for lubricating high-speed machine tool spindle bearings, one should be mainly guided by viscosity variation with increasing temperature. A too low-viscosity does not ensure the required strength of the lubricating film and results in early wear of the device, and even in its seizure. As the rotational speed of the bearings and the speed of travel of the carriage increase, so does the demand for lubricating medium volume delivered in a unit time. At the same time the sensitivity of the friction moment to the rate of flow of oil to the friction zone increases. This is illustrated for four bearings of a grinder spindle in Figure 3.15.

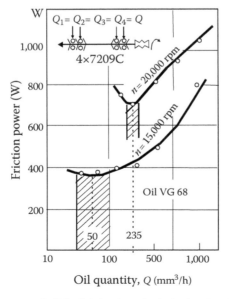

FIGURE 3.15 Optimum amount of oil for lubrication of grinder headstock bearings.

TABLE 3.3

Basic Ways of Lubricating Machine Tool Spindle Bearings

Kind of Lubrication	Application	Permissible Values of High-Speediness Coefficient ($n \times d_m$)	Disadvantages	Advantages
Grease	Slow-speed and highly loaded bearing sets	1.5×10^6	Reduced rotational speeds After assembly grease must be distributed according to special procedure	Intelligent lubricant, self-adjusting minimum optima lubricating film thickness ensuring minimum power losses Simple node design Generally no more lubricant needs to be added Low cost
Oil–air lubrication or with modified oil mist	High-speed bearing nodes	$2.3–3.0 \times 10^6$ upper value for ceramic rolling elements	Complex lubrication system Difficult to maintain oil wedge stability (at high speeds) Difficult optimum lubrication and dispensing and distribution of oil among individual bearings High cost	Ensures elastohydrodynamic lubricating film in each node bearing Lubricating oil does not escape from bearing No pollution environment
Oil injection lubrication	Very high-speed bearing nodes	$3.0–4.0 \times 10^6$ to be used only for ceramic rolling elements because of very high-rotational speeds	Wide oil Splash scatter Difficult to seal Difficult to drain off oil Oil cooling may be necessary Foam forms at very high speeds High costs	Ensures proper minimum lubrication at very high-rotational speeds Additional cooling effect of oil No adverse environmental impact

Grease is able to self-adjust the amount of base oil released in the friction zone. For this reason it is considered to be "intelligent" and finds a wide application. It is usually introduced into bearings for their whole lifetime. But prior to the normal operation of a machine tool, the lubricant introduced into the bearing's excess pocket during the assembly of the spindle unit must be distributed according to a proper time procedure. As the lubricant is distributed the bearing initially heats up very much but when the distribution process ends, the temperature stabilizes at an appropriately low level.

In order to ensure minimal oil lubrication of high-speed spindle bearings and other machine tool components exactly specified amounts of oil (in the form of a continuous streak produced by drawing out oil droplets with an air stream in a strictly controlled way) must be supplied. Such lubrication is called oil–air lubrication and is effected by means of special devices which via special hoses feed oil in the required amounts to each bearing individually. In this method oil droplets dispensed by a distributor at fixed time intervals are stretched by an air jet so as to ensure the required continuous oil delivery (from a few to tens of cubic millimeter per hour) to the bearing friction zone. Oil microstreams reach the particular bearings at a quasiconstant rate of flow required to ensure the minimal lubrication EHD film. Special nozzles at the hoses' ends accelerate the flow of oil, injecting (with a fixed force) microdroplets of oil into the friction zone.

A change in lubricating oil delivery is effected by appropriately setting the feeder of each hose. This system is fully automated and accurately monitored. The overpressure produced by compressed air in the lubricated bearing prevents pollution and seals the bearing. In the United States, oil mist is commonly used for lubrication meeting the minimal lubrication requirements. In this case, an aerosol oil–air mixture produced in a generator is fed at a constant rate of flow into all the bearings.

Both grease and oil–air lubrication is effective and ensures a proper elastohydrodynamic oil film but only up to a certain spindle speed specified by the spindle speediness coefficient: $n \cdot d_{\mathrm{m}}$, where n is the rotational speed and d_{m} is the average diameter of the bearing. As lubricating media are improved the value of this parameter constantly increases.

To very high-speed spindle bearings, oil is delivered by injection, individually to each bearing. Such lubrication is necessary in order to overcome the resistance of the layer of air adhering to the rotating rolling elements and the centrifugal force. Special injection lubrication devices are used for this purpose.

The above methods of lubricating spindle bearings and their limitations are described in Table 3.3. It follows from the table that higher rotational speeds are allowed for bearings with ceramic balls since the latter are lighter than steel balls whereby the centrifugal forces and the coefficient of friction are smaller, which in turn lowers somewhat the sensitivity of the bearings' thermal behavior to variation in the thickness of the lubricating film.

3.4 HEAT TRANSFER IN MACHINE TOOLS

3.4.1 SIGNIFICANCE OF HEAT TRANSFER

The thermal behavior of the machine tool in its operating conditions and the influence of this behavior on machining accuracy depends not only on the intensity of the heat sources, but also on the heat exchange between the machine tool and the environment and the heat transfer within the housings and the whole load-bearing structure of the machine tool. Heat accumulation in spaces closed in by the walls of the housings and by covers affects the machine tool components' heating up cycle and rate and thermal displacement variation. Accumulation of large quantities of heat close to intensive heat sources in the machine tool may very adversely affect its geometrical accuracy (thermal displacements of assemblies) and the lifetime of the friction nodes (particularly spindle bearings). This means that heat transfer conditions should be rationally shaped already during the design of the machine tool. Using the machine tool heat model and running numerical simulations of the heating

TABLE 3.4

Characteristics of Basic Materials Having Influence on Machine Tool Thermal Behavior

Parameters	Cast Iron	Resin Concrete	Steel for Welding	Aluminum	Glass
Density, ρ (kg/m³)	7340	2210	7830	2700	2700
Thermal conductivity coefficient, λ (W/m K)	75	2	43	170	0.8
Specific heat, C_p (kJ/kg K)	0.43	0.7	0.473	0.963	0.67
Thermal coefficient of expansion, β (μm/m K)	12	12	11	20	11

up and deformation of the machine tool, one can design optimum heat transfer by selecting suitable materials (with favorable thermal conductivity) and creating proper conditions of heat transfer through the joints between the machine tool components and housings, through natural convection or convection forced by the linear or rotary motion of elements, and through controlled heat abstraction by cooling overheated places. In forced cooling, the thermal properties of the fluids and gases used, e.g., oil, air, emulsion, water, etc., play a significant role.

In order to ensure that the machine tool's thermal behavior (and so its geometrical accuracy during the work cycle) is correct and replicable, one must properly match the materials of the mating elements so that any differences in the thermal expansion coefficients do not result in strains or stresses as the temperature changes since they could affect the machine tool's geometry. It is also important to shape the intensity of heat transfer through the joints between elements made of different or the same materials. Before a detailed analysis of heat transfer in machine tools it is worthwhile to have a close look at the thermal properties of the basic materials currently used for their construction. These are shown in Table 3.4.

3.4.2 NATURAL HEAT TRANSFER

In order to optimize the heat transfer within the machine tool structure and between the latter and the environment and thus minimize the heating up and thermal displacement of machine tool components, one must have a detailed knowledge of the heat exchange conditions which develop in a natural way through conduction, convection, and radiation. This knowledge is also needed to create a thermal model of the machine tool.

The following are used in the description and analysis of heat transfer:

- Specific heat of materials C_p, J/kg K
- Thermal conductivity λ, W/m K
- Surface film conductance α, W/m² K
- Heat flux density q, W/m²
- Radiation constant C_s, W/m² K⁴
- Temperature conduction coefficient $a = \lambda /(g \cdot C_p)$, m²/s ($g$ is the density)

3.4.2.1 Heat Conduction

Heat energy is transferred between the particles of solid bodies via the elastic vibration of the crystal lattice, and in metals through the flow of electrons. Heat transfer in liquids and gases results from the collision of their molecules. The density of the heat flux transferred through conduction toward a temperature drop is proportional to the temperature gradient and thermal conductivity λ.

Heat conduction processes in which temperature varies over time (whereby accumulation takes place) are described by Fourier's differential equation, which is the basis for modeling the nonstationary thermal processes which occur in machine tools:

$$\frac{\partial T}{\partial t} = a\left(\frac{\partial^2 T}{\partial^2 x} + \frac{\partial^2 T}{\partial^2 y} + \frac{\partial^2 T}{\partial^2 z}\right) \tag{3.21}$$

where

$$a = \frac{\lambda}{C_p \cdot g}$$

In the case of steady and one- and two-dimensional states the equation reduces itself to

$$\frac{\partial^2 T}{\partial^2 x} + \frac{\partial^2 T}{\partial^2 y} + \frac{\partial^2 T}{\partial^2 z} = 0 \tag{3.22}$$

On the surfaces of machine tool components heat transfer via conduction takes place through joints, gaps, coatings, and toward foundation. A drop in temperature in a material depends on the latter's heat transfer coefficient λ and in the contact layer on the thermal resistance of this layer (contact). For a contact layer the contact zone's equivalent heat transfer coefficient λe is used. An approximate value of the equivalent heat transfer coefficient can be calculated from the relation

$$\lambda e = \frac{R_{z1} + R_{z2}}{R} \times 10^{-6} \tag{3.23}$$

In the case of precise calculations good results are obtained when one uses contact resistance calculated from the relation

$$\frac{1}{R} = \frac{6.67 \times 10^4 \left(\dfrac{p}{\text{HB}}\right)^{0.75}}{\dfrac{R_{z1}}{\lambda_1} + \dfrac{R_{z2}}{\lambda_2}} + \frac{5.65 \times 10^4}{\left(\dfrac{R_{z1} + R_{z2}}{\lambda_p}\right)^{0.4} + \left[1 + 0.178\left(\dfrac{K}{\lambda_p}\right)^{0.25}\right]} \tag{3.24}$$

The above relation is suitable for analyzing the effect of intermediate layers whose function is to decrease or increase the heat flow to appropriate machine tool components or areas.

According to the relation, the heat resistance of the contact mainly depends on contact layer roughness R_z and on unit pressure P in the contact zone. Also the kind of contacting material plays a certain role.

The relation between thermal resistance and unit pressure used for matching steel and cast iron elements with milled and ground contact surfaces is shown in Figure 3.16.

It is apparent that under low unit pressures, the type of material and that of machining play a significant role, whereas under pressures higher than $150\,\text{N/cm}^2$ their influence is negligible. When a material with high heat conduction, e.g., aluminum, is introduced into the contact layer, the thermal resistance of the latter significantly decreases. Also a lubricant introduced into the contact zone significantly reduces the thermal resistance of the contact. Whereas ceramic materials and other insulators introduced into the contact zone may significantly reduce the contact's thermal conductivity and prevent the machine tool components separated by such a contact from components that heat up

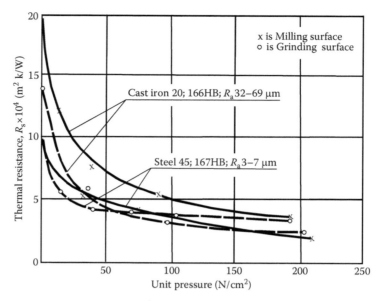

FIGURE 3.16 Thermal resistance versus unit pressure for steel or cast iron for milling and grinding contact surface.

very much from overheating and deforming. The effect of contact heat resistance on temperature distribution in the joint between two machine tool bodies, e.g., the headstock and the bed, is illustrated in Figure 3.17.

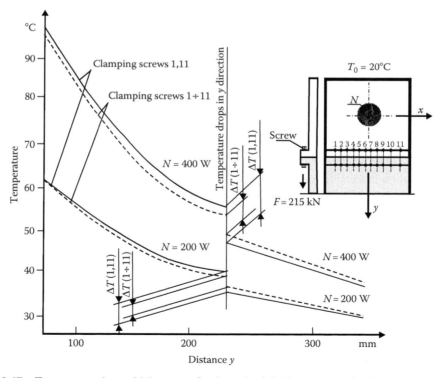

FIGURE 3.17 Temperature drop with heat transfer through a joint between two bodies.

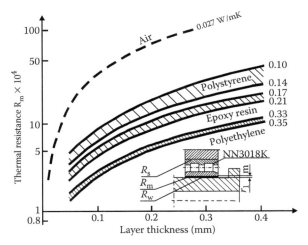

FIGURE 3.18 Thermal resistance of insulating layer versus thickness and conductivity. (From Jedrzejewski, J., *ASME*, 30, 174, 1998. With permission.)

A temperature jump (dependent on the unit pressure) appears in the screw joint between the walls of the two bodies. If the latter are joined by only the outermost screws the temperature jump is high but when all the screws are in place the jump is much lower.

Another case in which contact heat resistance may determine the machine tool's thermal properties is the resistance of the contact between the bearing and the spindle. The spindle's heat capacity is low and when it heats up and its diameter increases, the negative radial running clearance in the bearing (i.e., the latter's tension) increases. The bearing is a large source of heat which when transmitted to the spindle may produce such an adverse effect. If a material layer with high-heat resistance is introduced into the contact between the internal bearing ring and the spindle, then power losses, running clearance variation, and operating bearing temperature will be significantly lower and bearing lifetime will increase considerably.

Figure 3.18 shows the effect of different materials and interlayer thickness m on the heat resistance of the bearing/spindle contact. One can see that an effective increase in contact resistance R_m can be achieved already at an interlayer thickness of 0.2 mm. R_s and R_w are natural resistances of the bearing/housing/spindle contact.

Figure 3.19 shows possibilities of reducing the operating bearing temperature and running radial clearance for cylindrical bearings. If a natural contact with a heat resistance of 0.8×10^{-4} W/(m²·K) is replaced with a contact of 0.4-mm thick interlayer made of epoxy resin, the bearing's sensitivity to an increase in rotational speed greatly decreases and the limit rotational speed of the bearing increases significantly.

The curve representing the natural resistance of the direct contact between the internal ring of bearing NN3018K and the spindle clearly indicates the limit rotational speed at which the bearing is at risk of seizure. If the contact's thermal resistance is increased, the limit speed shifts appreciably toward high values. Ceramic materials, characterized by good adhesion to the base, are highly suitable for increasing contact heat resistance.

A similar but smaller (due to lower losses and a different bearing tension mechanism) effect of the heat resistance of the bearing/spindle contact occurs in the case of ball bearings.

3.4.2.2 Convection

Heat from the machine tool's walls to the environment is transferred through convection and radiation. The increase in wall surface temperature is directly proportional to heat flux density q

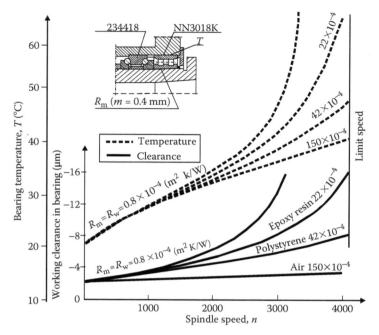

FIGURE 3.19 Influence of high-resistance contact layer on working clearance and heating up of spindle radial bearings. (From Jedrzejewski, J., *ASME*, 30, 174, 1998. With permission.)

and inversely proportional to the sum of convection (α_k) and radiation (α_r) heat exchange coefficients:

$$\Delta T = \frac{q}{\alpha_k + \alpha_r}, ^\circ C \tag{3.25}$$

In the case of convection, heat is transferred through heat conduction in a fluid, mainly with its flow. In forced convection the fluid flows owing to differences in external pressure, according to the Bernoulli law. The flow is then independent of the temperature distribution in the fluid and proceeds toward a pressure drop. In the case of free convection, the flow is caused by a temperature difference.

Heat transfer and friction processes at the housing material/fluid interface and the fluid/gas interface proceed only in a thin thermal or hydraulic near-wall layer. Apart from this layer the flow is undisturbed, free of friction and with a constant temperature distribution. The thickness of the layer increases with the flow (direction) and decreases as the temperature difference and the flow velocity increase. At the interface the velocity of the fluid is equal to zero.

The temperature gradient near the wall is approximately inversely proportional to the thickness of the near-wall layer. For convection which occurs in machine tools one can generally assume constant physical fluid values. As machine tool environment interference increases so does the natural convection coefficient, e.g.,

- In laboratory conditions, 0%–10%
- Due to air-conditioning operation, 10%–20%
- Due to draught, 20%–30%
- In production conditions, 30%

If the temperature of the heat transferring surface decreases, the above effect increases. A special case of convection occurs in vertical gaps. Heat transfer in closed spaces occurs as a result of natural vertical convection motions of the fluid. Because of the fluid's finite heat capacity its temperature tends toward the average of that of the surrounding walls. All the special cases of heat transfer through natural convection require a detailed mathematical description. In today's investigations of the thermal behavior of precision machine tools the highly complex heat transfer conditions which occur in them must be analyzed. It is also necessary to study the rate of cooling and heating up of machine tool assemblies.

A thermally ordered state can be expressed by a simple exponential relation describing the variation in temperature increments above ambient temperature T_0 at any point of a machine tool assembly over (cooling) time t.

$$\Delta T = T - T_0 = B \cdot e^{mt} \qquad (3.26)$$

where coefficient mt represents the cooling rate and it is the base quantity describing a thermally ordered state. The coefficient expresses the rate of heating up and cooling of an object according to the relation:

$$mt = \frac{d(\Delta T)/dt}{\Delta T} \qquad (3.27)$$

The cooling/heating up rate is constant over time, irrespective of the location within the object and is determined by

- Heat transferring surfaces
- Shape of the object
- Thermal conductivity coefficient λ
- Temperature conductance coefficient
- Heat transfer coefficient α
- Ambient temperature T_0

Owing to the above one can explain the thermal behavior of complex machine tool structures with many closed spaces.

Figure 3.20 shows an approximate distribution of heat transfer coefficient on variously located machine tool bodies, using the milling machine, as an example, for both natural and forced (by air movement produced by rotating elements) convection, as a function of temperature or the distance from the air movement forcing surface.

As the machine tool operates, the actual distribution of heat transfer coefficient values is highly complex and variable in time, since when forced convection occurs, the paths and velocities of air streams are variable. The complexity of heat transfer conditions in the case of forced convection through the lathe chuck can be illustrated with an example of the distribution of coefficient α on the frontal wall of the lathe headstock. This is shown in Figure 3.21 for a thermally steady state at a spindle speed of 1800 rpm.

One can see that the maximum coefficients α of 35 W/(m² K) occurred at a certain distance from the main heat source, i.e., the spindle bearing. The influence of an axially asymmetric temperature distribution caused by other heat sources and the presence of horizontal wall ribs running perpendicularly to the spindle axis is noticeable. Also the wall surface shape and the lathe bed reaction affected the distribution of α. It became apparent that the distribution of heat flux density q was similar to the distribution of heat transfer coefficient. Figure 3.22 shows the influence of spindle speed on temperature, coefficient, α and heat flux density q at a selected point of the headstock.

Figure 3.23 shows temperature, heat flux, and coefficient over time, after the spindle stopped. Immediately after spindle stoppage, as forced convection decayed and the heat flux transferred to the environment diminished, a characteristic rise in wall temperature occurred and later the dependence between α and q and wall point temperature is observed, which is typical of natural convection.

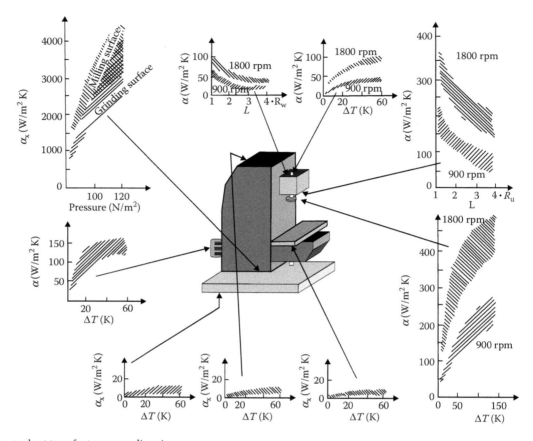

α—heat transfer to surrounding air
α_x—heat conduction to adjacent body or foundation
ΔT—temperature difference between body surface and environment
L—distance from rotating element
R_w—spindle radius
R_u—tool radius

FIGURE 3.20 Heat transfer coefficient values distribution within machine tool structure.

The higher the heat transfer coefficient α, the lower the temperature of the point at which the convection occurs. This means that convection shaping is an important tool for the designer of a machine tool in lowering the level to which it heats up and reducing its thermal deformations. In order to estimate coefficient α_w of convection produced by rotating smooth and ribbed parts, one can use the relations given in Figure 3.24. The relations can also be used to evaluate headstock wall-to-environment heat transfer caused by a rotating lathe chuck or a cutter-head.

Another special case of convection forced by the motion of machine tool parts is convection within the field of action of the belt transmission. This case, described for air motion forced close to the lathe headstock walls by V-belts, is shown in Figure 3.25 where it is illustrated with graphs of the dependence between coefficient α and the distance of the belts from the axis of rotation, the distance of the transmission gear housing from the outer surface of the belts in the pulley, and the distance from the headstock wall as a function of the pulley's rotational speed, the number of belts, and the distance from the axis of rotation.

As it can be seen, the closer one gets to the site where air movement is forced by the transmission gear components, the higher the convection coefficient. The highest convection coefficient values

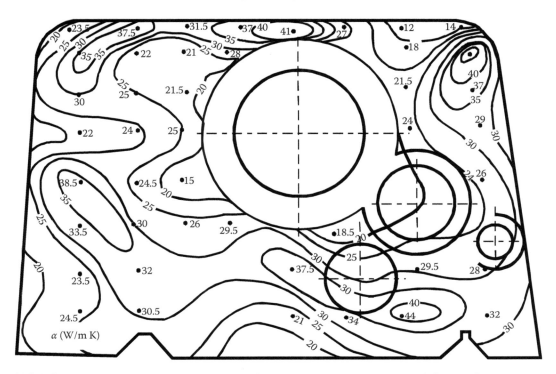

FIGURE 3.21 Distribution of heat transfer coefficient α values on a lathe headstock from wall.

occur for the maximum rotational speed at the shortest distance from the belt pulley. The figure also shows a relation for calculating the convection coefficient with all the needed data explained.

Another graph in this figure shows convection coefficient versus rotational speed. It follows from the graphs that the longer the distance from the belt pulley's diameter, the lower the value of convection coefficient α. The presented experimentally verified relations for coefficient α allow one to calculate the approximate value of the coefficient for any V-belt transmission design.

Recapitulating the discussion of the effect of convection on the heating up of machine tools, one should note that further theoretical and experimental research (using alpha-calorimeters) is needed to precisely describe convection on the surface of machine tool bodies and housings.

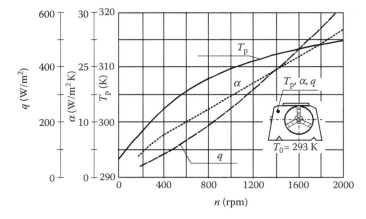

FIGURE 3.22 Influence of rotational speed on coefficient α, temperature T_p, and heat flux density q at selected headstock point.

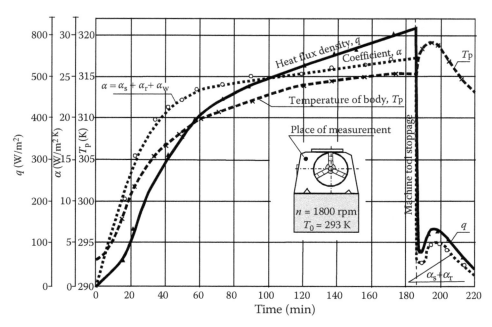

FIGURE 3.23 Changes in heat transfer coefficient α, heat flux density q, and body temperature T_p during machine operation and after its stoppage.

3.4.2.3 Radiation

Bodies whose temperature is higher than $0\,K$ are carriers of electromagnetic waves, which means that they radiate energy. The emitted/absorbed energy depends on the length and frequency of the waves and the absolute temperature of the body. One can assume that at temperatures typical for the surfaces of machine tool components almost all the emission is through radiation with a wavelength of 5–100 μm, i.e., in the infrared radiation range. This energy is called heat radiation.

Radiation processes are very complex and difficult to describe. In the case of machine tools the following simplifications are made:

- Solid bodies and liquids do not transmit radiation.
- Two-state gases, including air, totally transmit radiation (they are transparent).

If the heat emitting surface in different places is characterized by a different radiation coefficient, then $\alpha_r = \Sigma_1^n \alpha_{ri}$ and the following relation for transferred heat flux holds good:

$$q = \alpha_r \left(T_1^4 - T_n^4 \right) \tag{3.28}$$

Then α_{ri} stands for another nominal reference temperature. In practice, equivalent emissivity is used. To provide protection against radiation screens are put up between surfaces radiating at each other.

Emissivity may significantly affect the thermal condition of machine tools, especially when heat from the walls of the machine tool bodies is transferred via free and forced convection, but air movement close to the surface of the walls is less. The rate of radiation from the walls to the environment depends mainly on the emissivity of the heat imparting surface.

In the case of free convection, emissivity coefficient α_r assumes values equal to or higher than the coefficients α_k of this convection, as shown in Figure 3.26 where a solid line represents coefficient α_r for selected emissivities $\varepsilon = 0.5$, 0.7, 0.9, and 1 as a function of surface temperature and a broken line represents natural convection coefficient α_k.

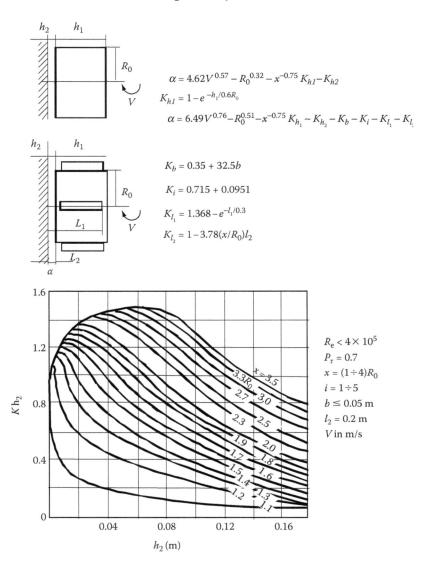

$$\alpha = 4.62V^{0.57} - R_0^{0.32} - x^{-0.75} K_{h1} - K_{h2}$$

$$K_{h1} = 1 - e^{-h_1/0.6R_0}$$

$$\alpha = 6.49V^{0.76} - R_0^{0.51} - x^{-0.75} K_{h_1} - K_{h_2} - K_b - K_i - K_{l_1} - K_{l_2}$$

$$K_b = 0.35 + 32.5b$$

$$K_i = 0.715 + 0.0951$$

$$K_{l_1} = 1.368 - e^{-l_1/0.3}$$

$$K_{l_2} = 1 - 3.78(x/R_0)l_2$$

$R_e < 4 \times 10^5$
$P_r = 0.7$
$x = (1 \div 4)R_0$
$i = 1 \div 5$
$b \le 0.05 \text{ m}$
$l_2 = 0.2 \text{ m}$
$V \text{ in m/s}$

FIGURE 3.24 Relationship for machine body surface heat transfer coefficient α due to ventilation effect produced by rotating elements. (From Jedrzejewski, J., *ASME*, 30, 177, 1998. With permission.)

It is apparent that emissivity coefficient α_r increases linearly while the convection coefficient increases nonlinearly with temperature. It is also clear that much greater quantities of heat can be transferred to the environment via radiation than via convection—emissivity is usually in an interval of 0.5–0.8.

This is corroborated by Figure 3.27 which shows the emissivity of the machine tool body surface (phthalic paint in different colors) versus the temperature of the heat emitting surface. The emissivity of ordinary paints, enamel paints, and varnishes used for coating the surface of machine tools ranges widely. It is specified in manufacturer catalogs (see Table 3.5).

For the particular kinds of paint coatings this coefficient generally is in a range of 0.1–0.2 and it depends on the color, temperature, and condition of the surface. The thickness of the paint coating has practically no effect on the emissivity of the surface.

It follows from the above considerations that the designer of a machine tool can to a large extent shape the intensity of heat transfer from the machine tool to the environment by selecting proper coatings.

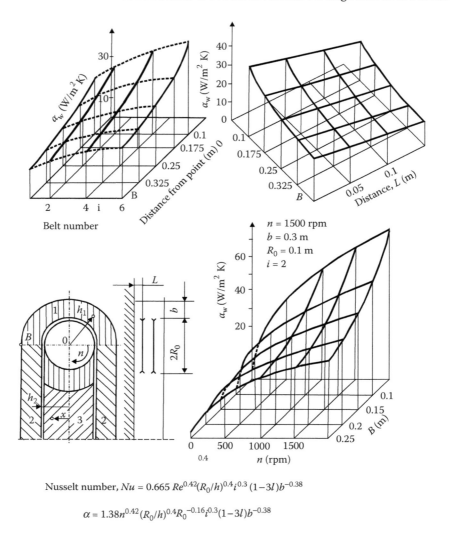

Nusselt number, $Nu = 0.665\, Re^{0.42}(R_0/h)^{0.4}i^{0.3}(1-3l)b^{-0.38}$

$$\alpha = 1.38n^{0.42}(R_0/h)^{0.4}R_0^{-0.16}i^{0.3}(1-3l)b^{-0.38}$$

h is distance from driving wheel axis 0 or symmetry axis X of transmission to point of α definition.

FIGURE 3.25 Influence of selected belt transmission parameters and working conditions on local value of forced convection coefficient α_w. (From Jedrzejewski, J., *ASME*, 30, 178, 1998. With permission.)

The emissivity of oil paints is very high while that of ordinary paints and aluminum enamel paints is low which makes the latter suitable for coating surfaces that should not transfer too much heat to the environment or from the environment to the machine tool. When selecting varnish coatings, one should consider the fact that heat transfer through radiation may balance the changes in the temperature of machine tool bodies due to the thermal resistance of the wall coating. In order to find out if this is the case, one must know the heat resistance value and the coefficients of natural convection and the convection forced by the air motion generated by the rotating parts producing the ventilation effect.

Figure 3.28 shows a complex case of heat transfer which occurs when the surface of the machine tool body is coated with a layer of putty with conductivity λp and then with a layer of varnish with conductivity λl. Temperature drops ΔT, and ΔT, respectively, in the coatings and the near-surface layer, due to the occurrence of α_r and α_k, are shown. The graph illustrates the relationship between

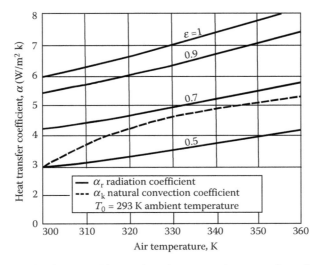

FIGURE 3.26 Heat transfer from machine tool surface to environment through radiation and natural convection.

temperature drop ΔT, and the thermal conductivity of the insulation layer for its different thicknesses S and different transferred heat flux densities q.

3.4.2.4 Forced Cooling of Machine Tools

Forced cooling of machine tool assemblies and components whose thermal deformations significantly affect the accuracy of machining is obviously costly and it is employed when other ways of reducing the temperature are ineffective. In precision machine tools, it is mainly spindle bearings and electrospindle stators, sometimes ball screw nuts and in special cases the screws themselves that are cooled. In very special cases also the bodies and housings are cooled. The required rate of cooling is generally determined using a thermal model and simulations of the thermal behavior of a given machine tool assembly or the whole machine tool. Cooling is costly because cooling units are expensive, the more so, the more precise the two-limit control of coolant temperature. Cooling is a highly effective method of reducing machine tool thermal errors, the more effective, the smaller the temperature control interval. In typical systems, the interval is 2°C and in precision system it is as small as 0.2°C–0.1°C. Oil is commonly used as the cooling medium, less often air, but the most desirable is water with anticorrosive additives.

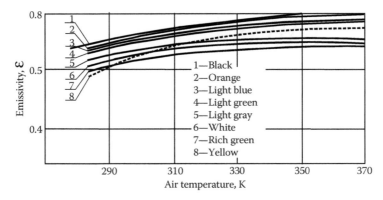

FIGURE 3.27 Dependence of emissivity of phthalate enamel from their temperature and color.

TABLE 3.5

Emissivity of Paints, Enamels, and Lacquers Used for Painting Machine Tools, in Air Temperature Range of 293–350 K

Sort	Emissivity
Aluminum paints	0.25–0.40
Oil paints	0.92–0.96
Phthalate paints	0.70–0.80
Oil-resins enamels	0.65–0.80
Phthalate enamels	0.62–0.78
Cellulose enamels	0.58–0.79
Chlorinated rubber enamels	0.60–0.70
Epoxide enamels	0.60–0.75
Polyvinyl enamels	0.65–0.75
Aluminum lacquers	0.26–0.39
Heat-resistance lacquers	0.90–0.95
Nitrocellulose lacquers	0.60–0.70

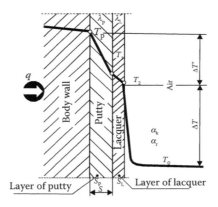

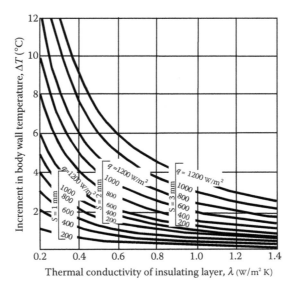

FIGURE 3.28　Influence of insulating layer on body temperature.

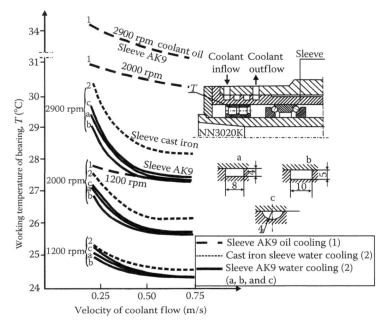

FIGURE 3.29 Effect of spindle unit cooling method on working temperature of bearings.

Figure 3.29 illustrates the effectiveness of cooling the spindle bearing assembly by means of different cooling media. The shape of the channels through which a coolant flows is of secondary importance. Heat reception in the vicinity of outer bearing rings markedly lowers not only their temperature but also the temperature of the headstock parts located further away. The effect of cooling milling machine spindle bearings with ceramic rolling elements on the temperature at selected headstock points is shown in Figure 3.30.

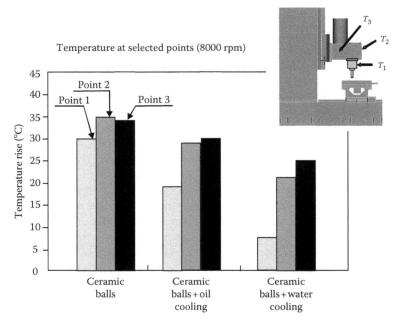

FIGURE 3.30 (See color insert following page 174.) Influence of cooling milling machine headstock spindle bearing on temperature at selected points.

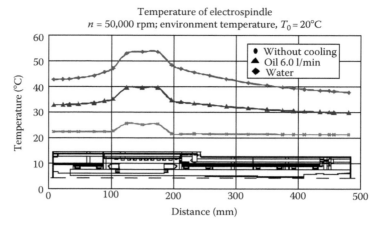

FIGURE 3.31 Effect of stator cooling conditions on temperature of fixed part of electrospindle.

It is apparent that water cooling has a clear advantage over oil cooling. The lowering of temperatures is reflected in the reduction of thermal deformations. In the case of oil, an increase in the rate of cooling—the rate of flow of oil through the cooling channels—causes a marked decrease in the temperature of the bearing housing. Figure 3.31 shows the cooling effectiveness of the high-speed electrospindle in a machining center.

In this particular case, thanks to the water cooling of the electrospindle, its lateral displacements diminished to almost zero while its axial displacements decreased from 100 to about 20 μm.

3.5 MODELING, SIMULATION, AND OPTIMIZATION OF MACHINE TOOL THERMAL BEHAVIOR

3.5.1 Machine Tool Thermal Model Design Fundamentals

The modeling and simulation of the thermal behavior of machine tools constitute a modern tool for analyzing their thermal properties at the design stage. This tool can be used to improve complex structures and minimize thermal errors, especially in the case of precision and high-speed machine tools. The effectiveness of modeling, simulation, and optimization depends on how precise the computational model is, i.e., how precisely it models the heat sources, the heat transfer, and the geometric structure in the machine tool's areas in which thermal phenomena impact the interrelationship between power losses, temperature distributions, thermal displacements, and the mutual displacements of machine tool assemblies.

In the past because of the low computing power of computers modeling was limited to thermally steady states whereby the analyses were simplified and did not take dynamic states into account. Today it is necessary to precisely model thermally unsteady states which predominate in the natural operating conditions of HSC machine tools. It is in thermally unsteady state that machine tool precision is most seriously disturbed. In order to accurately model dynamic states, it became necessary to integrate the modeling of heat sources and other structural machine tool components and take into account the variation of heat intensity and heat transfer conditions. Owing to the fact that the whole machine tool structure and all the heat processes which take place in it are modeled dynamically the model is closer to reality. The integrated modeling can be based on the finite element method (FEM) and a wide range of 3D elements. Another approach which ensures a high precision of modeling heat sources such as spindle and electrospindle bearing assemblies consists in the application of the finite difference method (FDM) to the axially symmetric assemblies and the FEM to the other structural machine tool components with an irregular shape. Such a hybrid model ensures a high precision

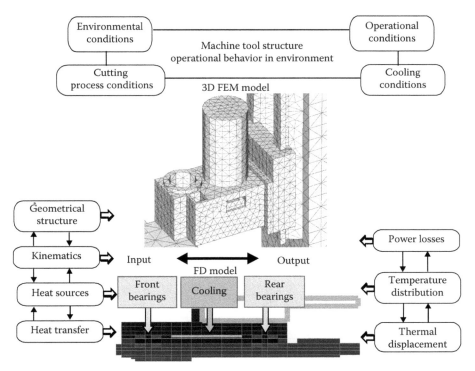

FIGURE 3.32 (See color insert following page 174.)   Scheme of hybrid model of machine tool headstock.

of modeling and of numerical identification of temperature and thermal displacement distributions under variable operating and heat transfer conditions.

A general structure of the hybrid model is shown in Figure 3.32 where the cylindrical components of the spindle assembly are modeled using the FDM and cylindrical elements, whereas the other part of the headstock is modeled using the FEM. Thanks to this approach, one can

- Easily model headstocks of any shape
- Omit complex bodies at certain stages of computational analyses, which greatly speeds up the computations
- Avoid very dense discretization in the field of action of heat sources
- Easily create models of typical bearing assemblies and store them in databases
- Easily analyze dimensional tolerances of bearing assemblies and their loads at high-rotational speeds
- Easily determine and analyze the lifetime of individual bearings

A general structure of the model and the interrelations between its components are shown in Figure 3.33. In this model power losses in the kinematic system components, which depend on the friction node design and the motion conditions, are variable over time and generate variable temperature and thermal displacement distributions.

A general diagram of the procedures used for calculating power losses, temperature matrices $\{T\}$, and displacement matrices $\{P\}$ is shown in Figure 3.34.

Having calculated loads P_1 from the model one can calculate the running clearance in the bearings and power losses Q, which depend on the distribution of temperatures T and loads P in any time interval of machine tool operation. The temperature distribution and the power losses in the links of the kinematic system are obtained by solving the system of equations:

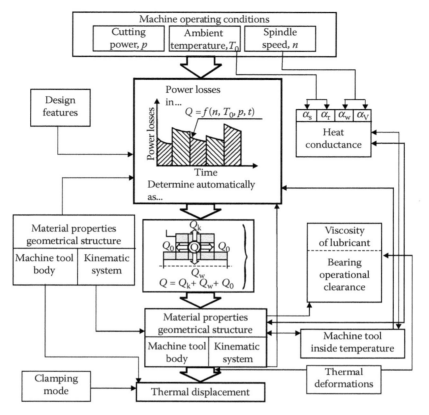

FIGURE 3.33 Schematic structure of model.

$$[K]\{T\} = B \qquad (3.29)$$

in which the matrix of coefficients $[K]$ has this form:

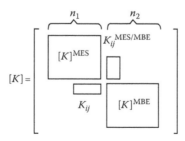

where

$[K]^{MES}$ is the system-of-equations coefficients formed according to the FEM

$[K]^{MBE}$ is the system-of-equations coefficients formed for cylindrical elements modeling the kinematic system

$K_{ij}^{MES/MBE}$ is the coefficients specifying heat transfer through joints between finite elements and cylindrical difference elements

In order to determine coefficients $[K]^{MBE}$, one should make heat fluxes balance for any element i in this form:

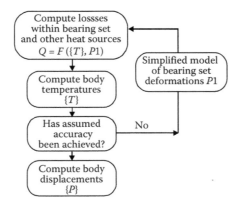

FIGURE 3.34 Procedure for determining temperatures and thermal displacements.

$$\frac{\sum_{j=1}^{p} Q_{ji} + \sum_{k=1}^{r} Q_{ok} + Q_i = cgV(T_i - T_i^{h-1})}{\Delta t}$$ (3.31)

where

Q_{ji} is a heat flux transferred by conduction between element i and the adjacent element j

p is the number of elements neighboring element i

Q_{ok} is a heat flux transferred between the environment and the outer surface of element i

r is the number of outer surfaces of element i

Q_i is the heat released in element i

c is the specific heat of element i

g is the density of element i

V is the volume of element i

T_i^{h-1} is the temperature of element i at instant $h-1$

T_i is the temperature of element i at instant h

Δt is a time difference between instants $h-1$ and h

The particular heat fluxes in the equation are defined by the relations

$$Q_{ji} = \frac{(T_j - T_i)}{R_{ji}}$$ (3.32)

$$Q_{ok} = (T_{ok} - T_i)\alpha_k S_k$$

where

T_j is the temperature of element j

T_i is the temperature of element i

T_{ok} is the temperature of the fluid surrounding element i from the surface k side

α_k is a coefficient of heat transfer on surface k

S_k is the surface area of k

R_{ji} is the resistance of flow between elements i and j

Then one determines

- Heat flow conditions
- Heat balance for the elements

and one performs other model completing and verifying activities. The matrix for cylindrical elements is solved in a similar way except that the heat sources are modeled together with the whole headstock structure.

3.5.2 SIMULATION AND ANALYSIS OF MACHINE TOOL THERMAL BEHAVIOR

A computing system based on this original model, supported with procedures aiding machine tool discretization and speeding up computation of complex geometric structures, is a tool which is highly suitable for the analysis of the thermal behavior of machine tools. An exemplary application of such a computing system is shown in Figure. 3.35.

Simulations of the thermal behavior of a machine tool or any of its assemblies can be run for any design assumptions, load conditions, and work time. Through an analysis—based on a precise thermal model—of the thermal behavior of a machine tool or its component one can discover the causes of the incorrect behavior of machine tool prototypes and easily rectify design and fabrication errors. The errors manifest themselves as differences between the simulated and measured temperature and displacement (deformation) distributions. The machine tool's load-bearing structure alone or together with covers and auxiliaries can be analyzed. Many of the figures shown here were the result of simulations and were used to examine the effect of different factors on the thermal state and displacements of individual structural nodes in order to improve the latter.

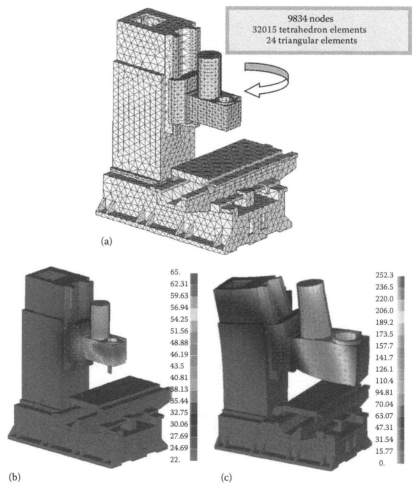

FIGURE 3.35 (See color insert following page 174.)   Example of computing system application to milling center thermal behavior analysis: (a) discretization of a milling center (spindle assembly discretization is not shown), (b) a numerically determined temperature distribution for a steady state at a spindle speed of 8000 rpm and an ambient temperature of 22°C, and (c) the deformations and displacements of the center.

3.5.3 MACHINE TOOL THERMAL BEHAVIOR OPTIMIZATION AND ITS EXAMPLES

Generally speaking, machine tool thermal behavior optimization consists in the minimization of machine tool power losses, heating up, and thermal displacements. In order to mathematically optimize the thermal behavior of a machine tool, one should formulate objective functions, optimization criteria, decision variables, and constraints. Through mathematical optimization based on a thermal model of the machine tool, one can get an insight into all the thermal processes taking place within the machine tool structure. Since it is necessary to repeatedly compute the objective functions of the thermal model, the latter must be simple enough to ensure that the time of a single objective function computation is short. Only then the optimization can be effective usually for the very large number of objective function computations. Such a simplified block diagram of computations is shown in Figure 3.36.

In most cases computations are limited to

- Thermally steady states
- Step-by-step optimization of individual assemblies

and the optimization procedures depend on how much time computations in the particular optimization steps take.

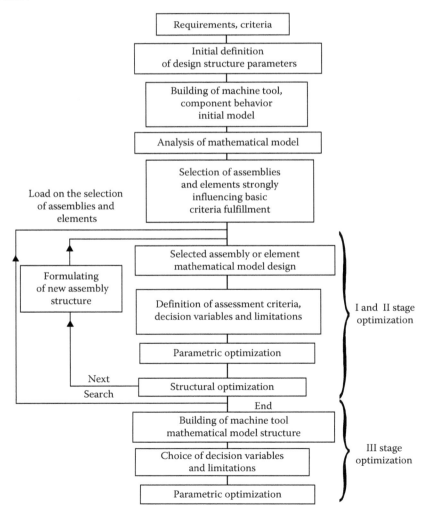

FIGURE 3.36 Optimization algorithm operations.

The search for the minimum heating up of machine tools is fast and does not impose any time constraints. The search for minimum thermal displacements is very time consuming since it must take into account the whole complexity of machine tool geometry and many variables dependent on the machine tool design parameters. The higher the required machine tool precision, the more the complex FEM model and longer the objective function computation times.

In the case of a very laborious optimization process, one can first minimize power losses, then the heating up and thermal deformation of the assemblies and finally perform optimization computations for the whole machine tool. A general example of this procedure is shown in Figure 3.37.

In the first step, parametric optimization, consisting in the minimization of power losses in spindle bearing assemblies, is carried out. For cylindrical bearings the constraint of power loss Q criterion is the permissible temperature rise ΔT and the assembly's stiffness C. The variable parameter is bearing

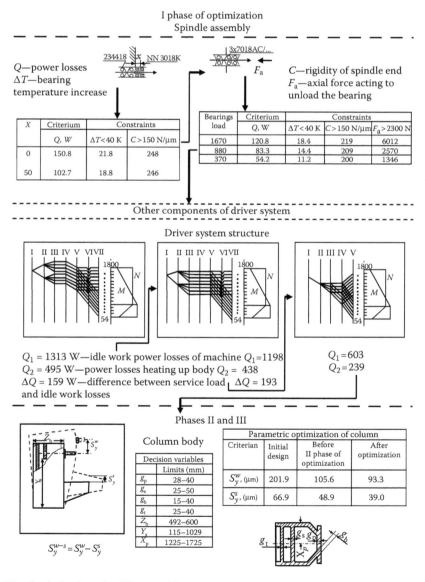

FIGURE 3.37 Optimization of milling machine.

position X. For ball bearings, the variable parameter is their tension while the constraints are temperature rise ΔT, stiffness C, and axial force F_o. In the second step, power losses in the headstock's toothed gear transmissions and other gear transmissions having a bearing on thermal displacements are reduced. In the third step, structural optimization covering the minimization of the thermal displacement of the main machine tool body through the optimization of the heat transfer conditions was carried out. The decision variables were the milling machine body design parameters shown in the figure. The parametric optimization criterion are spindle axis displacement S_y^w and table surface displacement S_y^S in three stages: in the initial state, before the second step of optimization, and obtained after the optimization. In the example, optimization led to a nearly double reduction in thermal displacements of spindle axis and table.

It is also possible to determine the most advantageous conditions for binding (fixing) machine tool bodies in such a way that the displacement of the tool relative to the workpiece is minimum. Other procedures can also be used for the optimization.

In cases when it is the headstock which most affects the thermal displacements of the machine tool it may be sufficient to optimize only the headstock. Then, it is usually enough to optimize the bearing nodes of the spindles with or without forced cooling. Power losses and temperatures can serve as evaluation criteria but thermal displacements are always the main criteria. The same applies to the optimization of the thermal behavior of machine tool bodies. A general chart of machine tool thermal optimization is shown in Figure 3.38.

As one can see, the optimization procedures are based on the sensitivity method, the gradient method, the random method, or experiments. The selection of decision parameters is usually aided by calculations or the use of the sensitivity method in which sensitivity coefficient values indicate the weight of the optimization parameter. The application of this method to the thermal optimization of the milling machine (column) body is illustrated in Figure 3.39.

It is apparent that thermal displacements were mainly determined by the body wall thickness g_s and g_b, spindle axis-base distance y_s, and body width z_b. The largest reduction was obtained in the

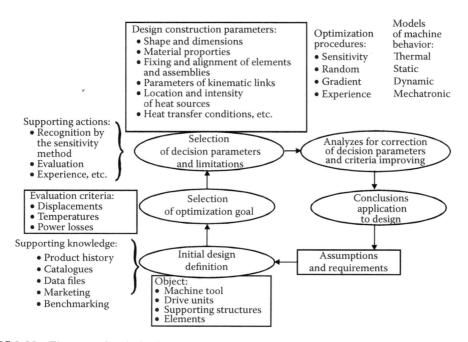

FIGURE 3.38 Elements of optimization process.

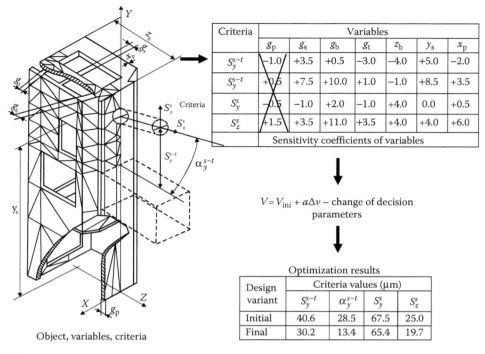

The table in the figure:

Criteria	Variables						
	g_p	g_s	g_b	g_t	z_b	y_s	x_p
S_y^{s-t}	−1.0	+3.5	+0.5	−3.0	−4.0	+5.0	−2.0
S_y^{s-t}	+0.5	+7.5	+10.0	+1.0	−1.0	+8.5	+3.5
S_y^s	−0.5	−1.0	+2.0	−1.0	+4.0	0.0	+0.5
S_z^s	+1.5	+3.5	+11.0	+3.5	+4.0	+4.0	+6.0
Sensitivity coefficients of variables							

$$V = V_{ini} + a\Delta v - \text{change of decision parameters}$$

Optimization results

Design variant	Criteria values (μm)			
	S_y^{s-t}	α_y^{s-t}	S_y^s	S_z^s
Initial	40.6	28.5	67.5	25.0
Final	30.2	13.4	65.4	19.7

Object, variables, criteria

FIGURE 3.39 Phases of milling body optimization supported by sensitivity.

angular displacement relative to the table (α_y^{s-t}) and in the displacement of the table relative to the spindle (S_y^{s-t}). It would be very difficult to achieve this effect by other methods.

Optimization based on the mathematical objective function is laborious and for this reason it is rather seldom used. Nevertheless, the method is very effective. It is anticipated that its use will widen with the development of computing systems.

3.6 DIAGNOSTICS OF MACHINE TOOL THERMAL BEHAVIOR

Diagnostics of the thermal behavior of machine tools and their assemblies generally is a part of the systemic diagnosis of machine tools and belongs to most of the kinds of diagnosis shown in Figure 3.40. The diagnostics can be limited to the identification of temperatures or temperature distributions at selected points in the machine tool structure and to the identification of thermal deformations and displacements of structural components and assemblies. It can be extended to cover identification combined with inference about the causes of given thermal behaviors. Generally, the task of the diagnostics is to

- Provide accurate information about the technical condition of a machine tool for evaluating the correctness of the operating conditions of the particular assemblies
- Enable the prediction of changes in machine tool thermal behavior
- Enable further diagnostic measures and their improvement
- Enable the evaluation of degradation of machine tool assemblies

It is also highly important to diagnose the thermal behavior of machine tools after its assembly during acceptance tests. Then temperature measurements are taken by temperature sensors located (attached to surfaces) within the machine tool structure or remotely by pyrometers.

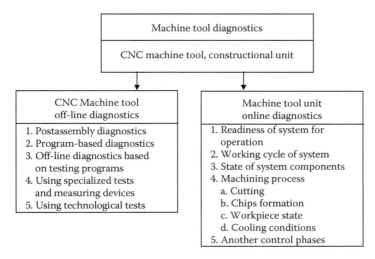

FIGURE 3.40 Kinds of machine tools diagnostics.

In special cases, thermovision cameras, enabling the identification of temperature distributions on the surfaces of machine tool components (idling or being in motion), are used. The results of such measurements must answer the question whether the temperature in a given machine tool point/area stays within the permissible range corresponding to the acceptable variation in the heat sources' output and in heat transfer or the question what exact value temperature reaches at a particular instant or in a given time interval. Precise temperature measurement may be needed to verify calculation models or for the precise compensation of thermal errors. Then accuracy of measurement and the efficiency of processing the temperature signal, so that it is free of excessive interference and has the required form, are important. Precise pointwise identification of temperature values is difficult and requires professional measuring systems and much practical experience. In order to measure temperature by means of touch sensors, one must have knowledge of touch measurement errors. When non-linear sensors are used the problem is much simpler since temperature increments are then measured. In very precision computer numerically controlled (CNC) machine tools and in computerized measuring machines (CMM) the required temperature measuring accuracy is usually 0.1 K. One should bear in mind, however, that the accuracy of temperature estimation is greatly affected by the variable temperature of the machine tool's environment, also in rooms with controlled temperature.

Thermal displacements are measured by means of a wide range of touch or touchless sensors. During such measurements it is important that the machine tool surface at measuring points (e.g., on a gauge plunger, a spindle tip, or the body itself) does not affect measuring accuracy. Under proper measurement conditions the available sensors' measuring accuracy is as high as 100–2 nm, which meets the requirements for precision and superprecision machine tools.

As regards laser interferometers used for measuring thermal displacements it is very important to know what can adversely affect their measuring accuracy in each individual case. When measuring temperatures and thermal displacements, it is extremely important to select a sampling frequency proper for the dynamics of the thermal phenomena. Also signal processing efficiency is critical. Specialized computerized diagnostic systems are helpful in external (acceptance and regular service) diagnostics.

In machine tool, self-diagnostics systems are integrated with CNC control systems and sensors are permanently located within the machine tool structure. This solution offers a possibility of remote diagnosis conducted from the service center of the factory using the machine tool or by the machine tool manufacturer.

Artificial intelligence tools are increasingly used in diagnostic processes, especially in intelligent inference.

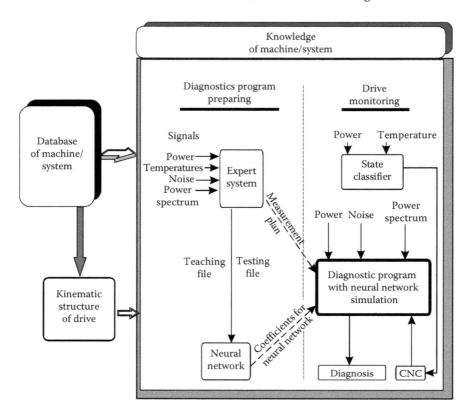

FIGURE 3.41   Generation of diagnostics program and its use in self-diagnostics of machine tool drive.

Intelligent diagnostics is generally integrated with intelligent prognostic monitoring of the machine tool thermal state. Also temperature monitoring by thermovision cameras (which provide a lot of information about the machine tool's thermal behavior since they monitor large areas of it during its everyday operation) is increasingly employed in diagnostics as CCD cameras monitor and record deformations and displacements within large machine tool areas. Using such cameras one can detect many irregularities in time intervals of any length.

Inferences about irregularities in the machine tool's thermal state are always based on a standard state determined experimentally or numerically. It is then necessary to specify the permissible variation in temperature and thermal displacement. Also the number of necessary measurements—the sampling frequency ensuring the required evaluation accuracy—should be specified for each diagnostic evaluation process.

Machine tool diagnosis strategies are constantly developed. More new advanced sensors, better signal processing, measurement automation, and more efficient procedures of inference based on AI and constantly updated knowledge appear. A typical structure of an intelligent machine tool drive diagnosis program based on an expert system and artificial neural networks is shown in Figure 3.41.

The ongoing development of efficient diagnosis strategies contributes greatly to the improvement of machine tool accuracy, the increase in machine tool life and machining accuracy monitoring precision, and the reduction in accuracy maintenance costs.

Diagnostics of the operating properties of machine tools, especially of their drives, needs a precise algorithm of diagnostic activities intelligently aided by knowledge. A general algorithm of this kind is shown in Figure 3.42. Such an algorithm requires detailed knowledge of drive and drive components diagnosis and ways of realizing drive paths and individual drive stages and their effect on the thermal condition. A diagnostic evaluation (including a thermal evaluation) of a drive is the

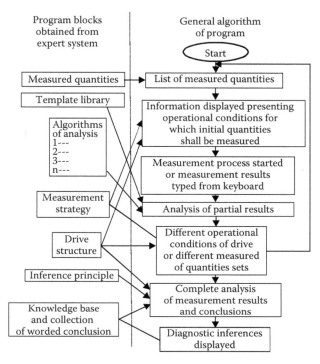

FIGURE 3.42   General algorithm of diagnostic program.

result of a multistage detailed analysis of data acquired from measurements taken in the course of normal operation of the drive in accordance with the adopted diagnosis strategy.

The general development of a diagnosis strategy is well illustrated in Figure 3.43. The first illustrated case (a) concerns diagnosis performed by an external diagnostic system by means of specialized measurement/diagnostic devices. In the second case (b) some of the diagnostic functions are assigned to a

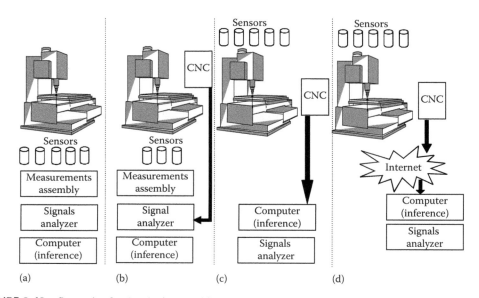

FIGURE 3.43   Strategies for developing machine tools diagnostics.

CNC system, which means that it is a combination of machine tool self-diagnostics and external diagnostics. In the third case (c) diagnostics is totally controlled by the CNC system. Remote diagnostics, which can be very deep and highly effective when machine manufacturer knowledge and expertise are required, is also possible here. In the fourth case (d), it can be realized, for example, through an Internet interface with the manufacturer of a machine or an assembly such as the electrospindle in which thermal processes are particularly complex and where disturbances, which a standard diagnostic system is unable to diagnose, may occur. In case of uncertainty, one should use the diagnostic and repair procedures which only the manufacturer can provide.

Recording and collecting data on the thermal processes taking place in machine tools enables quick diagnosis of the causes of irregularities, even when the machine tool unexpectedly stops working and sending reliable diagnostic signals. As the share of diagnostics and monitoring in machine tool operation increases, the operating cost and the cost of maintaining machining accuracy decrease and the machine tool becomes more sophisticated.

3.7 MODELING AND COMPENSATION OF THERMAL ERRORS OF MACHINE TOOLS

By improving the thermal properties of machine tools and minimizing their heating up through computer analysis and optimization, one can substantially reduce their thermal displacements. In the case of precision and HSC machine tools, however, also thermal error compensation is necessary. The compensation can be based on thermal displacement measurement in real time or the use of a volumetric error model. The former method would be very effective if a technique of accurate measurement of displacements during normal operation of the machine tool were available. Many factors, connected with the machining process, the machine tool's dynamics, and the lack of a necessary reference base for such measurements, make the latter rather ineffective.

Therefore, the only option left is an indirect method based on a model updated through periodic measurement of displacements or continuous real-time measurement of temperatures at machine tool points, representative of the model and located with the machine tool structure. Owing to the ease of measuring temperatures the latter method is now commonly used. A compensation procedure is usually based on the regression function or artificial neural networks. The two procedures produce a similar compensation effect. A thermal errors reduction of 50%–80% can be achieved in production conditions, which in many cases satisfies machine tool manufacturers. But this accuracy is not satisfactory for precision machine tools and better solutions need to be sought. Compensation is carried out using the error function and a compensation algorithm input into a CNC system with a structure open to such additional control tasks.

The problem with thermal errors compensation accuracy arises mainly from difficulties in sufficiently accurate mapping (by means of appropriate functions) of the actual thermal displacements. For real machine tool operating conditions at variable speeds of working motions and heat exchange conditions it is extremely difficult (in fact, still impossible) to mathematically describe dynamic changes in thermal displacements with an accuracy corresponding to the machining accuracy required after error compensation. Furthermore, the function describing an error must be a simplified function so that volumetric error values could be calculated and input into the compensation algorithm quickly enough to conduct compensation in real time. Hence such functions have a form peculiar to a given machine tool and are additionally corrected on the basis of temperature or displacement measurements. If such a function is to be effective, it must embrace not merely one controllable axis, but all the axes, i.e., the entire machine tool working space.

In special cases thermal errors compensation is limited to one axis—the one along which displacements (e.g., axial elongations of the spindle) are particularly large and most strongly affect machining accuracy. An example here can be the compensation of the thermal elongations of the

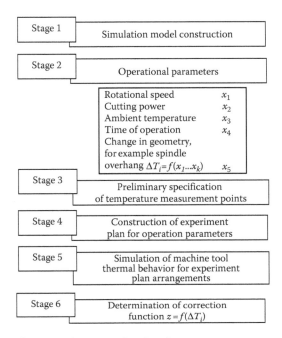

FIGURE 3.44   Algorithm of constructing correction function.

spindle of a horizontal milling–boring machine. Thermal displacements of spindles in such machine tools may exceed 200 μm. A demonstrative algorithm for creating a thermal error correction function is shown in Figure 3.44.

A typical arrangement of preselected (on the basis of a calculation model and simulations) temperature measuring points is shown in Figure 3.45.

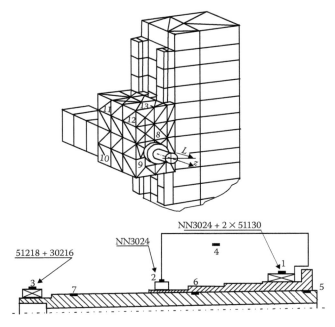

FIGURE 3.45   Boring–milling machine model with marked temperature monitoring points.

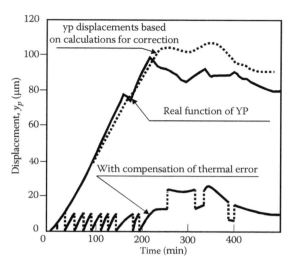

FIGURE 3.46   Effectiveness of correction for technological process.

The number of the points is minimized as a result of the application of the procedures. Typical curves of axial spindle displacements determined numerically and experimentally are shown in Figure 3.46.

The difference between the displacement values expressed by the two curves represents the error distribution after compensation. The figure shows how the error value changes during compensation in variable time intervals. It is apparent that in the time intervals in which the error is large, compensation accuracy significantly deteriorates. The real thermal error will be much larger due to the many disturbances already mentioned as well as errors in the realization of compensation by the CNC system.

It should be noted here that the identification of compensation accuracy (practically possible only through machining trials) is a separate problem. The CNC machine tool's measuring system itself introduces displacement measuring errors, which means that it cannot be used for the verification of compensation. Considering the above, the modeling of the thermal behavior of machine tools for the purpose of error compensation and the modeling of errors themselves must take into account the behavior of measuring systems (linear scales) made both of steel and quartz. In addition, the models should take into account the effect of the reciprocal motions of components and bodies on the distribution of temperatures and displacements. To recapitulate, one should say that very accurate mapping of errors in the model—the principle know-how of each machine tool manufacturer—is crucial for error compensation.

3.8   MAIN TRENDS IN IMPROVEMENT OF MACHINE TOOL THERMAL PROPERTIES

The current development of machine tools clearly indicates that their thermal behavior is and will continue to be very intensively improved. The output of heat sources will be reduced and heat transfer to the environment will be improved. In precision machine tools the precision of cooling the assemblies responsible for excessive thermal displacements will further increase. Online machine tool heating up monitoring and volumetric thermal error compensation based on intelligent procedures will play a major role. Compensation will be based in an increasingly larger scale on using a model of partial errors and of the total time-variable error as a function machine tool loading with machining forces and as a function of ambient temperature. It is anticipated that the virtualization of the machine tool thermal behavior, based on an online updated model will play an increasingly vital role. Improvement in the thermal properties of machine tools to a large extent will be effected through the use of new, more advantageous structural materials and further reduction of the influence of mechanical friction. It is anticipated that as the problems encountered in the design of magnetic bearings are overcome, rolling bearings and the associated rotational speed and lifetime limitations and errors (very difficult to compensate) will be eliminated from electrospindles.

However, the problem of effective and energy-efficient cooling will remain. It is anticipated that heat screens, which will be employed to control the flow of heat streams in a way optimal for the geometrical accuracy of machine tools, will find increasing use.

Further intensive development of precision machine tools with nanoaccuracy of machining will take place. Reduction of thermal errors in their case is critical and requires unconventional solutions with regard to both the design and path and control measurements. The computations necessary in error compensation conducted in the particular steerable axes will have to be supported with processors with ever greater computing power. The currently used hydrostatic water bearings (replacing hydrostatic oil spindle bearings), which being based on special ceramic sleeves with a porous structure, ensure (relative to the oil bearings)

- 70% increase in stiffness
- 40% reduction in temperature rise
- 40% reduction in power lost for friction
- Lower sensitivity to changes in the operating conditions

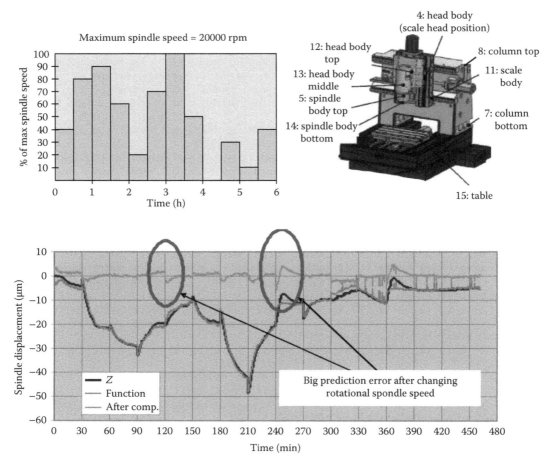

FIGURE 3.47 (See color insert following page 174.)   Experimentally identified and predicted machining center spindle displacement and theoretical compensation accuracy.

and will be further intensively improved. Their improvement sets the trends in the development of the new superprecision machine tool bearing technology. The contribution of basic research and practical applications of laser interferometry to the generation of similar new technologies will increase.

Concurrently with the R&D work on new innovative bearings, new generation HSC spindle bearing assemblies eliminating spindle drift during changes in rotational speed will be developed.

The complexity of thermal error compensation, considering temperature measurement distortions, sampling frequency, and radial spindle displacements (shift caused by spindle length changes and bearing balls contact angle with centrifugal force) for machining center with maximum rotational speed of 20,000 rpm, can be observed in the Figure 3.47. This figure shows displacement change during compound work cycle, in real thermal conditions present in the factory hall. In this figure, a numerically expressed displacement curve is shown, as well as predicted theoretical result of compensation.

In input temperature data (from measurements), distortions, introduced by unidentified measurement errors, appeared after about 300 min, which naturally result with the corresponding distortions in computed displacements. In real-life, temperature cycles are free from distortions, so they have no influence on spindle displacement, but such reaction of a very complex computation model shows high sensitivity. In presented displacement runs, sudden displacement changes and thermal displacement runs are clearly visible. Also, areas are marked, in which it would be very difficult, in numerical method, to achieve high precision of recreating measured displacements.

BIBLIOGRAPHY

1. Palmgren A., *Neue Unterzuchungen uber Energieverluste in Walzlagern*, VDI Berichte, Duesseldorf, Bd. 20, 1957.
2. Palmgren A., *Grundlagen der Walzlagertechnik*, Stuttgart, Frankh'sche Verlagsbuch handlung, 1964.
3. Jedrzejewski J., Kwasny W., and Potrykus J., *Beurteilung der Berechnungs—metoden fur die Bestimmung der Energiverluste in Walzlagern, Schmierungstechnik*, 20, VDI, Berlin, pp. 243–244, 1989.
4. Jedrzejewski J., Directions for improving the thermal stability of machine tools, thermal aspects in manufacturing, *ASME PED*, 30, 165–182, 1988.
5. Jedrzejewski J., Kaczmarek J., and Rejfur B., Description of the forced convection along the walls of machine tool structures, *Annals of the CIRP*, 37/1, 397–400, 1988.
6. Jedrzejewski J. and Buchman K., *Thermisches Verhalten von Weskzengmaschinen—Gestellen*, Industrie Auzeiger, 99, Ig. Nr 65, 1243–1245, 1977.
7. Jedrzejewski J., Kaczmarek J., Kowal Z., and Winiarski Z., Numerical optimization of thermal behaviour of machine tools, *Annals of the CIRP*, 39/1, 379–382, 1990.
8. Jedrzejewski J. and Kwasny W., Artificial intelligence tools in diagnostics of machine tool drives, *Annals of the CIRP*, 45(1), 411–414, 1996.
9. Jedrzejewski J., Effect of the thermal contact resistance on the thermal behaviour of the spindle radial bearings, *International Journal of Machine Tools and Manufacturing*, 28(4), 409–416, 1988.
10. Jedrzejewski J. and Kwasny W., Application of artificial intelligence to routine diagnostics of machine tools, *Proceedings of the International Conference On Computer Integrated Manufacturing*, ICCIM' 91, Singapore, pp. 609–612, 1991.
11. Jedrzejewski J. and Kwasny W., Multi-sensor system for diagnosing machine tools, *Proceedings of the International Conference,* IMTC 92, IEEE Instrumentation Measurement Technology Conference Metropolitan, New York, pp. 194–199, 1992.
12. Jedrzejewski J. and Modrzycki W., A new approach to modeling thermal behaviour of machine tool under service conditions, *Annals of the CIRP*, 41(1), 455–458, 1992.
13. Jedrzejewski J. and Strauchold S., Directions in improving thermal behaviour of spindle bearing assemblies in FMS modules, *Manufacturing Systems*, 23/4, 317–322, 1993.
14. Jedrzejewski J., Kwasny W., and Rodziewicz Z., Strategy and system for monitoring the machine tool main drive, *Manufacturing Systems*, 21(4), 273–276, 1992.
15. Jedrzejewski J. and Modrzycki W., High-speed machining centre modeling and simulation problems. W: Open and global manufacturing design, *Machine Engineering*, 2(2), 205–211, 2002.
16. Jedrzejewski J., Lorek F., and Modrzycki W., Agent-aided modeling of machine tool thermal behaviour, *Machine Engineering*, 3(2), 249–259, 2003.

17. Jedrzejewski J., Kowal Z., Kwasny W., and Modrzycki W., Hybrid model of high speed machining centre headstock, *Annals of the CIRP*, 53, 285–288, 2004.
18. Jedrzejewski J. and Modrzycki W., Numerical analyses and compensation of HSC machine tool thermal displacements, *7th International Conference and Exhibition on Laser Metrology, Machine Tool, CMM and Robotic Performance*. LAMDAMAP 2005. Ed. P. Shore, Bedford: Cranfield University, pp. 268–275, 2005.
19. Grochowski M. and Jedrzejewski J., Modeling headstock motion influence on machine tool thermal behaviour, *Journal of Machine Engineering*, 6(2), 124–133, 2006.
20. Winiarski Z. and Kowal Z., Investigation of the effects of heat generation in HSM electrohead stocks, *Journal of Machine Engineering*, 6(2), 107–115, 1996.
21. Chen J. S., Juan J. X., Ni J., and Wu S. M., Real-time compensation for time variant volumetric errors on machining center, *Transactions of the ASME*, 115, 472–479, 1993.
22. Attia M. H. and Kops L., Non linear thermoelastic behaviour of structural joints—solution to a missing link for prediction of thermal deformation of machine tools, *Transactions of ASME, Journal of Engineering for industry,* 101(3), 348–354, 1979.
23. Iwanicki R., Heat transfer through stationary joint in machine tools, PhD Thesis, Wroclaw University of Technology, 1980 (in Polish).
24. Kwasny W., The concept of diagnostics for machine tool service and supervising needs, *Inżynieria Maszyn*, 9(1), 122–136, 2004 (in Polish).
25. McKeown P. and Corbett J., Ultra precision machine tools. In: *Autonomie Produktion*, Klocke F. and Pritschow G. Eds., Springer-Verlag, Berlin, Heidelberg, New York, 2004.
26. Delbressine F. L. M., Florussen G. H. I., Schijvenars L. A., and Schellekens P. H. I., Modeling thermomechanical behaviour of multiaxis machine tools, *Precision Engineering,* 30, 47–53, 2006.
27. Bossmanns B. and Tu J. F., A power flow model for high speed motorized spindles – Heat generation characterization, *Journal of Manufacturing Science and Engineering*, 123(3), 494–505, 2001.
28. Bukowski A., Grochowski M., and Jedrzejewski J., Hybrid modeling of machine tool with integrated cooling system, *Journal of Machine Engineering*, 8(1), 84–93, 2008.

4 Design Strategies and Machine Key-Components

Samir Mekid

CONTENTS

…the experiment must be so arranged that the effects of disturbing agents on the phenomena to be investigated are as small as possible.

James C. Maxwell

4.1 INTRODUCTION

Precision machining has been continuously evolving over the last decade to reach subnanometer accuracy. Professor Taniguchi has predicted the evolution of the machine tool precision based on its past history as shown in Figure 4.1. It shows, for example, that components will reach a precision close to subnanometer in 2010. Machines have to be designed to achieve better precision than the expected precision for the products, and so they need to achieve better than subnanometer accuracy by 2020.

High-tech products will be able to deliver their full functionality if they are initially made accurately with high material integrity and surface integrity. Figure 4.1 shows a tendency towards too tight specifications on the new products thereby necessitating excellent design methodologies. Table 4.1 shows some examples of ultraprecision machining of optics requirement in current applications and products.

	1920	1940	1960	1980	2000	2020 (Extrapolated[1])
Normal machining	—	60	30	5	1	0.1
Precision machining	—	75	5	0.5	0.1	0.03
High-precision machining	75	5	0.5	0.05	0.01	>0.003
Ultrahigh-precision machining	5	0.5	0.05	0.005	0.001	<0.3 nm

————— Precision machining   ————— High-precision machining   ————— Ultrahigh-precision machining

FIGURE 4.1 (See color insert following page 174.) Evolution of machine required precision over the last century. Extrapolation from Prof. Tanigushi's graph, University of Tokyo. Values are indicated in micrometer.

4.2 DESIGN STRATEGY FOR STANDARD SIZE MACHINES

Various approaches to design mechanical systems are established and could be summarized as follows: deterministic approach by creating and modeling, axiomatic design, robust design known as Taguchi's method, systematic design, customer-centered product, and systematic design approaches. The aspects of virtual design could also be introduced to ease the design process and for economic reasons with less or no prototypes to check.

Virtual machine tool design encompasses the design, the test, optimization, control, and machine parts in a computer environment. The machine is usually designed in a CAD environment. The CAD model of the machine is exported to finite element analysis (FEA) for structural analysis, control loops depending on the speeds, and accelerations with the interaction between the CNC control model and machine itself.

TABLE 4.1
Optical Precision Requirements

	Wavelength λ (μm)	Contour Deviation λ/10 (μm)	Average Surface Roughness λ/100 (μm)
IR light	~3	~0.3	~0.03
Visible light	~0.5	~0.05	~0.005

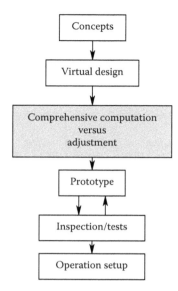

FIGURE 4.2   Virtual design strategies.

Figure 4.2 shows the step where virtual computation helps in readjusting the chosen design before the manufacturing of the prototype. The inspection tests will show how much adjustments are required on the prototype but expected to be minor.

4.2.1 DESIGN METHODOLOGY FOR STANDARD SIZE MACHINES

With the increasing demand for ultraprecision systems, the standard machine design methods are no longer appropriate to design precision machines and satisfy tight requirements. Hence, for a requirement of submicrometer precision in positioning, for example, the design strategy will not be based on standard methods. It may also appear that other key design parameters have to be identified and considered. The consideration of all physical phenomena usually assumed to be of second-order problems becomes extremely important. The design methodology will encompass a few critical steps such as the choice of design concepts with higher stiffness. It is important to investigate whether identified phenomena on the machine are affecting the overall accuracy separately or are coupled. The methodology will maximize the probability of success by reducing repeatable and nonrepeatable errors. Slide motion errors have to be measured and initially evaluated in terms of any Abbé* offset. The sequence design–analysis–performance is used as a continuous iteration with respect to the performance quality/cost ratio.

The main key design *principles* that will be required next are as follows:

1. High structural stiffness, damping, and stability of the machine structure [1]
2. Kinematic design between bearings and guideways
3. Minimization of known errors induced by the system such as Abbé errors
4. Use of direct displacement drives

---

* If errors in parallax are to be avoided, the measuring system must be placed coaxially with the axis along which the displacement is to be measured on the workpiece. (Abbé error definition.)

5. Suitable metrology requiring direct measurement to avoid Abbé errors and using sensors with appropriate resolution, repeatability, and time response
6. Control-securing high axis stiffness, high response, and high bandwidth
7. Error and/or force compensation [2]

The census of the factors that may affect the performance is necessary for an eventual minimization or compensation. The influence of the second-order physical phenomena such as microdynamics generated by rolling friction or electronic nonlinearities is dominant in high-precision behavior since they generate random errors that induce parasitic forces. Once these phenomena are identified, they have to be modeled and analyzed to determine whether they are acting separately or coupled. The unmodeled phenomena may be compensated for with a dedicated servo-control system. To reduce these types of errors as shown in Figure 4.4 that may affect the uncertainty in a machine, a design method was built and applied to designing several high-precision mechanical systems [3,4], and in other industrial applications whose reports are not yet available for the public domain. In this global design strategy, the output variables are to be set according to the specifications, while the input variables are actually related to the large number of parameters affecting the accuracy.

A design plan for precision machine design is presented in Figure 4.4. It shows logical steps and an attempt to address the majority of the errors described previously depending on the type of the machine to be designed. The conceptual design is an important step to satisfy all the required functionalities. The machine is divided into subsystems to be designed accurately by defining the design criteria derived from the main specifications. Each subsystem is evaluated and optimized to meet its own specifications. The designer has to impose meticulous criteria when selecting/designing actuators and bearings to generate precise motion. It is obvious that the performance of the whole system is reduced to the weakest element that contributes directly to the quality of the movement. The subsystems are linked together using congruent couplings (nonrigid and authorizing small lateral motion) at the interfaces to avoid static and dynamic interactions that may generate parasitic movements. In the next step, stress analysis and evaluation of all static errors are performed with the static determination for different source established with careful scrutiny followed by the evaluation of dynamic errors. The dynamic modeling of the whole system permits at this step to determine the eigen frequencies of both the parts involved in the generation of motion, or those parts of the machine structure. The first eigen frequencies have to be out of the servo-control bandwidth to avoid any excitation of the vibration mode that may degrade the accuracy of the considered axis. The next concern is to ensure that the machine working axis has a very high-axial stiffness. This means that any disturbing axial force (e.g., δF) such as friction will induce very little axial error (δx), hence, $x = \delta F/K$ where δx tends to be neglected if K is very large. The dynamic analysis will then be expanded to estimate the system response for the sake of correcting nonrepeatable errors, assuming that the dynamic decoupling has been achieved. At this stage, a budget of errors is required. It states how much uncertainty and nonrepeatability can be tolerated within the chain of components being manufactured. Finally, the validation of the precision system is then achieved applying procedures described in the international standards.

Homogenous transformation matrix is applied to simulate the influence of the actual errors on the required accuracy. Particular second-order errors that could be modeled would be added to the budget of errors, while in case of unmodeled errors, specific controllers [2,4] will be applied for eventual error compensation as detailed in the general design philosophy to maximize the probability of success by reducing repeatable and nonrepeatable errors and, hence, specifying tolerances of the machine components. Finally, the success of the machine servo control depends on the type of control applied and the interaction agility between subsystems. The validation of the precision system is then achieved by accuracy assessment using procedures described in the international standards such as ISO 230-1 to 230-5 for machine tools.

It is important to underline that a deterministic approach has to be systematic during the process of design. At this level of precision, any modification or change of properties in any element could have a mutual influence with another element and hence inducing an overall influence.

Figure 4.3a suggests a machine with some known source of errors that could be compensated for by various means. Figure 4.3b shows a machine with compensated errors. Compensations techniques are discussed later.

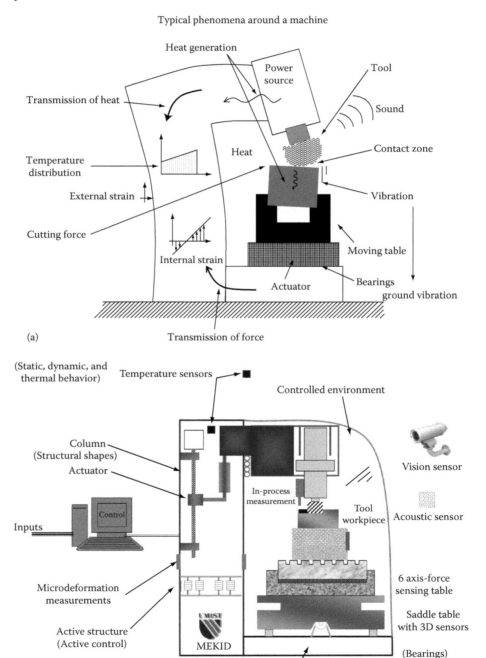

FIGURE 4.3 (a) Typical phenomena affecting the machine precision and (b) **(See color insert following page 174.)** possible suggestions to address a few problems.

4.3   REVIEW OF KEY PRECISION ENGINEERING PRINCIPLES EMPLOYED

In 1922, Pollard described the mechanical design of scientific instruments [5], and, 7 years later, introduced the kinematic design of couplings in instrument mechanisms [6]. It is only during the last three decades that enormous efforts have been made to introduce as complete as possible the multi-disciplinary activities that govern the precision engineering field. McKeown [1] has suggested 11 principles in the design of precision machines. Teague and Evans [7] have set patterns for precision instrument design. Since 1992, a few books have been published on precision machine design [8–11]. With all those principles, the main objective in designing precision systems is to secure a lower budget of the volumetric errors and minimize as much as possible errors that may affect the precision of the machine axes at all modes of operations.

4.3.1   DESIGN PRINCIPLES

Fundamentally, more attention has been implemented with the introduction of Abbé error by Ernst Abbé in 1890 with various aspects in machine tools and instruments introduced by Bryan in 1979. Many principles have been defined for the design of precision systems. With the advent of recent tight specifications and advanced requirements, these methods, although used as principles, become insufficient as systems require more verifications and comprehensive simulations. The following are most of the principles required in the design of a precision system.

Conceptual Design Analysis

- Kinematic semikinematic design, Abbé principle or options

Structural Analysis

- Structure Stiffness
- Balance of force, damping, and dynamic stability over machine axes
- Thermal drift stabilization and compensation

Machine Elements

- Carriage noninfluencing drive coupling and clamping
- Position of drives on axes of reaction
- Bearing averaging, friction, wear and effect of temperature
- Direct drive/actuation

Sensing and Metrology

- Sensing system independent from machine distortions

Control

- Servo-drive stiffness, position loop synchronization
- Error compensation techniques

4.3.2   MODELING AND SIMULATION

FEA is an efficient simulation technique used in engineering analysis. It uses numerical techniques called finite element methods (FEM). FEA should be considered with a number of conditions related to the definition and assumptions made for the elements used in the meshing (usually neglected by the users) to avoid inaccuracy and sometimes wrong computation results. The elements of FEA found in the software library are programmed to simulate physical governing behaviors in linear, nonlinear, thermal, and many other aspects. The elements written for two dimensions and three dimensions have successfully passed their respective benchmark tests; hence, there are precautions to consider when selecting elements and these are element numerical assumptions, condition of uses, boundary conditions, and meshing techniques. FEM may be considered for various physical aspects encountered in the design depending on the application.

4.4 STEPS IN THE DESIGN ROADMAP

The overall design procedure of a machine as shown in Figure 4.4 usually encompasses three-core steps that are concepts analysis, kinematic of motion including static, dynamic, and thermal analysis of the structure, and finally which suitable control strategy is to be used. The choice of an appropriate controller that could handle the specifications in positioning, speed, and acceleration will also depend on its ability to resolve precision motion and to handle error compensation routines. This will require appropriate sensors with chosen characteristics to enable the record of detailed information. The manufacturing of the machine components with specific materials and tolerances is an important task. The details of the first part covering the design roadmap is described with all steps required as follows. After the specifications are well understood, concepts are discussed in an attempt to satisfy the specifications.

4.4.1 CONCEPTUAL ANALYSIS

Initial investigation in the conceptual design is extremely important as this will enable to check the possible implementation functionality of the specifications. With a clear understanding of the functionalities required in the new machine, i.e., clear understanding of the specifications, various concepts could be drawn that may secure all functions. It is a step where all the aspects of the design are discussed before starting any modeling and simulation. Several variants may be available at this stage achieving the same objectives but a critical evaluation and comparison could be applied to decide for the best solution (variant) to adopt.

The house of quality is an evaluation tool for engineers and could be applied as a method of comparison between variants. A relationship matrix is constructed including design requirements and design factors. The design requirements are the characteristics required by the product. The design factors are the variables that would be considered to achieve an appropriate design solution. The initial design analysis starts with the use of the house of quality (Figure 4.5). The latter will determine the order of importance of the machine elements (design factors) with respect to the expected machine characteristics (design requirements). The designer will be aware of most possible interactions between design requirements and design factors. The following example shows a simple design analysis of a precision linear axis. The axis elements are material, bearing, guideways, and controller. Expected characteristics are stiffness, light weight, accuracy, and reliability. The relationship is defined with specific signs and weighting values are allocated from high-to-low relationship.

4.4.2 MATERIALS SELECTION

Materials are key-components in precision machines. Their properties have a great influence on the machine performance. The material properties are chosen to be relevant to the performance of the machine elements or the overall machine performance.

Granite for machine and metrology base is relevant because of its low coefficient of thermal expansion and its excellent secular stability. Mild steel could be very stable when properly stress relieved. Granite is preferable because of its damping coefficient (15 times better than steel). Zero-dur has low coefficient thermal expansion. To minimize thermal distortions, aluminum could be

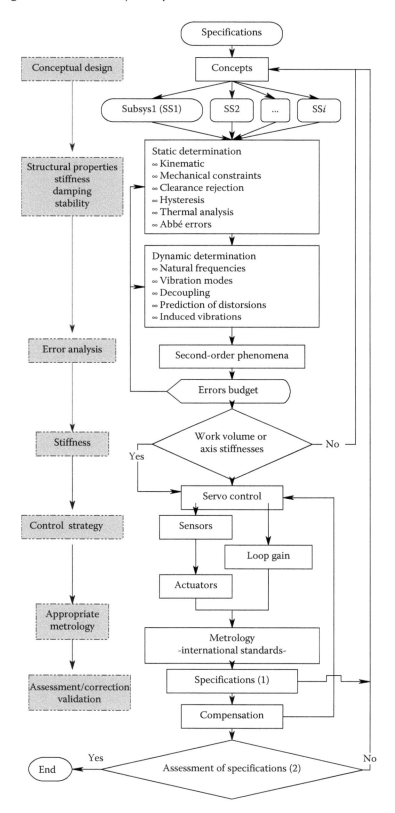

FIGURE 4.4 Design strategy for standard machine.

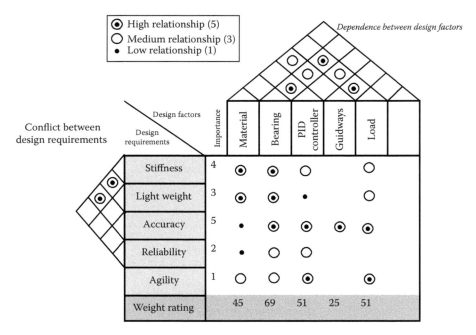

FIGURE 4.5   Example of use of the house of quality.

chosen with low heat capacity and high conductivity so that the temperature field in the machine reaches stability, for example, to avoid bending of machine elements. Concrete could also be used as a base material; it has generally poor stability. It needs a long-curing cycle where concrete is allowed to hydrate and improve its stability. Synthetic concretes are more expensive than granite and do not have concrete problems.

As an example for a structure used for measurements systems, the material is required to be

1. Inexpensive
2. Weightless
3. Extremely stiff
4. High damping capacity
5. Zero thermal expansion but high thermal conductivity
6. Low effect of moisture absorption
7. High material stability
8. High manufacturability

To give an idea about the key characteristics to look at when selecting materials, an overall view about key characteristics for metals, ceramics, and polymers are given in Table 4.2 and some material properties are given in Table 4.3.

4.4.3   Kinematic Design: Bearings and Guideways

This is a fundamental principle in the design of precise machines. In 1890, James Clerk Maxwell stated:

"In designing an experiment the agents and phenomena to be studied are marked off from all others and regarded as the field of investigation…the experiment must be so arranged that the effects of disturbing agents on the phenomena to be investigated are as small as possible." He also mentioned:

TABLE 4.2
General Characteristics of Structural Materials

Characteristic	Metals	Ceramics	Polymers
Density	Low–high	Low–high	Low
Hardness	Medium	High	Low
Tensile strength	High	Low–medium	Low
Compressive strength	Medium–high	High	Low–medium
Young's modulus	Low–high	Medium–high	Low
Dimensional stability	Low–medium	High	Low
Thermal expansion	Medium–high	Low–medium	High
Thermal conductivity	Medium–high	Medium	Low
Thermal shock	Medium–high	Low	High
Electrical resistance	Low	High	High
Machinability	Low	Medium	Medium

"The piece of our instrument are solid, but not rigid. If a solid piece is constrained in more than six ways it will be subject to internal stress, and will become strained or distorted, and this in a manner which, without the most exact micrometrical measurements, it would be impossible to specify."

He says: "There are certain primary requisites, however, which are common to all instruments, and which, therefore, are to be carefully considered in designing or selecting them. The fundamental principle is, that the construction of the instrument should be adapted to the use that is to be made of it, and in particular, that the parts intended to be fixed should not be liable to become displaced; that those which ought to be movable should not stick fast; that parts which have to be observed should not be covered up or kept in the dark; and that pieces intended to have a definite form should not be disfigured by warping, straining or wearing."

These apparently simple requisites of Maxwell may be regarded as the five fundamental axioms of design, one or other of which are frequently neglected, and sometimes are difficult to satisfy. A mechanical system generating motion is described as a kinematic system if it complies with the following five conditions.

TABLE 4.3
From Flexible Heavy to Light Stiff Materials

Materials	Density (kg/m³)	Young Modulus, E (GPa)	Thermal Expansion (10^{-6}°C)	Poisson Ratio	Stiffness/Mass Ratio (m²/s²) × 10^6
Epoxy resin	1200	10		0.26	8.3
Cast iron	7060	100	11	0.21	14.2
Invar	8130	140	0.36		17.2
Granite	2700	63	5.4	0.26	23.3
Aluminum	2780	70			25.2
Steel	7800	200	11	0.3	25.6
Zerodur	2500	81	0.05	0.32	32.4
Alumina (Al_2O_3)	3700	303	8.1	0.21	81.8
Carbon fiber epoxy	1580	150	0.1		95.0
Silicon carbide (SiC)	3100	365	4.3	0.19	118

Degrees of freedom: If a rigid body is not restrained in any way, then it has six Degrees of Freedom (DOF). If only a translation is allowed along X axis, then the other five DOF have to be locked.

Constraints: The previous five Degrees of Freedom (DOF) will be considered as constraints.

Closure force: The closure force maintains a constraint. It could be the own weight of the body, a clamping screw, or a loading spring.

Redundancy: Three contacts with the gravity closure of a rigid body are sufficient to enable the body to sit firmly on a horizontal surface. A fourth contact does not eliminate any of the remaining DOF. It is considered as "redundant." Hence, the body will rock on the surface. Any attempt to eliminate the rocking by increasing the weight will deform the structure. In a linear movement, the redundancy induces wear and distortions.

Conditioning: A movement is generated properly if the contact at the interface is properly conditioned.

Arrangement of constraints: The following cases are the six types of how a motion that could be generated. Constraints are indicated by letter C followed by the number of DOF.
    Kinematic couplings (Figure 4.6, C6) are highly repeatable fixtures because of exact constraints. Contact points are used between different types of surfaces. For very high repeatability, tough materials are used such as tungsten carbide or ceramics.

Direct actuation: After the guideways and the bearings have been identified, the motion generation will need a drive or an actuator for the movable stage. As high precision in positioning is required, a direct drive concept is necessary to secure a direct movement that is controllable and to avoid any loss or subtransformation of the movement that is sometimes difficult to estimate. Occasionally, the transformation of movement designed for certain systems is completely nonrepeatable.

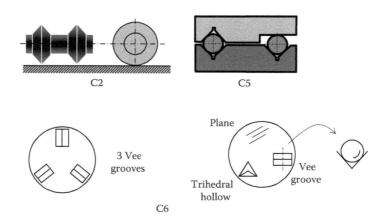

FIGURE 4.6 Example of technology solutions for kinematic concepts of linear motion.

4.4.4   STRUCTURAL ANALYSIS OF THE CHOSEN CONCEPT

In this critical step, the static, dynamic, and thermal analysis requires an accurate modeling of the structure to be able to determine static deformations, stiffness, and damping in key locations of the machine as well as thermal expansion depending on the identification of sources of heat.

4.4.4.1   Static Analysis

The census of possible forces that may affect the machine structure must be known depending on the machine modes of use. Deflections due to own weight (gravity effect), external forces, and either von Mises or maximum principal stresses are correctly computed. The resultant stress distribution must be checked against the material maximum stress failure characteristics. Safety factors could be used but kept low, for example, 1.15–1.5, as the incurred cost could increase if special materials are used. The simulation using FEM becomes more accurate in determining deflections and stresses.

A number of principles are involved in precision design such as Saint Venant principle, Maxwell reciprocity, center of masses, and permanent equilibrium. A brief introduction to these principles is given hereafter.

4.4.4.1.1   Maxwell Reciprocity and Betti's Interpretation

Maxwell's reciprocity theorem states that two identical forces applied at distinct points i and j on a linear structure, the displacement at i caused by the force at j is the same as the displacement at j caused by the force at i. Theoretically, the stiffness matrix and its inverse, the flexibility matrix, will be symmetric.

Betti's interpretation: The indirect or mutual work done by a loading system i during the application of a new loading system j is equal to the work done by the loading system j during the application of the loading system i. For more details, consult Den Hartog [12] and Timoshenko [13].

4.4.4.1.2   Saint Venant Principle

Saint Venant principle states that the localized effects caused by any load acting on the body will dissipate or reduce within regions that are sufficiently away from the location (point or small area) of the load. Another definition states that statically equivalent systems of forces produce the same stresses and strains within a body except in the immediate region (point or small area) where the loads are applied.

In engineering design application, one has to apply constraints over several characteristic dimensions of a component to secure a proper functioning.

4.4.4.1.3   Center of Mass and Permanent Equilibrium

The permanent equilibrium is usually determined by the position of the center of mass. If the center of mass is a point within the object's actual structure, then the object can be balanced at that point. Any moving slide should have its center of mass known with respect to the guideways.

4.4.4.2   Dynamic Analysis

The dynamic analysis includes the determination of the natural frequencies and their corresponding modes of vibration of the structure. The first analysis should be the comparison of the first frequencies with respect to the range of bandwidth used by the controller. It is also important to check whether the mode of vibration is degrading the quality of the generated motion. The user might have to solve a problem of the first mode that is critical to the motion axis and which natural frequency is excited by the controller or other sources of excitations.

The modeling of the effect of any varying force on the system and especially on the motion precision is important. The time response of the system is calculated depending on the application to observe the dynamics of motion.

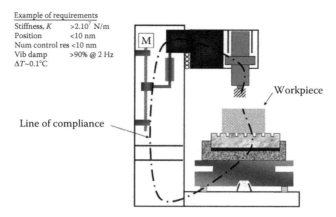

FIGURE 4.7   Compliance of a machine.

4.4.4.2.1   Stiffness of the Structure

To secure a very high accuracy in positioning, the axis movement should have a high stiffness as one of the conditions. Any external disturbance force to this axis will induce very low effects in terms of displacement as shown by Equation 4.1. An illustration of the stiffness loop (also called flexibility loop) in a machine tool is shown in Figure 4.7.

$$\delta x = \frac{F_{\text{dist}}}{K_{\text{axis}}} \tag{4.1}$$

where

F_{dist} is the disturbance force, for example cutting force
K_{axis} is the axis stiffness

The deformation behavior is described by the stiffness K or the compliance C (Figure 4.8).

$$K = \frac{\mathrm{d}F}{\mathrm{d}x}; \quad C = \frac{\mathrm{d}x}{\mathrm{d}F} \tag{4.2}$$

Therefore, in a mechanical system, all the components must be considered to characterize the compliance of the system. As a first estimate, an analogy could be made with springs mounted in serial or parallel mounting. An example is considered in Figure 4.9 in which the overall stiffness is estimated by

$$\frac{1}{K_{\text{total}}} = \frac{1}{k_1} + \frac{1}{k_2} + \frac{1}{k_3} + \frac{1}{k_4 + k_5} \tag{4.3}$$

The static behavior is mainly influenced by the shape and the form of the structure, location and design of joint, and the choice of the material. Some examples for typical shapes are compared in Table 4.4.

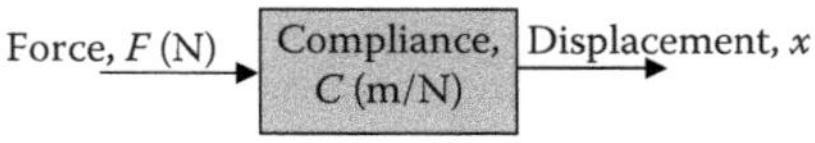

Force, F (N) ⟶ | Compliance, C (m/N) | ⟶ Displacement, x

FIGURE 4.8   Concept of compliance.

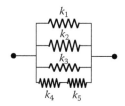

FIGURE 4.9 Model of stiffness.

Thermal drifts Thermal drift are treated separately in this book as these errors have been the focus of many researchers throughout the last three decades. The effects of thermal errors is not an easy task as the sources are multiple in a machine tool and sometimes complex to model (refer to Chapter 3).

4.5 MACHINE KEY-COMPONENTS

Precision system designers face myriad of choices when it comes to defining and specifying components of a high-precision motion system. With so many possibilities, it can be difficult to determine the optimum solution. Even with the selection of the best component, the resulting integrated system may not meet the application specifications unless careful engineering analysis is applied from a system perspective. In order to achieve optimum performance in such a system design, it is necessary to consider each of the component technologies individually. Then, the integration of the components into a working system must be carefully engineered. By breaking down a system into its building blocks, the design task can be approached systematically. Three main components are required to generate any movement; the guideways, the bearings, and the actuators. Without wishing to unnecessarily add to the plethora of available machine elements in most of the design books, supplier companies' Web sites, the author will discuss principles of known machine elements and a few ultrahigh-precision elements known to have high performance.

TABLE 4.4
Constraints Arrangement and DOF

Constraint	DOF	Concept	Constraint	DOF	Concept
C1	5		C4	2	
C2	4		C5	1	
C3	3		C6	0	

4.5.1 GUIDEWAYS

The guideways will support the moving parts of the machine. The translational or rotational movements have to be kinematically designed as described previously. However, in the design or selection of guideways, one has to comply with the following criteria depending on the design.

4.5.1.1 Criteria for Selection

1. Static and dynamic stiffness
2. Linearity or circularity
3. Wear resistance
4. Friction
5. Damping
6. Heat
7. Lifetime and maintenance
8. Cost

Sufficient stiffness of the guideways in the normal and lateral directions with respect to the main movement has to be secured once the kinematic design criteria are applied to design the guideways. It is also important to design guideways free of stresses and free from forces that may degrade the quality, for example linearity or circularity, of the expected movement.

Accurate rotational and translational motions are fundamental movement for machine tools; the accuracy of such machines is limited by the performance of one of these two movements.

Some conceptual designs of guideways become standard configurations for rotating and translating motion as shown in Figures 4.10 and 4.11. They were widely used during the last three decades for several applications that require accurate motion. The shaft and the slide are guided using the

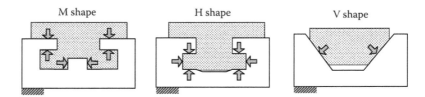

FIGURE 4.10 Configurations for linear movement.

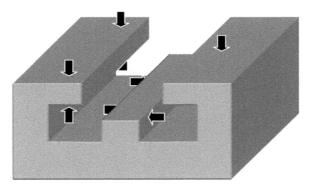

FIGURE 4.11 A typical guideway in previous generation of precision machines.

combination of several bearings achieving submicrometer precision. Stability and performance depend actively on the bearing type (orifice, porous media, magnetic, etc.).

4.5.1.2 Precision Linear Movement

Various concepts exist to generate a precise linear motion. It is important to secure a minimization of rigid body standard errors. A number of concepts are shown in Figure 4.10.

The damping of slideways plays an important role in the dynamic characteristics of a machine tool. The damping capacities were largely dependent on the sliding velocity of machine tables when lubricated slideways are used. For example, it was mentioned that the flexibility of the fixture at its natural frequency was greatly reduced with the increase of the sliding velocity of the table [14]. If sliding surfaces are normally loaded and subject to small tangential forces, then microdisplacement will appear at the interface. This is important for the stiffness and damping characteristics of the interface. Table 4.6 gives some characteristics depending on the type of the bearings and drives. Depending on the type of materials in sliding contact, shear deformation may show some cyclic hysteresis with submicrometer deformations. An example of long guideway for large-range machine tool is presented in Figure 4.12 requiring a stiff base on which the guideways are machined and finished.

Table 4.5 shows some characteristics comparison between three guideways used to generate linear motion. One type of guidance using roller bearing is shown in Figure 4.13. The guideways are added and assembled firmly to the base. The actuation is secured by recirculation lead screw mounted in the middle of the moving stage. This type of solution would generate large deviations and degrading quality of the linear motion.

Figure 4.14 uses similar principle as in previous figure but as this is a vertical axis, a hydraulic counter weight is added to help the motor dedicated to control this axis. Usually, such a motor is smaller than the standard one used with leadscrew. This is for a milling machine.

FIGURE 4.12 (See color insert following page 174.) Long guideway. (Courtesy of Goratu.)

TABLE 4.5

Influence of Form and Material Parameter on the Structural Stiffness

Standard Behavior	Corresponding Shape	Stiffness	Parametric Influence
Compression		$K_b = C \cdot E \cdot \dfrac{w \cdot a^3}{l^3}$	(+) a, E, w
			(−) l
Bending		$K_b = C \cdot E \cdot \dfrac{w \cdot a^3}{l^3}$	(+) a, E, w
			(−) l
	Open profiles closed profile		
	Torsion	$K_t = C \cdot G \cdot \dfrac{w \cdot a^3}{l}$ closed-profile	(+) a, G, w
		$K_t = C \cdot G \cdot \dfrac{w^3 \cdot a}{l}$ open-profile	(−) l
Shear		$K_s = C \cdot G \cdot \dfrac{w \cdot a}{l}$	(+) a, G, w
			(−) l

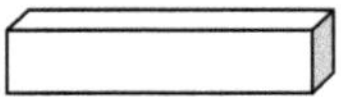

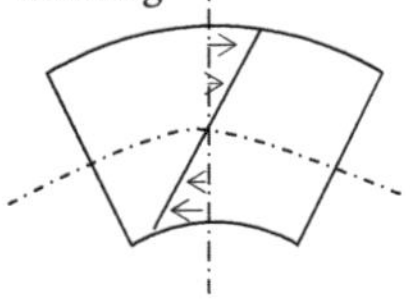

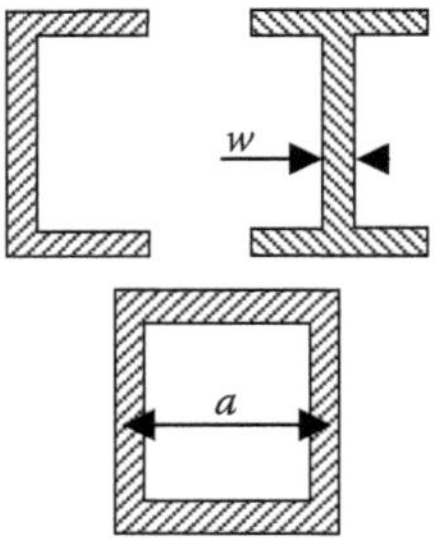

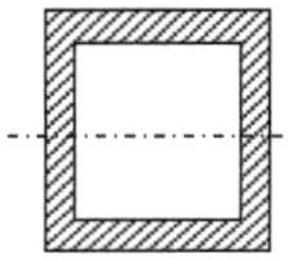

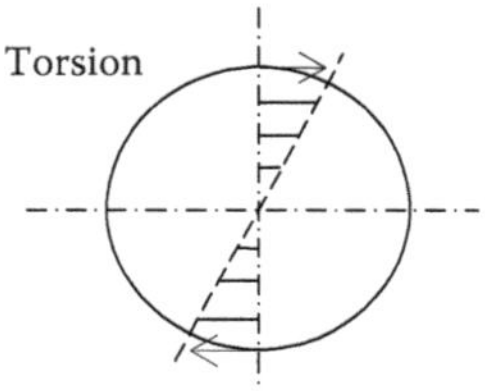

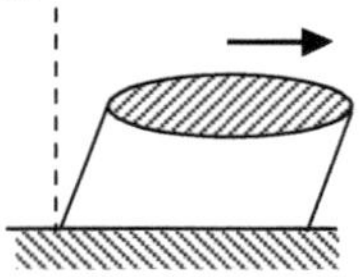

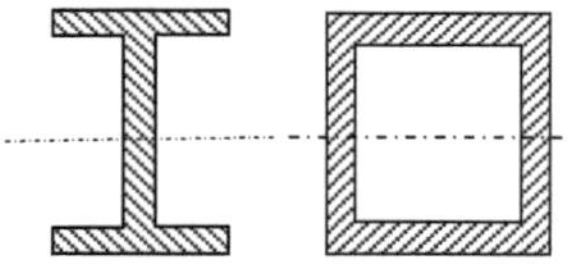

(+), positive influence; (−), negative influence.

TABLE 4.6

Characteristics of Linear Slides

Characteristics		Aerostatic Bearings	Hydrostatic Bearings	Friction Drive
Stroke (mm)		200	200	3000
Straightness (µm/100 mm)	Hor.	0.050	0.300 H	0.330 H
	Vert.	0.015	—	—
Stiffness (N/µm)		490	High	High
Load capacity (N)		3000		

FIGURE 4.13   Long guideway. (Courtesy of Goratu.)

The actuator could also be mounted on the side of a linear stage but this may create angular deviation (yaw) if the guideways are not designed with high precision as shown in Figure 4.15.

4.5.1.3   Precision Rotating Movement

Precision rotation concepts should be able to secure minimum geometric errors. Some concepts are shown in Figure 4.16.

Typical characteristics in precision rotative motion are used as criteria to compare spherical and circular air bearings (Tables 4.6 and 4.7).

FIGURE 4.14   Long guideway. (Courtesy of Goratu.)

FIGURE 4.15   Long guideway. (Courtesy of Goratu.)

4.5.2   BEARINGS

Without wishing to unnecessarily add to the plethora of available bearings in most design books, it was clear to the author that an introduction to bearings characteristics with some special bearings securing ultra high-precision motion is worth mentioning.

Linear or rotative motion can be provided with a wide variety of bearing types. The bearings are the key parts for the guideways. Thus, in case of incomplete analysis, the bearings become a source of vibrations. The appropriate choice depends on the requirements of the application.

The general requirements for bearings are

1. Precise displacement
2. Low level of vibration
3. No parasitic movement
4. Low friction
5. Easy implementation
6. Easy maintenance
7. Reasonable cost

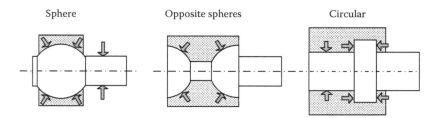

FIGURE 4.16   Configurations for rotating movement.

TABLE 4.7
Some Typical Characteristics of Spindles

Characteristics		Sphere Air Bearings	Circular Air Bearings
Rotational accuracy (μm)	Radial	0.05	0.05
	Thrust	0.05	0.03
Bearing stiffness (N/μm)	Radial	50	200
	Thrust	60	500
Load capacity (N)	Radial	300	1,500
	Thrust	360	3,500
Rotation speed (rpm)		10,000	3,600

4.5.2.1 Definition of the Different Types of Bearings

4.5.2.1.1 Linear Crossed Rollers

Crossed roller bearings exhibit high-load capacity due to their line contact (Figure 4.17). Linear ball bearings are lower cost but have point contact hence reducing load capacity and stiffness. Since linear rolling elements of the bearings do not recirculate, they require a longer length to accommodate the bearing motion. The bearings move half the distance the slide moves. Cross roller slide can support 8–10 times loads compared to standard rollers. Also, since the bearings move relative to the load, the slide can exhibit large cantilevers at the travel extremes. Overhung-load is better supported with cross rollers than ball rollers (see characteristics comparison in Table 4.8).

4.5.2.1.2 Aerostatic Bearings

Air bearings offer the highest-mechanical precision. However, they can be the most costly alternative. They have numerous advantages for very high precision: noncontact drive, which eliminates wear and friction, very high MTBF and requires no lubrication. Since the bearings essentially glide over a precision lapped reference surface, submicron geometric accuracies can be achieved. Also, since an air bearing can float in two directions, it enables single plane stage configurations, minimizing Abbé errors.

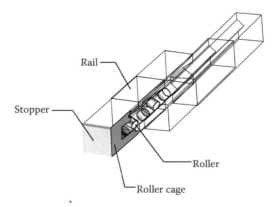

FIGURE 4.17 Crossed roller bearing.

TABLE 4.8

Characteristics Comparison between Cross Roller
Bearings and Ball Bearings

	Load Capacity	Axis Play	Friction
Ball bearings	↘	↗	↘
Cross rollers	↗	↘	↗

4.5.2.1.2.1 Orifice Bearings

The aerostatic bearings posses very low slide friction and thermal effects due to the low air dynamic viscosity. This will induce lower damping coefficient of air thickness compared to hydrostatic bearings (damping coefficient in the order of 0.2–0.5) (Figure 4.18).

The theoretical investigations and measurements achieved in the mechanical system division at Universite'de Technology de Compiegne (UTC) [15] show that the stiffness of the aerostatic bearings could rise till 10^8 N/m. The bearings designed with pockets show the known disadvantage of generating instability under certain conditions. This instability depends on the optimal choice of a few parameters such as supply pressure, design parameter depending on the intrinsic characteristic of the fluid and geometry of the orifice, and the volume of the pocket. The fabrication is very easy. The pivot, sometimes added to the bearing, allows an autoadjustment of the parallelism with the sliding surface, nevertheless this pivot is not recommended for high-precision guideways due to the induced instability. On the other hand, these bearings may be autopreloaded by creating a depression on certain part of the surface.

4.5.2.1.2.2 Porous Media Bearings

In case of porous media bearings, the air is brought to the sustentation gap through the porous media material such as a sintered alloy. The regular distribution of the pores allows an air supply on the whole active surface (loaded surface) and induces an improved relative stability (Figure 4.19).

This stability depends on the pressure supply, coefficient of compressibility, material porosity, and the bearings dimensions. The contact faces should have a good surface status, which permits a small film thickness and hence a high rigidity (till 10^8 N/m).

4.5.2.1.3 Hydrostatic Bearings

Hydrostatic bearings were used for their high stiffness and damping, the drawback with high-precision movement is a certain complexity of supply circuits (Figures 4.20 and 4.21).

The advantages of hydrostatic bearings are their small size, easier to design, and need of a minimal requirement for close tolerance: for example, a film clearance is typically 2.5 μm compared to 10 μm for aerostatic bearings.

4.5.2.1.4 Magnetic Bearings

Magnetic bearings allow contact-free levitation, high circumferential speeds at high loads; they do suffer neither friction nor wear, and therefore they offer a virtually unlimited lifetime while no maintenance is needed. Furthermore, the bearing force can be modulated, either for compensating unbalance forces or for deliberately exciting vibrations.

Active magnetic bearings with their control system inherently offer the possibility of continuously recording bearing forces and rotor displacements. This allows online monitoring of critical

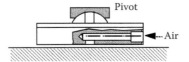

FIGURE 4.18   Aerostatic bearing (orifice bearing).

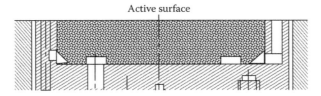

FIGURE 4.19   Porous media bearing concept.

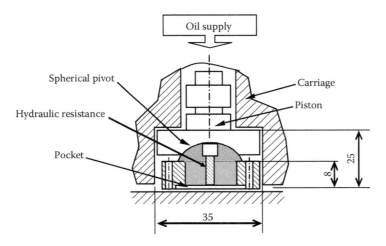

FIGURE 4.20   Preloading hydrostatic bearing.

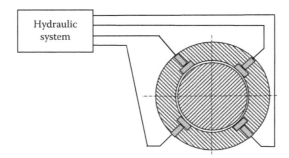

FIGURE 4.21   Application for rotative axle.

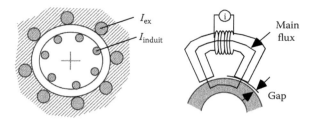

FIGURE 4.22   Magnetic bearings.

process parameters and early detection of incipient faults, such that reliability is increased. The self-sensing (sensorless) magnetic bearing is a special kind of magnetic bearing (ETH, Zurich), which needs no external position sensors. The position information is deduced from the air gap-dependent properties of the electromagnets.

The control schemes have succeeded in achieving micron level of precision [16] and very high static and dynamic stiffness. The research includes precision magnetic bearing design and vibration control in the micro-g level. In a recent development, the magnetic levitation stage for photolithography [17] operates with a positioning noise as low as 5 nm (Figure 4.22).

4.5.2.1.5   Advanced Bearing Design
4.5.2.1.5.1   Active Inherent Restrictor for Air-Bearings Spindles
The proposed method [18] employs an active inherent restrictor AIR, to control pneumatic pressure on the bearing surface. The AIR consists of piezoelectric actuator with a through hole, one end of which is small enough to function as an orifice when the actuator is embedded in the bearing to be controlled.

Thus, the stiffness and the rotational accuracy in both radial and thrust directions can be improved without the occurrence of pneumatic hammer. Instead of ordinary passive inherent restrictors, the AIR can be incorporated into most conventional aerostatic bearings. The run-out obtained for the AIR bearing is 70 nm using active control, with a pressure supply of five bars and a bandwidth in radial direction of 100 Hz.

4.5.2.1.5.2   Aerostatic Bearing Active Compensation
Error of form on the host surface could be compensated and external disturbances eliminated with an active compensation. The latter could control the air supply pressure film thickness and its parallelism. This will improve the static stiffness to virtually infinite stiffness, hence a much better control of the vertical motion or radial motion in case of a spindle (Figure 4.23).

4.5.2.1.5.3   Porous Ceramic Aerostatic Guideway Bearings
These bearings consist of porous ceramic material with a modified surface of sustentation. They offer higher-stiffness values than conventional air bearings. The commercially available graphite bearings also have a disadvantage in their permeability and therefore bearing performance is not consistent because during machining, the surface pores are partially closed which will affect bearing performance.

Thus, the permeability of the bearing is modified by the application of a surface layer to restrict air flow, using several techniques such as electron beam evaporation-ion plating, tape casting combined with slip casting (Figure 4.24) [19]. These bearings demonstrate maximum pneumatic stability and high stiffness required for high accuracy and low/high speed.

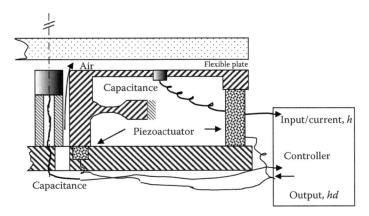

FIGURE 4.23   Smart bearing.

4.5.2.1.5.4 Dynamic Characteristics
The average stiffness value for a two-layered bearing was found to be in the order of 45 N/μm for bearing clearances below 4 μm, and with a supply pressure of four bars. This compares to a theoretical stiffness of 19 N/μm for a conventional "pocketed" orifice bearing of similar size, with a 2.5 μm bearing clearance, and pocket size of 4.6 mm diameter × 9 μm deep. To increase the pocket size, and thus stiffness, above this value increases the risk of pneumatic instability. For example, for a pocket size of 23 mm diameter × 9 μm deep the theoretical stiffness of a conventional bearing with a 2.5 μm clearance increases to 31 N/μm, but the bearing would almost certainly be unstable at five bar supply pressure.

The computation of porous media bearings using FEM shows that high stiffness is obtained when film thickness is around 8.5 μm for different air supply pressure (Figure 4.25). The required diameter for the porous media bearing is 100 mm.

4.5.2.2 Applications

4.5.2.2.1 Design of Porous Media Bearings for Delay Line
The porous media bearing is designed as shown in Figure 4.26, where parallelism of the surfaces is ensured by a pivot and a tilt control.

The very tight specification of level of vibration has been achieved using porous media bearings for the delay line interferometer [20]. The following measurements (Figure 4.27) show the attenuation of normal vibration at high frequencies (>300 Hz) of the upper stage when porous media bearings are active. The dynamic specification of $30 \, \mu m/s^2 \, \sqrt{Hz}$ is met.

4.5.2.2.2 Air Supply Equipment
The porous media bearings need a clean and stable air supply. An arrangement of air supply equipment is proposed in Figure 4.28.

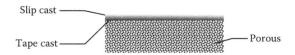

FIGURE 4.24   Combined slip and tape casting for ceramic porous media bearings.

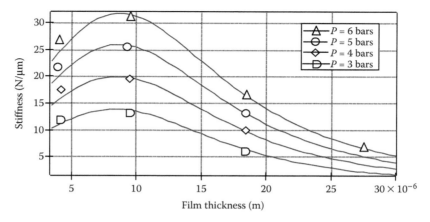

FIGURE 4.25 Stiffness of aerostatic bearing versus air film thickness.

4.5.2.2.3 Hydrostatic Bearings Design

Hydrostatic bearings are designed (35 mm diameter) for low speed with a minimum value film thickness (Figure 4.20), which should be compatible with surface flatness. For high-precision linear slide [21], other criteria have been added:

1. Pressure-regulation precision determined in such a way that the gap variation remains within $(\Delta h)_{\Delta p} < 0.05\,\mu m$ where Δp is the variation of the pressure in the hydraulic-bearing pocket.
2. The dissipated power should be under 20 W in order not to induce overheats.
3. The stiffness must be such that the displacement induced by an overload of 100 N is less than 1 µm.

To adjust the bearing rigidity, a second opposite pad bearing (Figure 4.20) was added as a preloading system. As the damping factor increases with fluid viscosity, bearing surface, capillary length, and alveolar pressure and decreases with capillary diameter and gap, the bearing stability is guaranteed. A dynamic study of the two-coupled pads (main and opposite) was carried out. The formulation of the whole system was written by using movement equation, mass conservation, and kinetic energy applied to each pad. The resulting damping coefficients of the main and preload pads are respectively

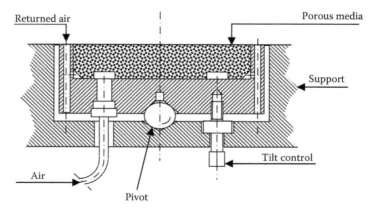

FIGURE 4.26 Concept of aerostatic porous media bearing.

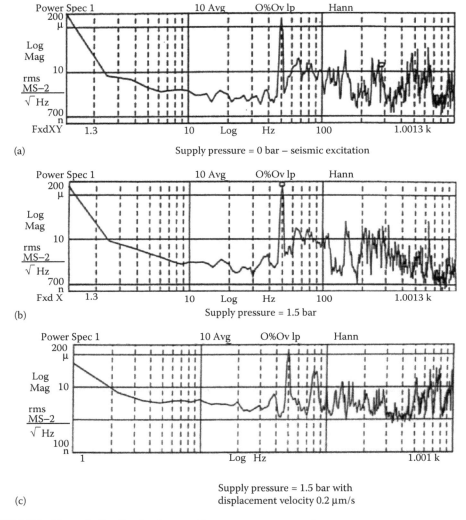

(a)   Supply pressure = 0 bar – seismic excitation

(b)   Supply pressure = 1.5 bar

(c)   Supply pressure = 1.5 bar with
                      displacement velocity 0.2 µm/s

FIGURE 4.27   Normal vibration level versus frequency measurements.

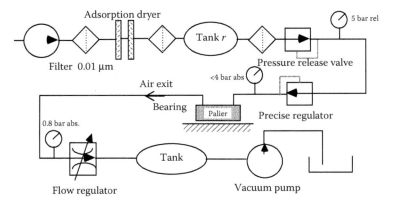

FIGURE 4.28   Air supply equipment for aerostatic bearings.

$$\xi_{\mathrm{m}} = \frac{\alpha_1}{2\left(M_{\mathrm{m}} \times \alpha_2\right)^{1/2}}; \quad \xi_{\mathrm{p}} = \frac{\beta_1}{2\left(M_{\mathrm{p}} \times \beta_2\right)^{1/2}} \tag{4.4}$$

where

α_1 is the fluid friction of the main pad (capillary restrictor + squeeze film)
α_2 is the hydrostatic stiffness of the main pad
β_1 is the fluid friction of the preloading pad (capillary restrictor + squeeze film + fluid viscosity)
β_2 is the hydrostatic stiffness of the preloading pad
M_{m} is the supported mass by the main pad
M_{p} is the preloading piston mass

The four parameters shown below are functions of the pad dimensions and flow characteristics. The natural frequencies are then defined respectively by

$$\omega_{\mathrm{nm}} = \sqrt{\frac{\alpha_2}{M_{\mathrm{m}}}}; \quad \omega_{\mathrm{np}} = \sqrt{\frac{\beta_2}{M_{\mathrm{p}}}} \tag{4.5}$$

As observed in Equation 4.5, the natural frequencies ω_{nm} of the preload pad are greater than those of the main pad because of the lower mass of the piston compared to that of the carriage. With the actual design, we have about $\omega_{\mathrm{np}}/\omega_{\mathrm{nm}} \approx 10$.

It is remarked from Equation 4.5 that the preload pad natural frequency ω_{np} is large enough to have no practical effect on the resultant natural frequency of the suspended carriage and consequently on movement precision. Because the coupled system is highly damped, a fluid with low viscosity must be used to reduce the response time of the system.

4.6 SECOND-ORDER PHENOMENA

The achievement of ultrahigh precision requires extremely advanced technology; therefore, all second-order physical phenomena affecting the accuracy have to be scrutinized and minimized.

However, some of them are unmodeled and the only way to compensate for them in a budget of errors will be by the use of adaptive or intelligent controllers. The influence of the second-order physical phenomena such as microdynamics generated by rolling friction, controllers via their frequency bandwidth, or electronic nonlinearities are dominant for high precision since they generate random errors that induce parasitic forces. Once these phenomena are identified, they have to be modeled and analyzed to determine whether they are acting separately or coupled. The unmodeled phenomena might be compensated for with a dedicated servo-control system.

The design methodology for precision engineering encompasses few critical steps such as the choice of concepts with higher stiffness, the influence of second-order phenomena with the parasitic forces, and the global interaction of mechanical dynamic properties with servo-controller parameters.

4.6.1 MODELING

At the ultrahigh-precision level, the modeling and simulation get difficult due to parameters affecting the system instantaneously. As the second-order phenomena affect the precision at this scale, measurement techniques have to be defined at an early stage of the design to check the results provided by the proposed model. For any considered mechanical system, there are various physical phenomena that could affect the accuracy at nanoscale level. External parameters will interact at this level, and it is important to define the contribution of each phenomenon.

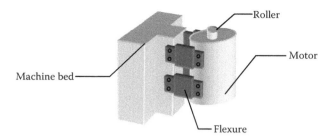

FIGURE 4.29   Motor attached to machine bed via flexures.

The following methods have the characteristic to compensate for unmodeled second-order phenomena where the servo system for positioning within a translational movement could be described simply by Equation 4.6 as

$$M\ddot{x}^2 + C\dot{x} + Kx = F(x) + \Delta(x, \dot{x}) \tag{4.6}$$

where
   M, C, and K denote the mechanical parameters inertia, viscosity, and stiffness, respectively
   F are the explicit known applied forces
   $\Delta(x, \dot{x})$ represents the implicit forces

The second-order phenomena are represented by an implicit nonlinear function $\Delta(x, \dot{x})$ added to the explicit servo system. This function is treated by the radial basis function compensation, which is used as neural networks to model nonlinear functions that are not explicitly defined.

Other examples are given in the following sections. Modeling is not an exact science when computing with accuracies at the scale of the submicrometer. Experimental tests and inspections are extremely important when the prototype needs to be adjusted according to specifications. The following section will elaborate an erratic modeling encountered in dynamic analysis of a coupling of an actuator with a slide. A number of mechanical behaviors remain partially addressed by FEM specialists. The designer in this case is advised to have a good background on the physical problems addressed by various elements and their assumptions and to be comfortable with best engineering practices. The following example shows a typical problem faced in FEM.

A direct current (DC) motor is held with four flexures (Figure 4.29). The purpose of the work is to model and analyze the dynamic characteristics of the whole motor attached to the bed of the machine. The motor and flexure were then modeled by using FEM and various types of elements were tried (Table 4.9). The background theory related to each element was consulted to check the assumptions made and possibly conditions of use.

4.6.2   PARTICULAR PARASITIC ERROR TO STRAIGHTNESS AND LINEARITY

Other possible errors such as form errors may occur from the geometric errors of the manufactured actuator components, but also from errors induced by the design concept itself. Figure 4.17 shows a comparison between a standard design and a precision design. The latter has shown far better precision performance than the standard one [3]. Other errors may result from external constraints such as thermal expansion. This could shift the whole roller assembly with respect to bar. The following example (Table 4.10) shows how a lateral-induced force F_t could generate a small yaw angle. It is important to observe that the magnitude of this angle θ depends on the type of the precision ball bearings arrangement.

The measurements performed with the collaboration of SKF in France show in Table 4.2, the behavior of this angle for "X" and "O" arrangements. Face-to-face arrangement "X" is more suitable for reducing the angle θ when the tangential force F_t increases (Figure 4.30).

TABLE 4.9

Comparison of Various Elements in Vibration Modeling

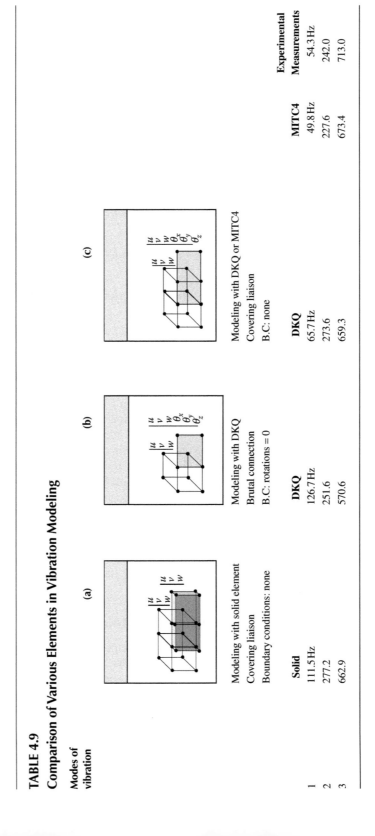

Modes of vibration	(a)		(b)		(c)		Experimental Measurements
	Solid		**DKQ**		**DKQ**	**MITC4**	
1	111.5 Hz		126.7 Hz		65.7 Hz	49.8 Hz	54.3 Hz
2	277.2		251.6		273.6	227.6	242.0
3	662.9		570.6		659.3	673.4	713.0

(a) Modeling with solid element
Covering liaison
Boundary conditions: none

(b) Modeling with DKQ
Brutal connection
B.C: rotations = 0

(c) Modeling with DKQ or MITC4
Covering liaison
B.C: none

TABLE 4.10

Variation of the Angle with Respect to the Tangential Force

Bearings Arrangement	$F_t = 5$ N	$F_t = 10$ N	$F_t = 50$ N
"X"	$\theta = 11.6\,\mu\mathrm{rad}$	$\theta = 8.7$	$\theta = 2.9$
"O"	$\theta = 14.5$	$\theta = 23.3$	$\theta = 37.8$

Example of Concepts

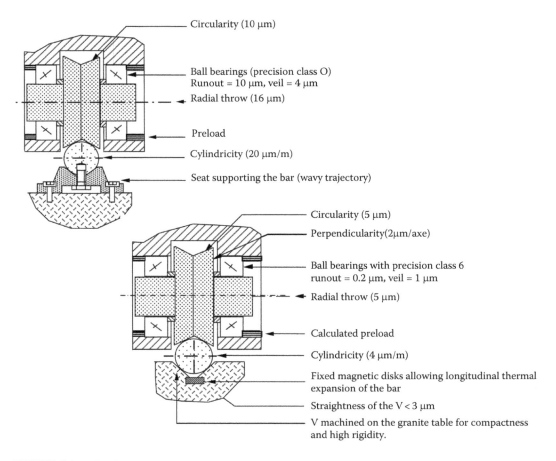

FIGURE 4.30 Design concept comparison between standard and precision design.

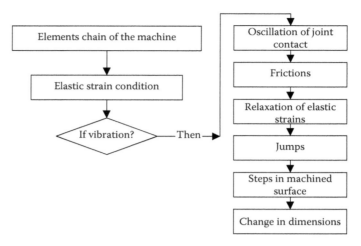

FIGURE 4.31 Effects of vibration on the chain elements.

4.6.3 NOISE

The measurements at the nanoscale are related directly to the noise level present in the system. In a wide range frequencies spectrum, noise from air-conditioning at low frequencies, pick-up due to the power supply, processor clocks or external interferences, and electromagnetic interferences at high frequencies could be easily identified. The noise observed at nanoscale manifests itself as a variation of a measured quantity around a mean value, especially when the system is "held on." A simple fast Fourier transform applied to the signal spectrum could reveal the frequency distribution and hence the physical mechanism responsible for its generation. This noise affects the accuracy of the controller's feedback and hence only partial compensation could be achieved within an acceptable lag time.

4.7 VIBRATION ISOLATION

In many cases, vibration isolation specifications for high-precision objects may not be as stringent as usually assumed [22]. The main reason for this is that sensitivity to external vibrations does not necessarily increase with increasing accuracy requirements, because improvements in accuracy of precision machines and apparatus are usually accompanied by better designs. Figure 4.31 shows in general cases the effects of vibration on the elements of the machine.

To satisfy the obvious need of isolation, a wide inventory of passive vibration isolation mounts was developed and is commercially available. Recently, this inventory was complemented by actively controlled isolators that can provide even better vibration protection and/or maintain a constant level of the device mounted on soft isolating mounts, regardless of changes of mass distribution within the machine [23].

For microelectronic manufacturing, maximum level of vibration velocity is 1.5–5 μm/s in the range of 10–60 Hz with the corresponding displacement amplitude not exceeding 0.05 μm. The stringent criterion for displacement amplitude is 0.012 μm for some specific applications [22]. Properly selected passive isolators having high damping can satisfy isolation requirement. In case of very tight specifications, active isolator systems will solve the problem but with higher cost.

4.7.1 DESIGN AND INSPECTION EXAMPLES

Example 1

Precision in positioning would require examination of the accuracy and rigidity of the mechanical drive. It is important to estimate the torsion of the threaded shaft as this can add a parasitic axial motion and degrade position accuracy. But are not sufficient, sizing the motor with better selection

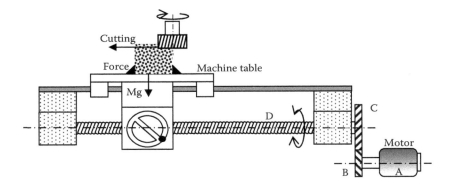

of the drive and controls to reduce cogging help in securing good motion accuracy. Brief calculation notes are introduced hereafter.

Frictional torque due to external load

$$T_r = \frac{F_a l}{2\pi \cdot \eta}$$

where

F_a is the axial load including cutting force and friction force due to total weight
l is the lead
η is the efficiency

Frictional torque due to preload

$$T_p = k\frac{F_{ao} \cdot l}{2\pi}$$

where

k is the internal coefficient of friction of the preload in the nut (leadscrew)
F_{ao} is the friction torque due to preload

Load torque due to acceleration, T_r

$$T_r = I_M \cdot \dot{\omega}$$

where I_M is the moment of inertia on the motor and the angular acceleration $\dot{\omega}=(2\pi N /(60 \cdot t))$

The moment of inertia acting on the motor could be

$$I_M = I_B + \left(\frac{N_B}{N_C}\right)^2 (I_B + I_C) + I_A + M\left(\frac{l}{2\pi} \cdot \frac{N_B}{N_C}\right)^2$$

where

N_B is the number of teeth of gear B
I_B is its moment of inertia
M is the total mass of the table

The powers are calculated for brush or brushless motors using the following:

$$P_{out} = \gamma \cdot N \cdot T$$

where N is the speed and T is the torque.

Efficiency:

Eff = 100 · (P_{out}/P_{input}), usually for brush motors Eff is between 55% and 70%, while for brushless it is between 70% and 90%, the motor become expensive with high efficiency.

The current could also be calculated as $i = P_{input}/V$

You can continue this example to identify a real motor and its characteristics.

Example 2

A special precision machine is designed to inspect large-scale workpieces. The sliding table holds the workpiece and moves along y-axis for 5 m. Three rollers from which two have Ve shape to support it.

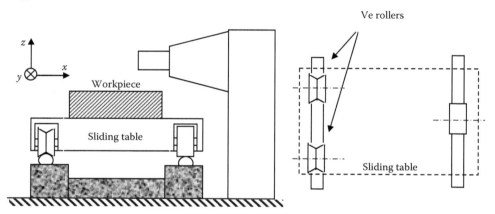

The slide has a weight of 100 kg and is made of steel. The guideway is made of granite. During the inspection, the temperature increases by 5°C.

1. Estimate the relative material dilatation of this system in the critical direction if the dilatations coefficients are $\lambda_{steel} = 13\,\mu m/m/°C$ and $\lambda_{granite} = 5\,\mu m/m/°C$.
2. Calculate the rolling resistance force induced by the three rollers.
3. Calculate the parasitic force in x-axis. Could you explain how to estimate the resulting displacement in this direction?
4. Comment on the current concept used for inspection and highlight any possible weakness.
5. The pitch angle has been measured along y-axis as well as the positioning accuracy in y-axis. The Abbe' offset is 190 mm. Calculate the Abbe' error and the compensated position. Explain how this compensation could be implemented?

Position Measurement (μm)	Forward Movement (μm)	Pitch Angle (arcsec)	Abbe Error (μm)	Position Compensated (μm)
0	3.15	1.5		
30	4.90	3.6		
60	6.31	5.9		
90	9.26	6.5		
120	11.12	8.4		
150	15.65	11.5		
180	16.32	12.5		
210	11.99	8.6		
240	8.00	7.3		

4.8 COMMERCIAL PRODUCTS

Some commercial machine elements with high performance exist already in the market. When designing a new system for a specific application, it may be good to envisage implementing such elements after performance inspection and suitability within the overall design.

Aerotech Ltd. has produced a number of linear and rotative stages with their independent controllers (A3200) that could achieve submicron accuracy over limited strokes. Some of them are described hereafter.

4.8.1 AIRBEARING ABL1000 SERIES

ABL1000 is a miniature linear positioner with noncontact design. It is a fully preloaded stage driven by a noncontact linear brushless servo motor. This stage proves the ultimate solution whether the application requires small, accurate steps, or constant smooth velocity. It offers superior servo performance with robust and perturbation-free cable management system. It is available in ranges with maximum axis travel of 25, 50, 100, and 150 mm. Generally, it is reported to posses an achievable accuracy of ±0.20 μm, repeatability of ±0.05 μm, straightness of ±0.25 μm, flatness of ±0.25 μm, and resolution of 1 nm. The Pitch and Yaw are each ±0.25 arcsec. The *XY* stacking of the linear stage is shown in Figure 4.32.

4.8.2 AIRBEARING ABL9000 SERIES

The ABL/ABLH9000 incorporates an active preload on both vertical and horizontal surfaces. The opposing thin-film pressure maintains the bearing nominal gap tolerance. This design, in addition to the large air-bearing surface that distributes the load over a large surface area, results in a stage with outstanding stiffness that is ideal for heavy or offset loading. The air bearing has an inherent averaging effect that maximizes performance. The thin film will fill small surface voids and allow for other irregularities. This characteristic yields superior pitch, roll, yaw, straightness, and flatness specifications. Because there is no mechanical contact between moving elements, the ABL/ABLH9000 experiences no wear or reduction in performance over time. It is claimed that the linear stage ensures years of maintenance-free operation at the high-performance levels. Service life is virtually unlimited. The ranges of the *XY* travel of the

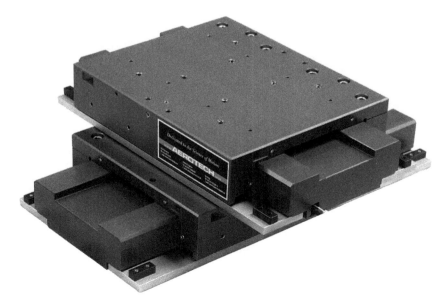

FIGURE 4.32 ABL1000 series. (Courtesy of Aerotech.)

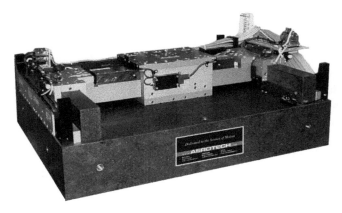

FIGURE 4.33　ABL9000. (Courtesy of Aerotech.)

stage are 300×300 mm, 500×500 mm, 750×750 mm, 1000×1000 mm, and 1200×1200 mm. Over such a long range suitable for standard machine tool, the stage is capable of achieving ±0.5 μm accuracy. The repeatability, straightness, and flatness are ±0.1 μm, ±0.5 μm, and ±0.5 μm, respectively while the pitch, roll, and yaw are each 2 arcsec. The stage is as shown in Figure 4.33.

4.8.3　LMA Actuators

The LMA and LMAC series actuators harness the speed, acceleration, and accuracy capability of a linear motor for the latest in high-throughput linear actuator technology. This modular stage is ideal for pick-and-place machines, gantry axes, shuttle stages, assembly machines, or as a general-purpose positioned. The mechanical and electrical components of the linear actuators (Figure 4.34) are integrated thereby eliminating the guesswork involved in choosing bearings, motors, encoders, and the

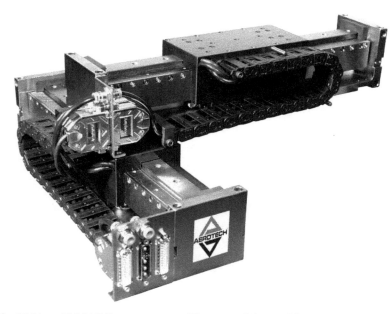

FIGURE 4.34　LMA and LMAC linear actuators. (Courtesy of Aerotech.)

other components required for a linear motor system. With a no-load acceleration of 3 g and a top speed of 5 m/s, the LMA and LMAC series actuators are the ideal solution to increasing throughput. A high-power brushless linear servo motor drives the LMA and LMAC actuators to speeds and resolutions that are impossible with a ball screw. The feedback device is a directly coupled, high-accuracy linear encoder. The encoder is mounted in close proximity to the work surface; hence accuracy problems associated with pitch, roll, and yaw are greatly reduced. The actuators are available in travel ranges between 100 and 1000 mm at an interval of 100 mm. Repeatability is ±0.5 μm over the length of travel. Resolution ranges between 5 nm and 1 μm. The LMA/LMAC linear motor drive system consists of two noncontacting parts, making the actuators virtually maintenance-free. As a result, there is no backlash, windup, wear, or maintenance that is normally associated with contacting-type systems such as ball screws or belts.

4.8.4 ANT ACTUATORS

The ANT-25LA (Figure 4.35) is a direct-drive linear actuator, is designed, and manufactured to overcome the inherent shortcomings of existing screw-based designs. It employs linear motion guide bearings and a glass linear encoder equipped with home marker and optical limits. These elements are connected directly to the output spindle assembly which is constrained in low-friction bearings. The directly coupled mechanical configuration allows backlash-free operation, 10 nm resolution, excellent repeatability, in-position stability. The tip of the actuator output shaft is tapped with a standard 2 mm thread to facilitate the acceptance of threaded-ball or clip assemblies. The travel limit is 25 mm.

4.8.5 GANTRY

The ABG10000 incorporates an active preload on both vertical and horizontal surfaces. The opposing thin-film pressure maintains the bearing nominal gap tolerance. This design, in addition to the large air-bearing surface that distributes the load over a large surface area, results in a stage with outstanding

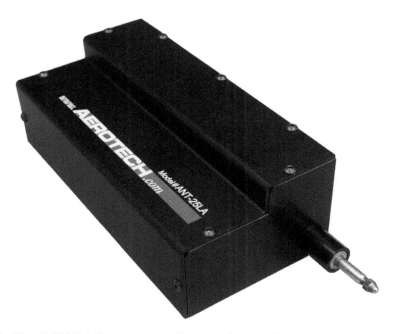

FIGURE 4.35   The ANT-25LA linear actuator. (Courtesy of Aerotech.)

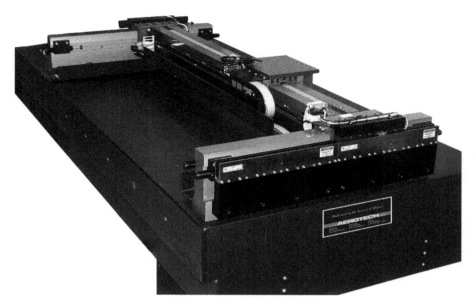

FIGURE 4.36   ABG1000 air-bearing gantry system. (Courtesy of Aerotech.)

stiffness that is ideal for heavy or offset loading. The systems are used in applications such as high-speed pick-and-place, automated assembly, vision inspection, dispensing stations, and high-accuracy inspection. The linear motor is a noncontact device with no backlash, wear, or maintenance. The air bearings are magnetically preloaded and assembled to provide optimized stiffness and load distribution. Large bend radii and highflex cables ensure that the air-bearing gantry (Figure 4.36) provides millions of cycles of maintenance-free operation. In the unlikely event of a component failure, a modular design ensures that part replacement is fast and easy. The gantry system is available in 250×250 mm, 500×500 mm, 750×750 mm, and 1000×1000 mm travel limit ranges. The accuracy of the system is $\pm 2 \mu$m, repeatability is $\pm 0.5 \mu$m, and the minimum resolution within all the travel ranges is $\pm 0.005 \mu$m.

4.8.6 ALS135 LINEAR MOTOR STAGE

The ALS135 employs a center-driven, noncogging linear motor as the driving element. Since the linear motor is a direct-drive device, there is no backlash, windup, or "stiction" that is normally associated with a lead screw or ball screw drive. The linear motor drive also offers the advantage of higher speeds and accelerations. The compact yet powerful linear motor drives the ALS135 to a peak unloaded acceleration of 1g and a maximum velocity of 300 mm/s. The result is a high-accuracy device with outstanding throughput that significantly outperforms comparable high-accuracy screw-driven stages. It is the ultimate solution for high-accuracy alignment and inspection stations. The ALS135 (Figure 4.37) is capable of 10 nm resolution, when used with Aerotech controller, due to the direct-drive linear motor that allows the ALS135 to make precise, small resolution steps. Straightness and flatness for the standard stage is less than or $\pm 2 \mu$m over the entire travel. It exhibits the outstanding ripple-free motion required for scanning and inspection applications. It is available in travel limit of 25, 50, 100, 150, and 200 mm. Accuracy of the smallest travel limit stage is $\pm 2 \mu$m while the repeatability and resolution is generally ± 100 nm and 0.0025μm to 1μm respectively. They are easily configured as *XY* assemblies.

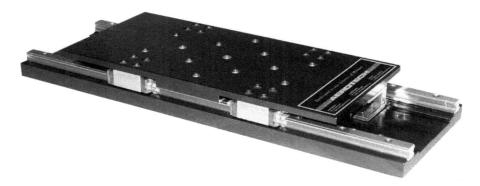

FIGURE 4.37 ALS135 linear motor stage. (Courtesy of Aerotech.)

4.8.7 ALS130 Linear Motor Stage

The ALS130 (Figure 4.38) is basically similar in design with the ALS135 with the same repeatability, resolution, and accuracy capability. The major difference is that the ALS130 are available in 25, 50, 100, and 150 mm only. They are also easily configured as *XY* assemblies.

4.8.7.1 Test

Forward and backward step motion test was done for the ALS 130H ant stage and is shown in Figure 4.39. The motion seems to be well resolved for steps of 25 nm but with a slight delay in response time.

4.8.8 Rotary Axis: ADR240

ADR series (Figure 4.40) direct-drive rotary stages provide superior angular positioning and velocity control. Applications range from high-speed laser machining to precision wafer inspection. Large diameter, matched-set ABEC-7 angular contact bearings are used to maximize performance with

FIGURE 4.38 ALS130 linear motor stage. (Courtesy of Aerotech.)

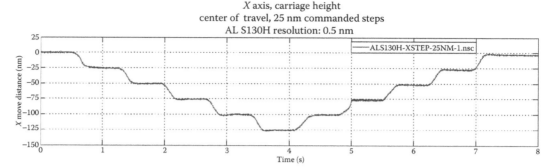

FIGURE 4.39   Forward and reverse step test of ALS 130H.

respect to wobble, moment stiffness, and rotating friction. A thick-walled, precision-ground shaft further minimizes wobble. The large diameter bearings permit large payloads without compromising performance.

To maximize positioning performance, the ADR series utilizes Aerotech's S-series brushless, slotless motor. This motor has all the advantages of a brushless direct-drive motor—no brushes to wear, no gear trains to maintain, and high acceleration and high speeds. With its low-inherent inertia and high-power output, the ADR is capable of speeds and accelerations that are an order of magnitude greater than typical direct-drive devices or worm-driven stages. The low inertia and zero backlash makes the ADR the ideal solution for applications requiring frequent directional changes. Typical line counts range from 3,600 to 36,000 lines per revolution and resolutions can be as fine as 0.036 arcsec.

4.9   MICROMACHINES

4.9.1   RESEARCH NEED FOR MICROMACHINES

The miniaturization of machine components is currently perceived as a core requirement for the future technological development of the growing markets and its future challenges in biotechnology, life science, telecommunication, and mobility. This technology promises enhanced health care, quality of life, and economic growth by providing microsystems such as microchannels for lab-on-chips and medical devices.

It requires mesoscopic parts with complex microscopic features (part size on the order of $10\,cm^3$ with machining accuracy of less than $1\,\mu m$). Figure 4.41 shows the position of mesoscale with feature

FIGURE 4.40   ADR175/240 series. (Courtesy of Aerotech.)

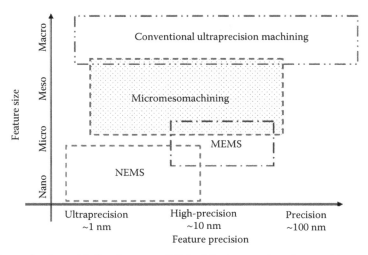

FIGURE 4.41 (See color insert following page 174.) Missing gap for micromachines.

precision compared to other known scales. Traditional fabrication methods of microstructures are often very expensive, have poor repeatability, are limited to a few silicon-based materials, and are unable to fabricate complex 3D shapes. Therefore, precise micromachining becomes an important requirement in the new concept of the microfactory allowing for the use of a wide range of materials (metals, ceramics, or polymer), an advantage over silicon-based microfabrication processes. The arising products are characterized by miniaturization and integration of mechanics, electronics, and information processing. The required mesoscale parts have delicate features that will require five axis milling and tight tolerances. It is inefficient to produce parts such as molds and dies with electrical discharge machining or investment casting techniques, as is the current practice. The actual production of microparts is performed by standard machine tools consuming high energy, occupying large space, and requiring large air-conditioning room; therefore, it is important to save driving energy, space, and resources in controlling the environment, for example temperature by applying the concept of micro-factory and robust designing micromachines with suitable volume-ratio of machine/workpiece.

Hence, it becomes essential to develop flexible reconfigurable and cost-effective miniaturized production units. This is an emerging new research activity with limited knowledge on barriers limiting the accuracy on the existing few examples. An innovative design approach will be considered with more integrated design objectives and encompassing new key identified issues limiting the current performance of such machines. The desktop machines have an average overall size of $300 \times 300 \times 300 \, \text{mm}^3$ with a high stiffness and superior dexterity in the entire working volume.

These types of machines will secure a low-cost production for on demand micromachining of applications in electronics, MEMS, biomedical, automotive, and micromachines industry. Low-cost and time saving will be achieved by high-speed machining due to reduction in distance and mass with an increase of precision by smaller forces and high-natural frequencies of the structure. The hybridization reduces machine inertias, thus allowing the control of higher accelerations compared to those permitted by conventional machines. The vibration amplitudes of small machine tools are lower than those of larger machine tools because the inertial forces decrease as the fourth power of the scaling factor and the elastic forces decrease as the second power of the scaling factor. But, these remain significant effects inducing errors at low scale.

4.9.2 International Context of the Research in this Field

The conventional large-scale equipment consumes much energy, space, environmental conditions, and material, while the second method (MEMS) seems to be more advanced but has some limitations in size and dimension (2D and 2.5D). The micromachining refers to the mechanical material removal using a tool. Most of current micromachining is provided by machine having standard size. A great interest is for portable machines. The Japanese people have realized the first portable machining microfactory [24–26], which consists of two or three micromachines: a lathe, milling, and/or press machine with a manipulator. It uses 3 CCD miniature cameras to ease the handling of microparts. This microfactory has produced ball bearings with a diameter of 0.9 mm [27]. In single process machines, Kitahara et al. [28] have developed a microlathe with an XY driving unit comprising a slider and a V-shaped guide. A brass rod of 2 mm diameter was obtained with $Ra = 1.5 \mu m$ and Roundness = 2.5 μm. With the suitability of the inherent excellent characteristics on the micromachines, the existing machines did reach the required level of accuracy. Mishima [29] stated that design and optimization were not studied enough. Ito [30] has demonstrated that with an improved model based on the previous lathe, the new microlathe could achieve on a 10 mm stainless steel an estimated $Ra = 60$ nm and a circularity of 50 nm. When the small machine tools are made with expensive components and materials, it is expected to make savings in a long-term run. On the other side, savings in energy and materials could be achieved rapidly with the concept of designing a simple micromachine, whose size is equivalent to the workpiece as it was proposed by researchers in Mexico [31]. For approximately the same size as standard machines, Fanuc has recently commercialized "Robonano" achieving ultrahigh-precision micromachining with milling, shaping, and grinding and capable of producing mirrors, gratings, lenses, etc. The linear axes have a resolution of 1 nm while rotary axes have 0.0001°. The trend in machining accuracy should reach subnanometer scale in the next decade. Table 4.11 gives an overview about past achievements in terms of resolution and uncertainty achieved with respect to machine workspace.

4.9.3 REASONS FOR MINIATURIZATION

Many companies offer services to produce microcomponents to precision accuracies from processes using contemporary CNC machines, electrodischarge machining (EDM), and laser machining to industries such as biomedical applications, components of microdevices, and part of the MEMS production. Logically, if one requires a small component to be manufactured, a suitable solution would be to have a proportionally sized machine to perform the manufacturing tasks. The relevant reasons are given hereafter:

TABLE 4.11

Micromachine Tool Key Characteristics Comparison

Machines	N Axes	Processes	Workspace (mm³)	Resolution (μm)	Uncertainty (μm)
Nanowave MTS5 [40]	3	Milling	55 × 55 × 16	0.1	0.5 (positioning) 0.02 (roughness)
MTS2–4	2	Turning	—	0.1	1
Kussul et al. [31]	3	Milling	20 × 35 × 20	1.87	20
Kitahara et al. [28]	2	Turning	Hold 2 mm diameter maximum	0.05	0.5
Lu et al. [41]	2	Turning	—	0.004	1
KERN [44] (Germany)	5	Milling	160 × 100 × 200	0.1	± 1 (positioning) ± 2.5 (cutting)
Ashida et al. [42]	2	Turning	—	—	1.5 (roughness) 2.5 (roundness)
Ashida et al.	3	Milling	—	—	—
Okazaki et al. [43]	3	Milling	—	0.1	2 (surface roughness achieved)

1. *Negligible thermal drift*:   Thermal drifts that are generated by the machining process causing deformations effect directly the accuracy of standard machines. These effects are reduced in micromachines due to the miniature nature of the components, but could have similar scale effect.

2. *Reduced vibration amplitude with high natural frequencies*:   Reasonable decrease in vibration amplitude is due to reductions in mass of moving parts and therefore larger natural frequencies of the machine. As an estimation, the stiffness $K \approx L$ and the mass $m \approx L^3$, where L is the size of the component, therefore if the machine is miniaturized with a scaling factor of s, the natural frequencies will be s time larger. This will depend, of course, on the mode of vibration.

$$f_\mu = s f_o \qquad (4.7)$$

3. *Reduction of material consumption*:   As micromachine tools are dedicated to manufacturing components of microdimensions, they are capable of using billets of comparable size to the end product. This is not the case in standard size machines where limitations on work-holding devices may impose a restricting minimum diameter that a billet can be of.

4. *Magnitude of induced forces*:   The eccentric loading of a shaft induces a deflection δ due to the centrifugal force F_i [26]. Table 4.12 summarizes the analysis of a miniaturized shaft whose deflection is reduced by a power of the scaling factor s depending on the scale of speed. Three different rotating speeds scaling were considered.

    If the rotating speed of the shaft is increased or decreased, δ is reduced by a factor of s^5 or s, respectively. The eccentric force is given by $F_i = m\omega^2\varepsilon$ and the deflection $\delta = c(L^3 F_i/Ed^4)$ the standard axis is down scaled as $L_o = sL_\mu$, $d_o = sd_\mu$, $\varepsilon_o = s\varepsilon_\mu$ and $m_o = s^3 m_\mu$ where the indices are o for standard scale and μ for micromachine.

5. *Enhanced machining accuracy*:   It is improved by the inherent reductions of machine component inertia, negligible thermal drift, and larger eigen frequencies. As an example, any eccentricity in a lathe machine will be reduced by s, s^3, or s^5, where s is the scaling factor. The effect of miniaturization upon inertial, elastic, electromagnetic, and electrostatic forces can be quantified as represented by equations in Table 4.9. As can be seen, miniaturization can play a beneficial role by reducing these forces in varying magnitudes.

6. *Agility*:   As the inertia of moving parts decreases, it indices larger accelerations to enhance the agility of such small machines

7. *Space and energy consumption*:   Conventional machine tools can occupy up approximately 5 m² (including service area), whereas a single micromachine from the Japanese microfactory

TABLE 4.12

Magnitude of Eccentric Deflection

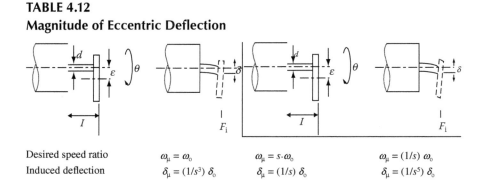

Desired speed ratio	$\omega_\mu = \omega_o$	$\omega_\mu = s \cdot \omega_o$	$\omega_\mu = (1/s)\, \omega_o$
Induced deflection	$\delta_\mu = (1/s^3)\, \delta_o$	$\delta_\mu = (1/s)\, \delta_o$	$\delta_\mu = (1/s^5)\, \delta_o$

uses only $0.09\,m^2$ [29], thus providing considerable spatial savings. This translates into economical advantages in terms of the costs of rent of premises. This provides a driving force for the replacement of conventional machine tools with miniature machines. Issues of portability of such a machine can also be addressed, in that the costs of relocation and installation of large machines are considerable.

On the other side, the saving in energy is of 10% in microlathe [32], whereas others have claimed 60% [33]. But, precision manufacturing often requires special working environments (e.g., controlled temperature). The reduction of the working volume of the manufacturing system would diminish the complexity of the maintenance of such operational conditions.

4.9.4 DESIGN APPROACH

The optimal parts manufacturing at mesoscale will not simply be achieved by scaling-down versions of conventional manufacturing devices. As the micromachine has larger natural frequencies and the thermal effect is insignificant, a direct robust design method, implementing, for example, Taguchi's approach will be set up to directly identify the design parameters, which significantly affect the machining tolerances. After the specific concept with all functionalities is defined, it is necessary to ensure that the working volume of the machine is located within the region of the highest stiffness of the mechanical system. The machine is then divided into subsystems to be designed accurately by defining design criteria derived from the main specifications. Each subsystem will be evaluated and optimized to meet the latter criteria. Needless to mention that issues related to friction and induced microdynamics is addressed at this scale. The overall steps of the design methodology are described as follows:

- Problem identification and concept development.
- The functional requirements of the machine will be defined based on the main specifications, followed by an outline viable concept, which includes combined axes, motion generation, and minimization of errors.
- The system is split into critical subsystems with detailed subspecifications.

Fundamentally, the two strategies that are most likely to succeed are one design that is most mechanically simple, but would require more controls effort, and one design that is simple to control, but may require more mechanical complexity [34]. The selection will be based on risk assessment of preliminary calculations of accuracy.

4.9.5 DESIGN CHALLENGES

The stringent specifications in quality, precision, and time are directing efforts towards more integrated machines as a general concept even for mesoscale workpieces. A machine that can fabricate a component with a variety of operations without the need to transfer the artefact in any way would provide advantages in terms of positional accuracies and set up times. The design challenges are revealed by the subsystems of the micromachine and are succinctly described hereafter.

4.9.5.1 Kinematics

Machines with parallel kinematics have difficulties in achieving stiffness [35] and avoiding singularities in the workspace where some of the DOF are lost. Serial machines accumulate errors from each axis. Stacked axes lead to larger machines generally with small working volumes; hence, a compromise in kinematic concept is acceptable for hybrid configurations. The structure of the machine must be robust in order to obtain high-machining accuracy and repeatability especially in line productions.

4.9.5.2 Interactive Forces

With microscale components, interactive forces (e.g., van der Waals, surface tension, and electrostatic) will exist between components, which instigate difficulties in manipulation and control. To overcome these problems, contact type manipulators such as ultrasonic traveling waves, or mechanical grippers; or noncontact type manipulators, for example, magnetic fields, aerostatic levitation, or optical trapping could be used in place of conventional solutions [36].

4.9.5.3 Actuators

When scaling-down actuators, it is important to keep an acceptable range of forces required to cut materials or to move and position devices (e.g., the tool). A rapid comparison between different types of forces shows that electrostatic force remains exactly the same if it is scaled down.

1. *Inertial force*: The miniaturization of actuator delivering such type of force will result in a reduction of the force by s^4 where s is the scaling factor.

$$F = m \cdot a = m \cdot \frac{dx^2}{dt^2} \Rightarrow [L]^3 \cdot [L] = [L]^4 \qquad (4.8)$$

2. *Elastic force*: The reduction of the force is about s^2.

$$F = \frac{E \cdot A}{L} \delta x \Rightarrow [L] \cdot [L] = [L]^2 \qquad (4.9)$$

3. *Electromagnetic force*: The reduction of the force is about s^2.

$$F = \frac{\beta^2}{2\mu} \cdot A \Rightarrow [L]^2 \qquad (4.10)$$

4. *Electro static force*: The force remains constant.

$$F = \frac{\varepsilon V^2}{2} \cdot \frac{A}{d^2} \Rightarrow [L]^2[L]^{-2} = [L]^0 \qquad (4.11)$$

The forces incurred in the forming process must be evaluated so that suitable systems can be designed. The forces in a metal turning operation can be evaluated as described in Figure 4.42. The forces are F_s, axial force; F_a, radial force; and F_v, tangential force. The forces are dependent on cutting speed (V), feed rate (S), and depth of cut (a).

Governing equations for these forces on steel work are $F_v = 1750\ S^{0.75}a$, $F_s = 650\ S^{0.35}a$, and $F_a = 250\ S^{0.25}a$.

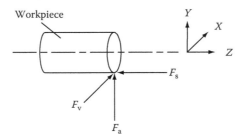

FIGURE 4.42 Forces involved in metal turning operation.

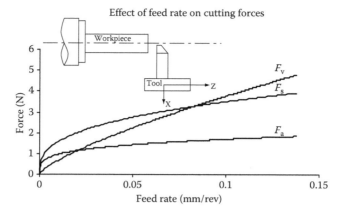

FIGURE 4.43   Effect of feed rate on cutting forces.

Figure 4.43 shows the effect that varying the feed rate has upon the cutting forces incurred in the forming process. Cutting speeds for turning operations may also cause problems at miniature scale; this is due to the reduction of the size of products to be machined.

$$N = 1000V/\pi D \tag{4.12}$$

If spindle speed N is obtained via Equation 4.11, where V is cutting speed and D the work diameter, then it can be seen that the required speed is about 300,000 rpm for a diameter of 1 mm using Al-alloys with TiN-coated carbide tool. The speed decreases in hyperbola with the diameter. Selection of such motors for spindles is obviously problematic but one has to design direct drive motors for better torques and acceptable speeds.

4.9.6   Miniaturized Controller

An important aspect to be considered is the system of control, which is increasingly being required to perform a wide variety of complicated tasks under varying operating conditions and in different environments, while at the same time achieving higher levels of precision, accuracy, repeatability, and reliability [37].

Modern CNC approaches employ PC-based solutions to incorporate extensive functionality in order to combine high quality and flexibility, with reduced processing time. The controller is required to satisfy several key issues set out in the "Technical Committee of Open Systems" of the IEEE 1003.0. These criteria address portability, extendibility, interoperability, and scalability in order to specify flexibility of the controller to varying applications [38]. To achieve interoperability, one must ensure that the controller is vendor neutral, achievable by employing a modular approach; however, there is a compromise between the degree of openness and the cost of integration [39]. One must also consider the processing power of the controller hardware, as too great a modularity can result in deterioration in the real-time performance of the system.

Integration of the PC-based control system into a CAD/CAM manufacturing system is a fascinating area of research. The CAD/CAM system incorporates a feature recognition program which links directly to a computer-aided process planning software tool; hence, the manufacturing process can become totally automatic thus improving efficiency. A general proposed design strategy for micromachines is shown in Figure 4.44. It encompasses various aspects that may affect the performance of micro-machines.

4.9.7   Concluding Remarks

Miniaturization of machine tools for mesoscale components is very tempting due to the reasons exposed earlier. For high efficiency and robustness, the challenges discussed earlier have to be

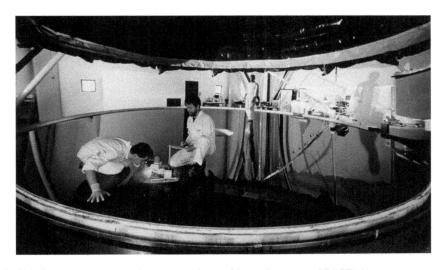

COLOR FIGURE 1.2 Inspection of the VLT mirror of 8 m. (Courtesy of SAGEM.)

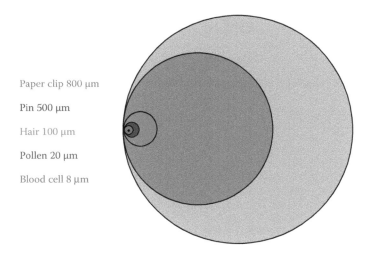

Paper clip 800 μm

Pin 500 μm

Hair 100 μm

Pollen 20 μm

Blood cell 8 μm

COLOR FIGURE 2.1 Size comparison of five items.

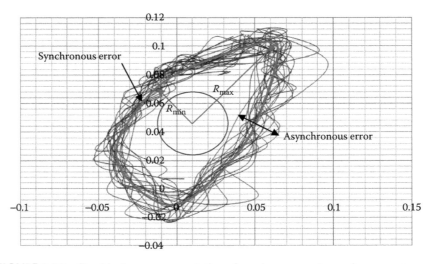

COLOR FIGURE 2.28 Graphical error representation of synchronous and asynchronous errors.

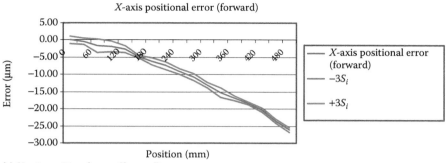

(a) *X*-axis positional error (forward)

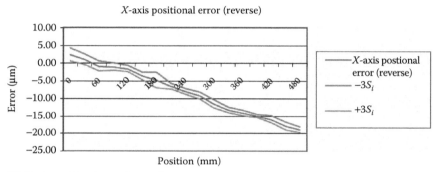

(b) *X*-axis positional error (reverse)

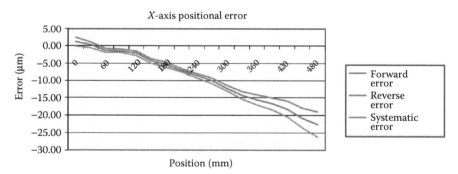

(c) *X*-axis positional error (bidirectional)

COLOR FIGURE 2.33   Positional error for (a) forward, (b) reverse, and (c) bidirectional.

TABLE 2.17
Diagrams Resulting from Contouring Assessments

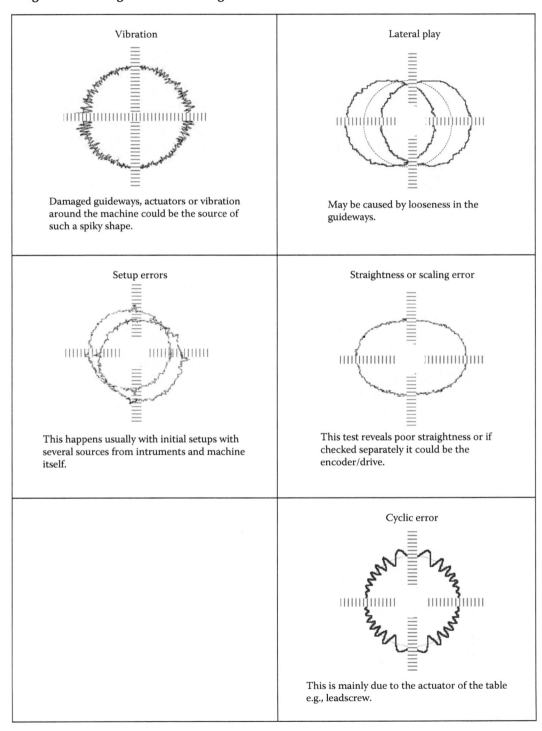

Vibration

Damaged guideways, actuators or vibration around the machine could be the source of such a spiky shape.

Lateral play

May be caused by looseness in the guideways.

Setup errors

This happens usually with initial setups with several sources from intruments and machine itself.

Straightness or scaling error

This test reveals poor straightness or if checked separately it could be the encoder/drive.

Cyclic error

This is mainly due to the actuator of the table e.g., leadscrew.

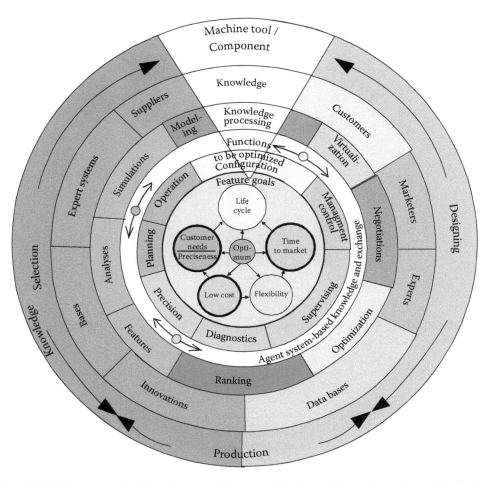

COLOR FIGURE 3.1 Levels of knowledge and data processing by machine tools creation and optimization.

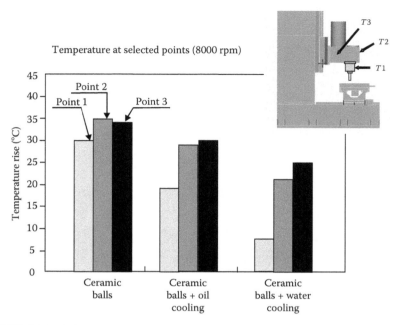

COLOR FIGURE 3.30 Influence of cooling milling machine headstock spindle bearing on temperature at selected points.

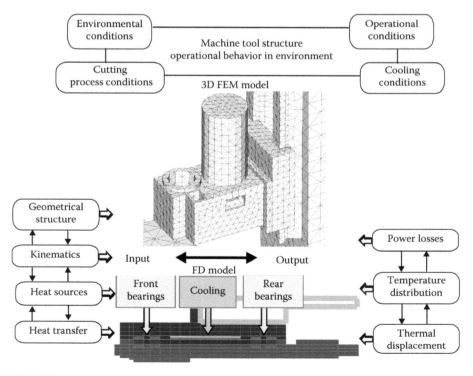

COLOR FIGURE 3.32 Scheme of hybrid model of machine tool headstock.

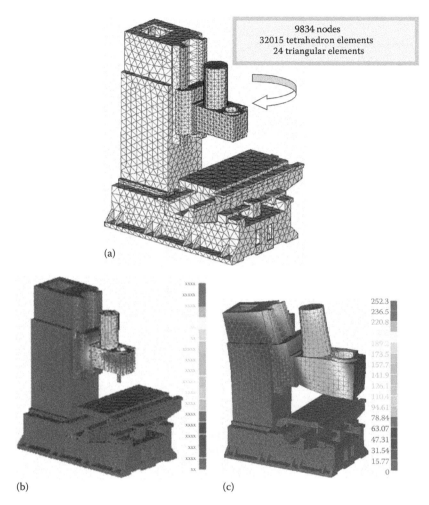

9834 nodes
32015 tetrahedron elements
24 triangular elements

(a)

(b)

(c)

252.3
236.5
220.8
189.2
173.5
157.7
141.9
126.1
110.4
94.61
78.84
63.07
47.31
31.54
15.77
0

COLOR FIGURE 3.35 Example of computing system application to milling center thermal behavior analysis: (a) discretization of a milling center (spindle assembly discretization is not shown), (b) a numerically determined temperature distribution for a steady state at a spindle speed of 8000 rpm and an ambient temperature of 22°C, and (c) the deformations and displacements of the center.

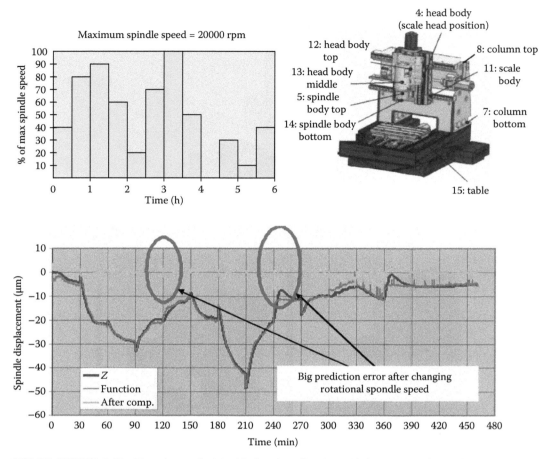

COLOR FIGURE 3.47 Experimentally identified and predicted machining center spindle displacement and theoretical compensation accuracy.

	1920	1940	1960	1980	2000	2020 (Extrapolated[1])
Normal machining	—	60	30	5	1	0.1
Precision machining	—	75	5	0.5	0.1	0.03
High-precision machining	75	5	0.5	0.05	0.01	>0.003
Ultrahigh-precision machining	5	0.5	0.05	0.005	0.001	<0.3 nm

━━━ Precision machining   ━━━ High-precision machining   ━━━ Ultrahigh-precision machining

COLOR FIGURE 4.1 Evolution of machine required precision over the last century. Extrapolation from Prof. Tanigushi's graph, University of Tokyo. Values are indicated in micrometer.

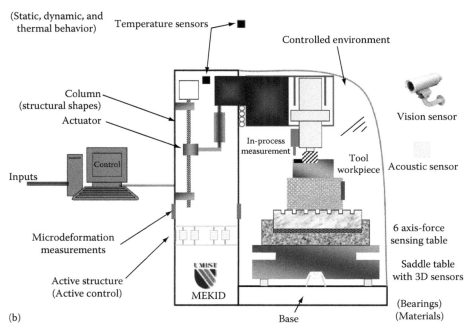

(b)

COLOR FIGURE 4.3b Possible suggestions to address a few problems.

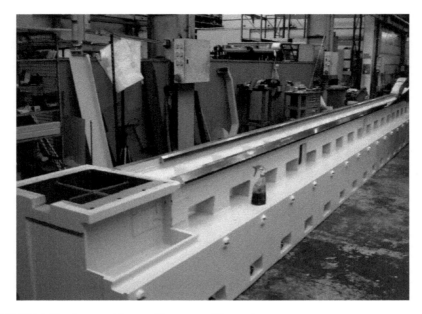

COLOR FIGURE 4.12   Long guideway. (Courtesy of Goratu.)

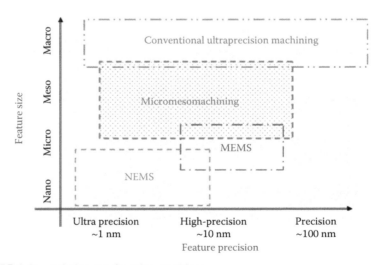

COLOR FIGURE 4.41   Missing gap for micromachines.

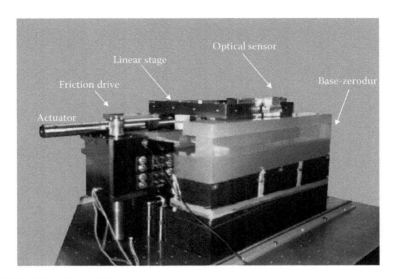

COLOR FIGURE 4.45   Linear slide.

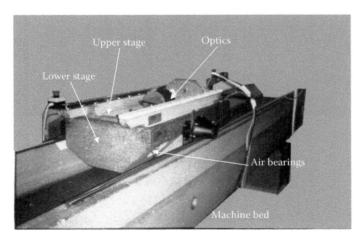

COLOR FIGURE 4.47   Optical delay line.

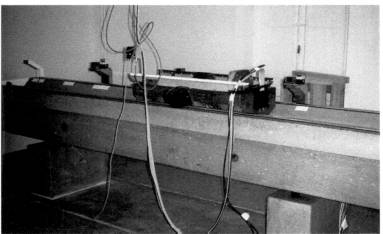

COLOR FIGURE 4.50   Overall optical delay line structure.

COLOR FIGURE 4.52   Magnetic guideway system. (Courtesy of IFW, Hannover.)

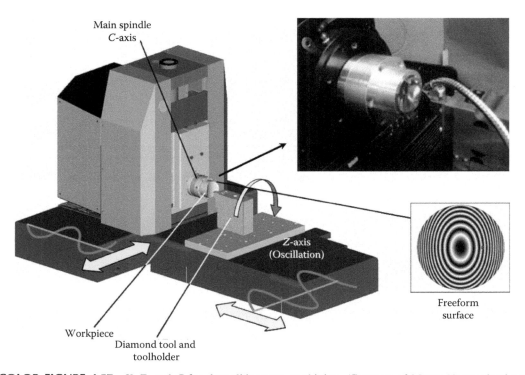

Main spindle
C-axis

Z-axis
(Oscillation)

Freeform
surface

Workpiece

Diamond tool and
toolholder

COLOR FIGURE 4.57   *X*, *Z*, and *C* for slow slide servo machining. (Courtesy of Moore Nanotechnology Systems.)

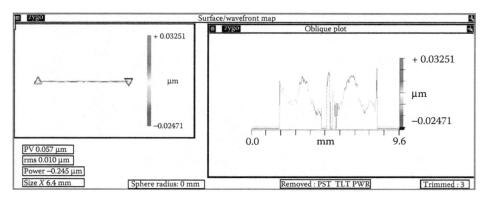

COLOR FIGURE 4.59 Form accuracy: PV = 0.057 μm, 8.6 mm radius as measured on Zygo HeNe interferometer.

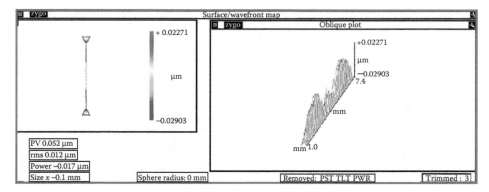

COLOR FIGURE 4.60 Form accuracy: PV = 0.052 μm, 13 mm radius as measured on Zygo HeNe interferometer.

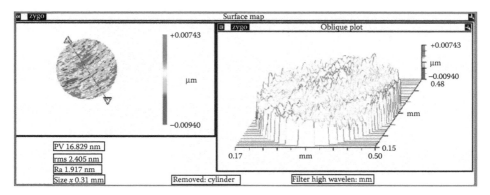

COLOR FIGURE 4.61 Surface finish: Ra = 1.917 nm as measured on Zygo Scanning White Light Interferometer.

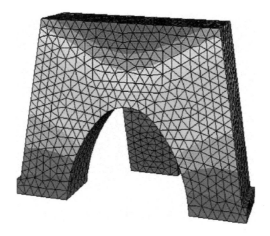

COLOR FIGURE 4.62 Finite element modeling of optimal Three legged structure.

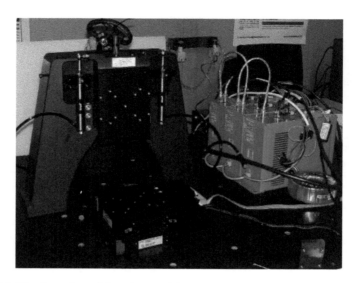

COLOR FIGURE 4.63   Overview of the micromachine.

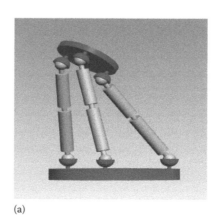

(a)

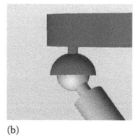

(b)

COLOR FIGURE 5.1   (a) PKM machine and (b) spherical joint.

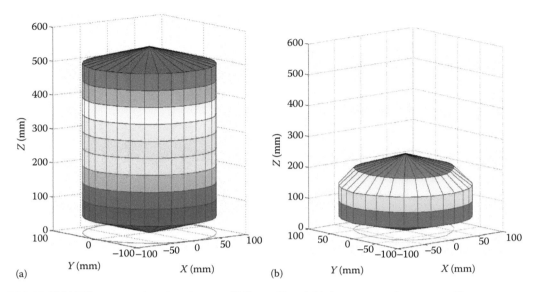

COLOR FIGURE 5.19 (a) Total workspace (SPS), $\gamma = 0°$ and (b) dexterous workspace, $\gamma = 0°$.

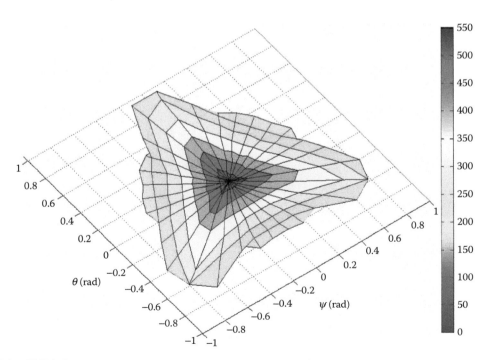

COLOR FIGURE 5.20 Dexterity corresponding to the total w/s for SPS using forward and pseudoinverse approach of conventional Jacobian.

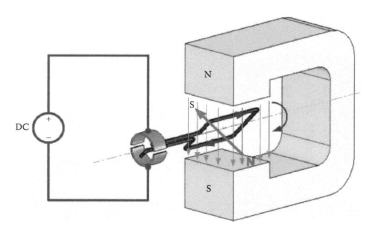

COLOR FIGURE 7.3 Functional diagram of an electric motor.

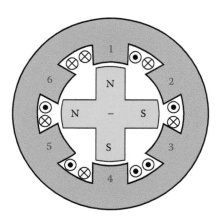

COLOR FIGURE 7.4 Functional diagram of a stepper motor.

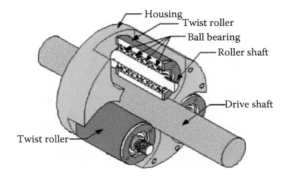

COLOR FIGURE 7.21 Twist friction drive. (From Mizumoto, H., Yabuya, M., Shimizu, T., and Kami, Y., *Prec. Eng.*, 17, 57, 1995.)

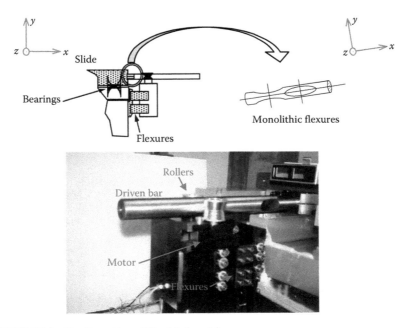

COLOR FIGURE 7.24 Configuration of the friction drive.

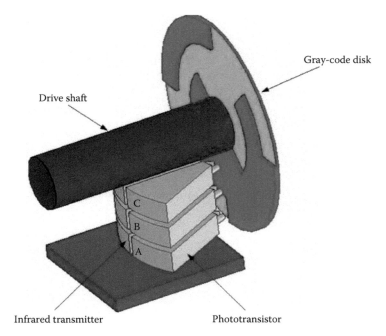

COLOR FIGURE 7.38 Optotransducer.

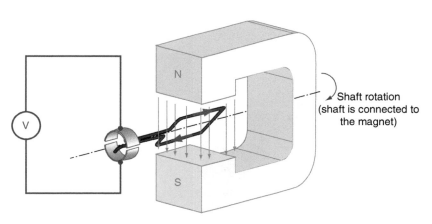

COLOR FIGURE 7.42 Tachogenerator structure.

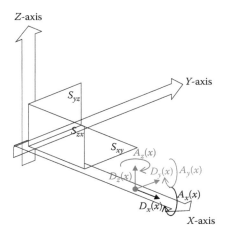

COLOR FIGURE 8.2   Squareness errors between axes.

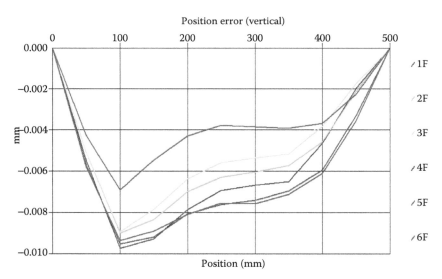

COLOR FIGURE 8.7   Measured x-axis straightness errors $D_y(x)$ at six different thermal conditions, Run #1 to Run #6.

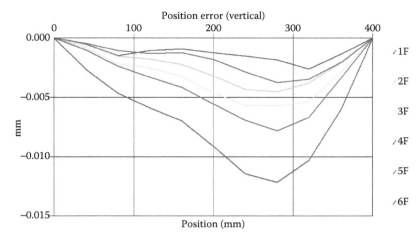

COLOR FIGURE 8.8   Measured y-axis straightness errors $D_x(y)$ at six different thermal conditions, Run #1 to Run #6.

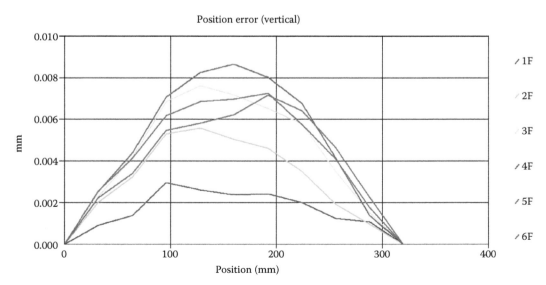

COLOR FIGURE 8.9   Measured z-axis straightness errors $D_x(z)$ at six different thermal conditions, Run #1 to Run #6.

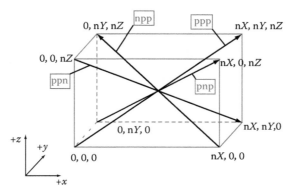

COLOR FIGURE 8.13 Four body diagonal directions.

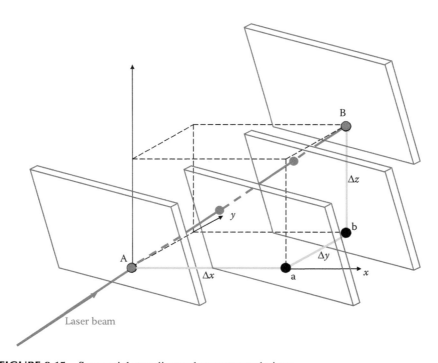

COLOR FIGURE 8.15 Sequential step diagonal or vector technique.

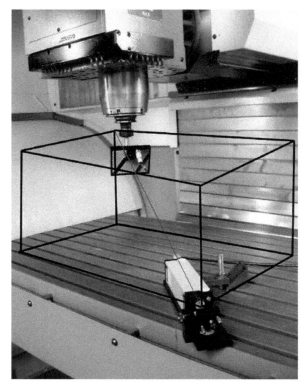

COLOR FIGURE 8.16 A photo of actual laser setup for the vector measurement.

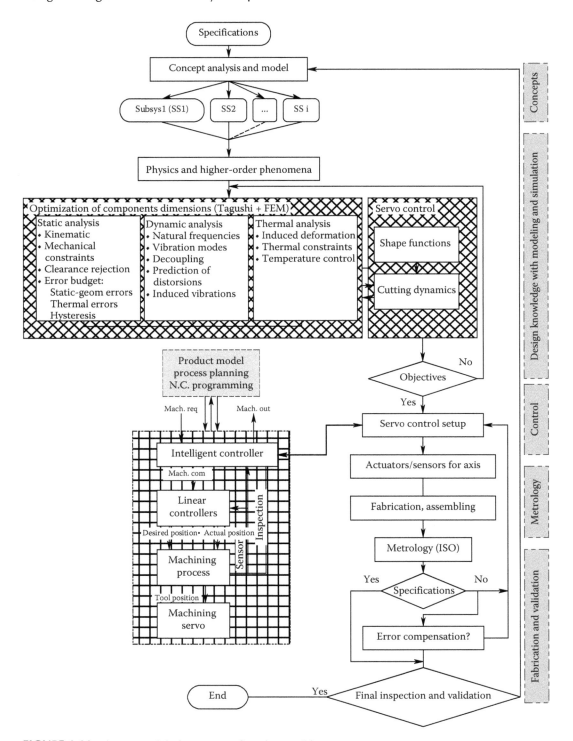

FIGURE 4.44 A proposed design strategy for micromachines.

addressed for hybrid configurations. Exploitation of existing development in microsystems will tremendously help the implementation of specified devices to ensure superior performance in their functionalities. Performance assessment of micromachines is required today to evaluate the real performance and limitations as no results have been published so far.

4.9.8 Current Limitations and Challenges

Currently, the primary technologies used in industry for miniaturization are the microelectronic fabrication techniques (at mesoscale level), such as lithography, LIGA, and their spin-offs to produce MEMS. The principal shortcomings associated with such technologies for MEMS (up to 99% of commercial MEMS production uses LIGA technology) are related to the inability of this technology to produce arbitrary 3D-sculptured form features in not only electronics (Silicon-based material) but also in a wide range of metallic and nonmetallic materials (stainless steel, brass, gold, titanium, plastic, and glass). The new generation of MEMS will have enhanced functions (mechanical, physical, and control) and will be strongly based on 3D shapes micromechanical components fully embedded in intelligent electronic devices to create the so-called micromechatronics devices. Therefore, it is intended to go with further machining miniaturization to reach MEMS scale.

Moreover, while conventional micromachining techniques (e.g., LIGA) can achieve excellent absolute tolerance, relative tolerance is rather poor compared to those achieved by more traditional technique at macroscale level through metal cutting machining. In traditional macromachining, relative tolerance of 0.0001% of an accuracy/part-size ratio is becoming standard, whereas in the Electronics/Integrated Circuit industry a 1% relative tolerance is considered to be good. Then, parts with relative tolerance of 0.01% do actually exclude LIGA, where absolute size in one or more dimensions is in the micrometer range. Alternative processes like micro-EDM can be suitable just for workpieces presenting 2D symmetry; but such a process introduces environment problems as a consequence of using electrodischarge fluids. The current developments are often driven by the philosophy of using regular size (and high costs) equipment and conventional processes to produce miniature components and devices. This, in turn, has resulted in emphasizing ultraprecision machine tools instead of looking towards the possibility of developing new cost-effective and ecoefficient manufacturing processes to achieve the required levels of precision, accuracy, and low-cost productivity in 3D complex shapes component machining. Apart from current research, big-size machines are often used in industry to fabricate microparts where cutting forces are in the milli/micro-Newton range. Indeed, hexapod-based machining centers grabbed a lot of attention in the machine tool world for potential use in large applications; however, they are actually not well suited for large applications. They do, however, scale exceptionally well for small applications (e.g., micromanipulators).

A small variation in the manufacturing process caused by material or cutting tool characteristics, thermal variations in the machine, vibration, and any second-order physical phenomenon would have a direct impact on the ability to produce the required features in mass production.

Handling microparts if not robust would have an impact on the repeatability of a process that has a desired tolerance of less than a micron. Adapting metrology to inspect workpieces and systematic machine accuracy maintenance is another challenge.

4.9.9 Justification: Evaluation of the Cost/Accuracy Ratio in Standard Machines

The evolution of the machining accuracy becomes very tight according to Taniguchi's accuracy expectations in the next two decades. The corresponding gap to be filled by the micromachine is for mesoscale parts up to MEMS with a machining accuracy in the range of submicrometers (Figure 4.41). The driving parameters that justify the design of micromachines are

1. Reduction of energy consumption
2. Reduction of space with an easy control of environment
3. Decrease in material consumption (especially expensive materials)
4. Decrease of the effect of heat deformation
5. Reduced vibration amplitude with large natural frequencies
6. Quality increase in machining accuracy

4.10 EXAMPLES OF PRECISION MACHINE STRUCTURES AND CONCEPTS

A number of examples of precision machines are discussed hereafter. The examples will include initial specifications, description of the key design aspects used, key elements and their characteristics, and current achieved performance.

4.10.1 DESIGN OF THE LINEAR SLIDE

This was the first learning project initiated in 1990s at the University of Technology at Compiegne in France where the author was a PhD student. The aim of the project was to design a linear stage with 16 nm positioning precision and an overall motion error within a virtual cylinder having a length of the slide stroke of 220 mm and a diameter of 1 μm. A steel slide with a mass of 100 kg is fully floated by three hydrostatic bearings. Very high-axis stiffnesses are achieved. The linear axis is driven with a maximal translational speed of 10 mm/s using a friction drive actuator. A closed loop servo-position control, i.e., internal model controller (IMC) is used to compensate automatically for unmodeled mechanical behaviors such as prerolling phenomena (Figure 4.45).

Specifications

- Mass: 100 kg
- Stroke: 220 mm
- Position accuracy: 16 nm
- Travel speed: 10 mm/s

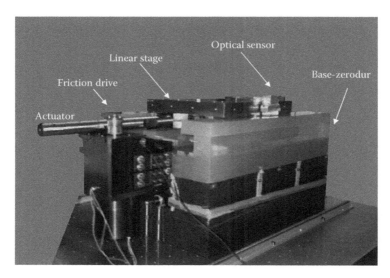

FIGURE 4.45 (See color insert following page 174.) Linear slide.

- Overall geometric error within 1 μm
- Straightness <1 μm
- Pitch and Roll angle <1 arcsec

Subsystems

The subsystems considered in this system were the carriage, the guideways, the hydrostatic bearings, and the actuators.

Identified Sources of Error

- Room temperature
- Yaw angle and lateral stiffness of the slide
- Straightness of guideways
- Stiffness of the hydrostatic bearings
- Coupled pads in the hydrostatic bearings
- Pressure regulation of oil
- Dissipation of power
- Tangential stiffness of the actuator
- Geometric errors of the rollers and the rails
- Rolling frictions
- Electronic nonlinearities of controllers

Modeling and Analysis

- Static analysis for Von-Mises stresses and gravity effects
- Dynamic analysis for natural frequencies and modes of vibrations

First natural frequency: 122 Hz, flexural mode of vibration
Second natural frequency: 249 Hz, pitch mode of vibration

Control Strategy

Internal model control to achieve positioning, speed control, and compensation for errors induced by rolling friction and electronic nonlinearities.

Control bandwidth: 50 Hz.

Measurements (to be Compared to Specifications)

The measurements have been achieved with IMC controller:

- 16 nm resolution (limited by the available sensor resolution)
- Straightness ±0.6 μm
- Variation of pitch angle 0.55 arcsec
- Variation of roll angle 1.3 arcsec
- Well resolved motion at 5 nm. (Figure 4.46)

Optical Delay Line

Two challenges have to be addressed in this case study compared to the previous case. The first one is the long stroke requiring precision positioning; and the second is the tight requirement on the low level of vibration required at positioning. Hence, the concept chosen for long stroke precision positioning is based on master–slave concept, for example a double-stage system. The slave stage (lower)

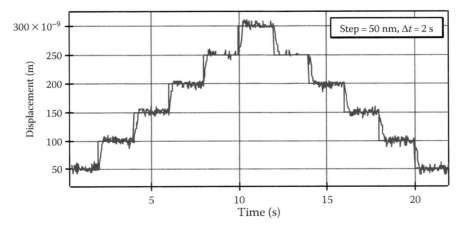

FIGURE 4.46   Continuous step response.

will take the system to a rough position with friction drive actuator and the master (upper stage) will refine the position of the optics using voice coil actuator. The level of vibration needs a dynamic modeling of the whole system. These two particular constraints are discussed next. To secure a long and continuous stroke of 3 m, reinforced concrete was cast on the mountain Massif at the Observatory of Cote d'Azur (France) to construct a support for the delay line.

Purpose of the Design

This is an optical delay line required by ESO for a prototype for Chile telescopes. The recent concept of stellar interferometry consists of collecting light from several telescopes aiming at the same star to build a fringe pattern. To ensure that observation of the image can be pursued independent of the star motion relative to the observation site, the path differences between the light beams coming through different telescopes must be canceled. This compensation is done by optical delay lines consisting of mobile cat's eye retroreflectors where the movement is controlled to follow the sidereal motion, so that the optical paths between the two arms of the interferometer are continuously maintained equal (Figures 4.47 through 4.50).

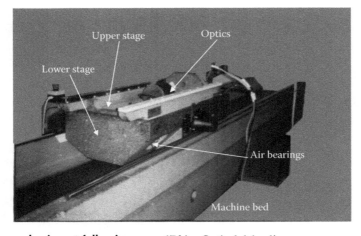

FIGURE 4.47 (See color insert following page 174.)   Optical delay line.

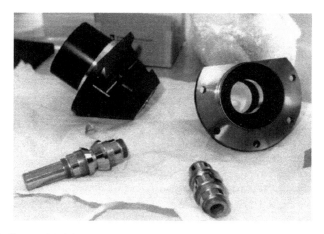

FIGURE 4.48   Ve Rollers and holders.

Specifications

- Mass: 190 kg
- Stroke: 3 m
- Velocity 0.2 km/s to 3 mm/s for tracking mode and up to 40 mm/s in positioning mode
- Straightness ±50 μm
- Pitch and yaw angles ±103 arcsec
- Level of normal vibration
- ±12.5 nm rms for 15 ms exposure time

Subsystems

- Conceptual analysis and comparison (four concepts)
- Upper stage
- Lower stage
- Guideways
- Ground foundation

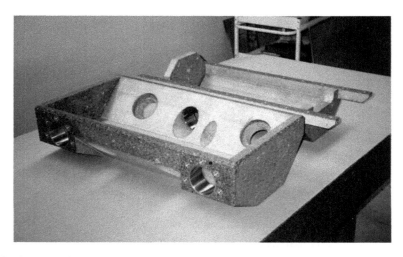

FIGURE 4.49   Lower and upper stages.

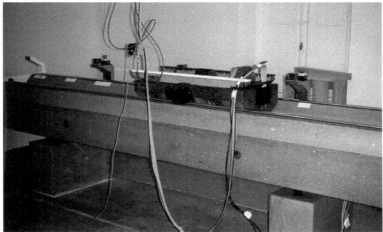

FIGURE 4.50 (See color insert following page 174.) Overall optical delay line structure.

Identified Sources of Error

- Gravity center not located in the guideways plane
- Excitation transmitted from lower to upper stage
- Mass of lower stage (lower to ease control)
- Abbé errors
- Kinematic support for both stages
- Parasitic movement induced by railways waviness
- Air-bearing stiffnesses
- Torsional deflections
- Locations of position and speed sensors

Modeling and Analysis

- Static analysis to determine gravity effects and stresses
- Level of vibration: normal and longitudinal
- Thermal drifts
- Dynamic analysis for modes of vibrations and natural frequencies

The modeling using FEM of the lower stage on the rollers contact stiffnesses with the upper stage on air-bearing stiffnesses shows that the first natural frequency is 81.2 Hz, which corresponds to the pitching mode of the upper stage. The second and the third modes are 138.9 and 242.3 Hz related respectively to the yawing mode and rolling mode of the upper stage. The above first natural frequencies of the two stages are chosen to be out of the control bandwidth. The previous computed frequencies are obtained by using a bearing stiffness of 50 N/μm. The air-bearing stiffness is adjustable to secure high stability during the experimental measurements.

Control Strategy

Two modes of control: positioning and tracking with rejection of low-frequency perturbations.
  Master (upper stage) and slave (lower stage) strategy.
  Control bandwidth: 50 Hz.

Measurements (to be Compared to Specifications)

After establishing the budget of errors and compensating for the majority of errors:

- Lateral straightness: 7.5 μm
- Vertical straightness: 22.1 μm
- Pitch angle: 103 arcsec
- Yaw angle: 61.8 arcsec

Level on vibrations:

- 12.4 nm rms for 15 ms exposure time

4.10.2 THE SCHNELLE MACHINE

4.10.2.1 Overview

The prototype center consists of three machining axes driven by linear direct drives. The x- and y-axis host the workpiece while the z-axis hosts the tool. The workspace of the machine is estimated to be $500 \times 500 \times 300$ mm³. The x- and z-slide of the machine are driven by two parallel drives to provide higher-acceleration forces. Both the x- and y-axes are guided by conventional ball type guides, whereas for the z-axis of the machine a new type of active magnetic guide has been developed. The drives were selected for a maximum feed acceleration of 5 g in all three axes, while the HSC work spindle has a cutting power of 15 kW at a maximum speed of 60,000 rpm (Figures 4.51 through 4.53).

4.10.2.2 Machine Function and Purpose of the Design

High-speed and -dynamic machining strategies require specific solutions for the components. The limits for speed and accuracy depend mainly on the machine structure, the drives, and the guides. High-performance machine frames based on granite, cement concrete, or polymer concrete provide a significant higher damping ratio compared to conventional frames made of welded steel or cast iron. Therefore, the base frame of the machine prototype is made of Hydropol, a modern composite consisting of a steel reinforced sheet metal cover filled with cement concrete. Besides the high damping of the composite due to friction effects, the frame provides enough mass to absorb the large acceleration forces of the axes. The mass of the machine frame is approximately 15.85 t. Like the frame, the machine slides are excited by driving and processing forces. Although being lightweight,

FIGURE 4.51 Side view of the schnelle machine. (Courtesy of Leibniz University Hannover, IFW, Hannover.)

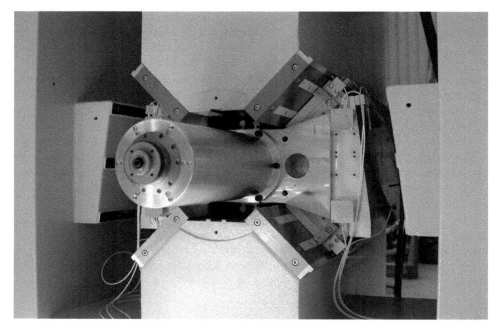

FIGURE 4.52 (See color insert following page 174.) Magnetic guideway system. (Courtesy of Leibniz University Hannover, IFW, Hannover.)

FIGURE 4.53   Work space of the machine. (Courtesy of Leibniz University Hannover, IFW, Hannover.)

they need to provide high stiffness and damping. Sandwich structures fulfill these requirements best, so for the gantry x-axis of the machine a lightweight honeycomb steel structure was chosen. With conventional roller guideways, the positioning accuracy and speed of a machine axis is limited due to friction and low damping. In high-speed machining, these effects are responsible for increasing contour errors. Additionally, wear of the guideways will degrade the machining accuracy. This will reduce the lifetime of the machine. Hence, alternative guiding solutions based on new concepts have been investigated. Active magnetic guides represent a new and contactless alternative for high-end applications. They avoid the undesired effects of roller bearings and can achieve extremely high damping ratios. Additionally, the dynamics of the machine axes, namely, vibrations and path errors perpendicular to the feed direction, can be actively influenced by appropriate feed back and feed forward control. The contactless axis is also suited for fine positioning in the x- and y-direction. This gives the possibility to compensate contour errors of the linear drives based on the higher-dynamic bandwidth of the guides.

Positioning Uncertainty

Uncompensated positioning uncertainties caused by the machine frame do not exceed ±6 μm. Using the magnetic guideway system, these uncertainties have been successfully compensated to stay below ±1 μm. The improvement of the dynamic positioning behavior is currently under investigation.

Natural Frequencies

Table 4.13 shows calculated and measured eigenfrequencies of the x-slide.

TABLE 4.13
FEM Calculated and Measured Normal Modes

No.	Frequency (Simulation) (Hz)	Frequency (Experimental Measurements) (Hz)	Difference (%)	Damping (Experimental Measurements) (%)
1	156	150	3.8	0.61
2	305	301	1.3	1.10
3	306	299	2.3	0.68

Thermal Problems

Currently, no problems due to thermal issues have occurred. Possible displacements caused by thermal effects can be compensated by the magnetic guideway system. The power inserted into the system by the magnetic guideway itself is very low and does not affect the machine precision. As a benefit from the frictionless guide, the z-axis does not show wear or thermal influences due to friction.

Sample Component

A first sample component is shown in Figure 4.54. This component has been milled without any active compensation. The resulting accuracy of the test piece is shown in Figure 4.55.

Smart Features

As a prominent component of the machine prototype, the magnetic guideway system does not only feature a friction and wearless operation, but is also capable of implementing further possibilities to increase precision and to actively compensate vibrations. Using the magnetically guided z-axis as a highly dynamic additional drive, positioning deviations can be compensated online. Furthermore, the integrated sensor network can be employed to monitor process forces and stability. This enables the machine to perform a continuous process-force measurement and adjust to undesired operating conditions as well as detect disturbances such as a tool breakage.

4.10.3 Machine Configurations with Moore Nanotechnology Systems

Single point diamond turning machines dedicated to ultrahigh-precision machining are designed with up to five axes (Figure 4.56). The combination of axes is used to generate various types of surface shapes as explained hereafter.

- Axes X and Z are to be used to machine rotationally symmetrical surfaces like flats, cones, spheres, aspheres, and diffractives.
- Combining axes X, Z, and C allows also certain types of freeform shapes to be generated by "slow slide servo" machining. This technique offers hardly any limitation on stroke depth.
- Combining axes X, Z, and Y allows another class of freeform shapes to be generated by "Raster" machining, such as certain microgrooved components and microcorner cubes.
- The B axis can be utilized in addition to any of the above configurations. It is mainly used for "tool normal" machining and wheel normal-grinding applications.

Various types of attachments can be added on the Z-slide of the machines for fly-cutting, micro-grinding, or micromilling. These could be rotary indexing table, motorized Y-axis positioning stage, flexible spindle attachments for vertical, horizontal, or 45° tilted grinding or milling applications.

Cutting velocity = 380 m/min
Feedrate = 3 m/min
Feed per tooth = 0.1 mm

FIGURE 4.54   Test workpiece manufactured on the machine.

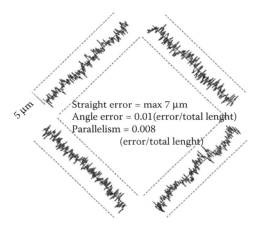

5 μm

Straight error = max 7 μm
Angle error = 0.01(error/total lenght)
Parallelism = 0.008
                (error/total lenght)

FIGURE 4.55   Exemplary accuracy measurements of test workpiece.

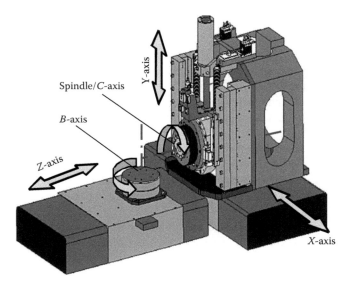

Y-axis

Spindle/*C*-axis

B-axis

Z-axis

X-axis

FIGURE 4.56   3D design concept for 5 axis machine. (Courtesy of Moore Nanotechnology Systems.)

Some design features characterize such type of machines and are briefly introduced as follows:

- The machine base is made with a metallic robust frame that holds a synthetic epoxy granite base for maximum stability. It is also secured by four air bags for vibration isolation, two of which are tied together for kinematic support. The base incorporates significant ballasting to lower the center of gravity of the machine relative to the air isolation system.
- The slideways are oil hydrostatic and follow the kinematic design discussed earlier in this chapter.
- The Y axis with integral work spindle is counter balanced with an adaptive pneumatic cylinder which senses and adapts to changes in workpiece mass.
- Brushless DC linear motor drives are used on all linear axes.

4.10.3.1 Slow Slide Servo Machining (S3)

To generate freeform optical surfaces, Moore Nanotechnology Systems (Keene, NH) has introduced an alternative method via a novel slow slide servo technique. The slow slide servo technique is similar to the fast tool servo (FTS), in that the part is mounted on the spindle and as the spindle rotates, the tool oscillates (Figure 4.57). Unlike the FTS method, this system does not use any additional axes for oscillating the tool; the Z-axis slide generates the oscillations. Another difference is the spindle position control (or C-axis). In an FTS setup, the spindle has an encoder that feeds the position to the FTS unit without putting the spindle in position control. In a slow slide servo, all axes are under fully coordinated position control. The slow slide servo can oscillate at ranges up to 35 mm and beyond, is easy to setup, inexpensive, and allows the manufacturing of highly accurate freeform parts.

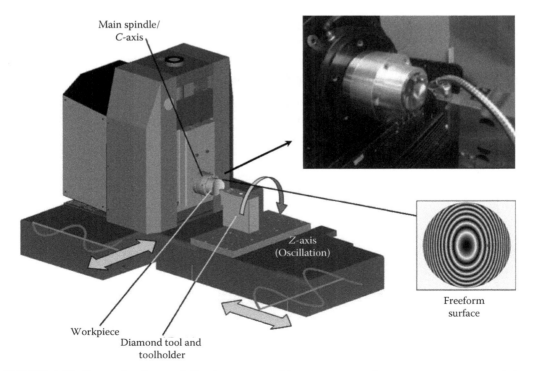

Main spindle/
C-axis

Z-axis
(Oscillation)

Freeform
surface

Workpiece

Diamond tool and
toolholder

FIGURE 4.57 (See color insert following page 174.) X, Z, and C for slow slide servo machining. (Courtesy of Moore Nanotechnology Systems.)

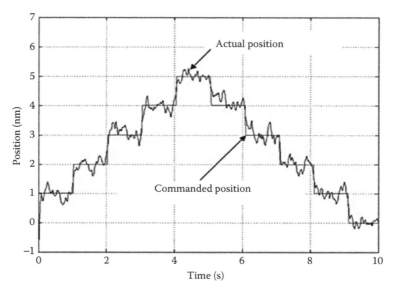

FIGURE 4.58 Nanometer step motion with previous machines. (Courtesy of Moore Nanotechnology Systems.)

To implement the slow slide servo method, several key features must be available on a diamond turning lathe. Most of these features are the same for both the linear and the rotary axes. They include friction-free oil bearings that generate little heat, direct drive motors with no mechanical compliance between the motor and the feedback system, high-resolution encoders, and minimal structural dynamics in the control loop. Another key feature is the control system or the CNC. The CNC must have high-speed data processing, look-ahead capability, a high-resolution data acquisition system, and high-order trajectory generation.

It is aimed to achieve ultraprecision motion well resolved as shown in Figure 4.58.

4.10.3.2 Example of Inspected Machined Surfaces

A slow slide servo diamond machining of a toric surface shows from the inspection of the machined surface a submicron accuracy in form and nanometre level surface finish. Data related to the workpiece are given (Figures 4.59 through 4.61)

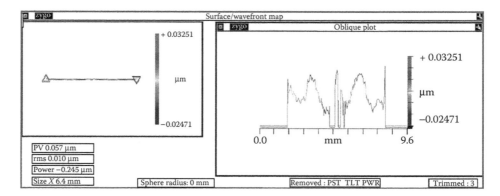

FIGURE 4.59 (See color insert following page 174.)   Form accuracy: PV = 0.057 μm, 8.6 mm radius as measured on Zygo HeNe interferometer.

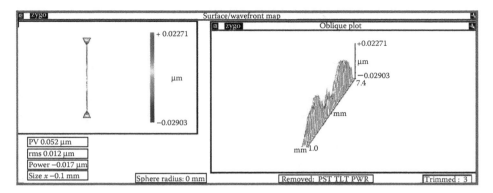

FIGURE 4.60 (See color insert following page 174.)   Form accuracy: PV = 0.052 µm, 13 mm radius as measured on Zygo HeNe interferometer.

- Diameter: 10 mm
- Radius (R1): 8.6 mm
- Radius (R2): 13 mm

4.10.4 MANCHESTER MICROMACHINE

A desktop three axis micromachine has been designed and manufactured at the University of Manchester as part of the research development in the area of micromachining. A large investigation has taken place prior to the design of such a machine [40–48]. The following is a brief description of on-going project to build a micromachine.

For the current machine, three important design parameters of the tapered egg bridge design were studied to determine how each of them influences the performance characteristics of the bridge. The design parameters are width of each leg of the column, height of column, and its thickness. The observed characteristics of the vertical column are natural frequency and stiffness/deflection. While the highest-possible first natural frequency is preferred, the deflection should be ideally zero. Using FEM and Taguchi technique, a reduced set of finite element simulation was conducted in terms of Taguchi orthogonal array. Classical analysis of variance was applied to the simulation responses to determine the influence of design parameters on the product performance. Result shows that the

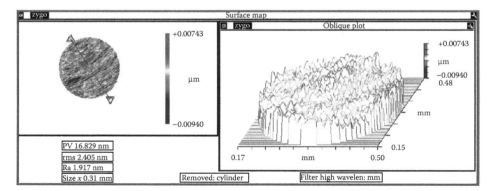

FIGURE 4.61 (See color insert following page 174.)   Surface finish: Ra = 1.917 nm as measured on Zygo Scanning White Light Interferometer.

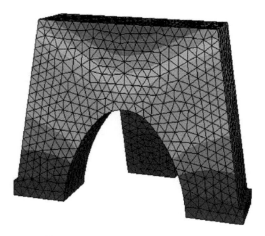

FIGURE 4.62 (See color insert following page 174.) Finite element modeling of optimal three legged structure.

thickness of the column has the highest contribution to both natural frequency and deflection of the column. Also, it shows that the width of each leg influences the stiffness while the height influences the natural frequency.

The equally sized three legged (Figure 4.62) tapered egg bridge design revealed that the desirable characteristics of the structure is greatly improved with the optimized three legged tapered egg bridge design showing a remarkable improvement with $0.0572\,\mu m$ deflection and $1025.5\,Hz$ natural frequency. Further studies are related to dynamic stiffness in all directions.

The three axes of the machine are composed of ALS 130 H from Aerotech assembled with an orthogonality of around 5 arcsec. Positioning could easily resolve 10 nm steps as shown previously. The vertical axis uses a counterbalance. The controller is A3200 with open architecture planned for compensation and in-process inspection (Figure 4.63).

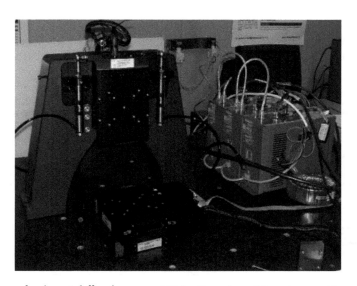

FIGURE 4.63 (See color insert following page 174.) Overview of the micromachine.

REFERENCES

1. McKeown, P. A. 1988. The development of high precision CNC machine tools, *Journal of Mechanical Working Technology*, 17, 225–236.
2. Mekid, S. and Olejniczak, O. 2000. High precision linear slide. Part II: Control and measurements, *International Journal of Machine Tools and Manufacture*, 40(7), 1051–1064.
3. Mekid, S. and Bonis, M. 1997. Conceptual design and study of high precision translational stages: Application to an optical delay, *Precision Engineering*, 21(1), 29–35.
4. Tan, K. K. et al. 2001. *Precision Control*, Springer, London.
5. Pollard, A. F. C. 1922. *The Mechanical Design of Scientific Instruments,* Cantor Lectures, Imperial College of Science and Technology, Parsons Ltd., Hastings.
6. Pollard, 1929. *The Kinematic Design of couplings in Instrument Mechanisms*, Imperial College of Science and Technology, Hilger and Watts Ltd., London, United Kingdom.
7. Teague, E. C. and Evans, C. 1989. *Patterns for Precision Instrument Design (Mechanical Aspects)*, National Institute of Standards and Technology, Tutorial Notes, ASPE Annual Meeting, Norfolk, Virginia.
8. Slocum, A. 1992. *Precision Machine Design*, SME, U.S.
9. Smith, S. T. and Chetwynd, D. G. 1992. *Foundations of Ultraprecision Mechanism Design*, Gordan and Breach Science Publishers, U.K.
10. Nakazawa, H. 1994. *Principles of Precision Engineering*, Oxford University press, United Kingdom, ISBN 0-19-856266-7.
11. Hicks, T. and Atherton, P. 1997. *The Nanopositioning Book*, Qeensgate Instruments.
12. Den Hartog, J. P. 1987. *Advanced Strength of Materials*, Dover, New York.
13. Timoshenko, S. P. 1982. *History of Strength of Materials*, Dover, New York.
14. Kobayashi, T. et al. 1994. The effects of the damping characteristics of slideways on the dynamic characteristics of workpiece fixtures mounted on machine tool tables, Proceedings of the Institution of Mechanical Engineers. Part B. *Journal of Engineering Manufacture*, 208, 245–251.
15. Fourka, M., Tian, Y., and Bonis, M. 1996. Prediction of the stability of air thrust bearings by numerical analytical and experimental methods, *Wear*, 198(1–2), 1–6.
16. Zhang, X., Shinshi, T., Li, L., and Shimokohbe, A. 2003. A combined repetitive control for precision rotation of magnetic bearing, *Precision Engineering*, 27(3), 273–282.
17. Kim, W. and Trumper, D. L. 1998. High-precision magnetic levitation stage for photolithography, *Precision Engineering*, 22(2), 66–77.
18. Mizumoto, H., Arii, S., Yabuta, Y., Kami, Y., and Tazoe, Y. 2004. Active aerostatic bearings for ultraprecision applications, *Proceedings of the 35th International MATADOR Conference*, Springer, London, pp. 289–292.
19. Corbett, J. 1999. The Development of 2-layered porous ceramic aerostatic guideway bearings for ultra precision applications, *EPSRC Research Grant Report GR/K 89801*, Cranfield University, London.
20. Mekid, S., Bonis, M., Glentzlin, A., and Sghedoni, M. 1995. High precision optical delay line for stellar interferometers, *International Precision Engineering Seminar 8th Compiegne*, France, May 95, Elsevier, pp. 495–499.
21. Mekid, S. 2000. High precision linear slide. Part 1: Design and construction, *International Journal of Machine Tools and Manufacture*, 40(7), 1039–1050.
22. Rivin, E. I. 1995. Vibration isolation of precision equipment, *Precision Engineering*, 17, 41–56.
23. DeBra, D. B. 1992. Vibration isolation of precision machine tools and instruments, *Annal of the CIRP*, 41(2), 711–718.
24. Naotaki, O. et al. 2000. Desktop machining micro-factory, *Proceedings of 2nd Internarnational Workshop on Micro-Factories*, Switzerland, pp. 14–17.
25. Okazaki, Y., Mishima, N., and Ashida, K. 2002. Micro factory and micro machine tools, *The 1st Korea-Japan Conference on Positioning Technology*, Daejeon, Korea.
26. Tanaka, M. 2001. Development of desktop machining microfactory, *Riken Review*, 34, 46–49.
27. Maekawa, H. and Komoriya, K. 2001. Development of a micro-transfer arm for a micro-factory, *Proceedings of IEEE International Conference on Robotic Sand Automation*, Seoul, Korea.
28. Kitahara, T et al. 1996. Development of micro-lathe, *Journal of Mechanical Engineering Laboratory*, 50(5), 117–123.
29. Mishima, N., Ashida, K., Tanikawa, T., and Maekawa, H. 2000. Development of desktop machining microfactory, *Japan–USA Flexible Automation Conference*, Michigan.

30. Ito, S. et al. 2004. Precision turning on a desk- micro turning system, *EUSPEN Conference*, Glasgow, United Kingdom, May–June.

31. Kussul, E. et al. 2002. Development of micromachine tool prototype for microfactories, *Journal of Micromechanics and Micro Engineering*, 12, 795–812.

32. Naotake, O. et al. 2000. Desktop machine microfactory, *Proceedings of 2nd International Workshop on Microfactories*, Switzerland 9th and 10th October, pp. 14–7.

33. Kitahara, T, et al. 1996. Development of micro-lathe, *Journal of Mechanical Engineering Laboratory*, 50(5), 117–123.

34. Mekid, S. 2005. Design strategy for precision engineering: Second order phenomena, *Journal of Engineering Design*, 16(1), 63–74.

35. Zhang, D., Xi, F., Mechefske, C. M., and Lang, S. Y. T. 2004. Analysis of parallel kinematic machine with kinetostatic modelling method, *Robotics and Computer-Integrated Manufacturing*, 20(2).

36. Alting, L., Kimura, F., Hansen, H. N., and Bissacco, G. 2003. *Micro Engineering*, CIRP 52, 635–657.

37. Smith, M. H., Annaswamy, A. M., and Slocum, A. H. 1995 Adaptive control strategies for a precision machine tool axis, *Precision Engineering*, 17(3), 192–206.

38. Pritschow, G., Altintas, Y., Jovane, F., Koren, Y., Mitsuishi, M., Takata, S., van Brussel, H., Weck, M., and Yamazaki, K. 2001. Open control architecture—past, present and future, *CIRP Annals—Manufacturing Technology*, 50(2), 463–470.

39. Koren, Y. 1998. *Open Architecture Controllers for Manufacturing Systems*, Open Architecture Controllers, ITIA series.

40. www.nanowave.co.jp.

41. Lu, Z. and Yoneyama, T. 1999. Micro cutting in the micro lathe turning system, *International Journal of Machine Tool and Manufacture*, 39, 1171–1183.

42. Ashida, K., Mishima, N, Maekawa, H., Tanikawa, T., Kaneko, K., and Tanaka, M. 2000. Development of desktop machining microfactory, *Proceedings of J–USA Symposium on Flexible Automation*, pp. 175–178.

43. Okazaki, Y., Mori, T., and Morita, N. 2001. Desk-top NC milling machine with 200 krpm spindle, *Proceedings of 2001 ASPE Annual Meeting*, 192–195, 6/6 Reported in The 1st Korea-Japan Conference on Positioning Technology, Daejeon, Korea, 2002.

44. www.kern-microtechnic.com.

45. Mekid, S., Gordon, A., and Nicholson, P. 2004. Challenges and rationale in the design of a miniaturised machine tool, *International MATADOR, Conference*, UMIST.

46. Khalid, A. and Mekid, S. 2006. Design of precision desktop machine tools for meso-machining, *Proceedings of the 2nd Virtual International Conference on Intelligent Production Machines and Systems*, Elsevier, Oxford.

47. Khalid, A. and Mekid, S. 2006. Design and optimization of a 3-axis micro milling machine, *6th International Conference on European Society for Precision Engineering and Nanotechnology*, Baden, Austria, May 2006.

48. Mekid, S. and Khalid, A. 2006. Robust design with error optimization analysis of a 3-axis CNC micro milling machine, *5th CIRP International Seminar on Intelligent Computation in Manufacturing Engineering*, CIRP ICME '06, July 25–28, 2006, Ischia, Naples, Italy.

5 Introduction to Parallel Kinematic Machines

Samir Mekid

CONTENTS

Do parallel lines meet in infinity?

5.1   DEFINITION

A parallel kinematic machine (PKM) is a closed loop mechanism of which the end-effector, e.g., mobile platform or tool tip, is connected to the machine base by various kinematic chains. The parallel links are actuators employed with prismatic or rotating actuation elements.

PKM is found in industry in various aspects. It is used mainly in machine tools, medical applications for ophthalmic surgeries, fast packaging, flight simulators for both fixed wing and rotary wing aircrafts, precise positioning for very large telescopes, receiving antennas, and as satellite platforms for maneuvering in space.

Usually, a serial kinematic machine (SKM) presents an isotropic behavior with respect to the workspace, while a PKM presents an anisotropic behavior. However, with a parametric design approach, isotropy in workspace and stiffness could be achieved [1]. A system is completely anisotropic in a singularity position defined by the determinant of the Jacobian equal to zero.

In singular configuration, a PKM could gain one or more degrees of freedom (DOF); hence, the tool tip, for example, cannot resist forces or torques applied to it even if the actuators and the joints are locked. Precision PKMs requiring accurate motion and high stiffness are mainly affected by this aspect. The following are some useful definitions required for PKMs:

1. Total workspace: Total reachability of the end-effector, located at the moving platform.
2. Dexterous workspace: It is defined as the real workspace (x, y, z) in 3D space of the manipulator having superior capability of maneuvering.
3. Dexterity: It is defined within the workspace as a superior capability of PKM to achieve full orientations in 3D space.

5.2   INTRODUCTION

Theoretical works related to parallel mechanisms particularly hexapods, date back to centuries ago, when the geometricians were obsessed with polyhedra.* One of the first devices, a spherical parallel mechanism, was designed by James E. Gwinnett in the United States. He applied for a patent in 1928 to be used as an amusement device. In 1947, a new parallel robot was invented in United Kingdom by Dr. Eric Gough, the one that became the most popular, the variable-length-strut octahedral hexapod. The universal tire-testing machine, or the universal rig, as Dr. Gough called his brainchild, was invented in order to respond to problems of aero-landing loads. Gough platform was the arrangement of the six struts. The machine was fully operational in 1954 and a decade later, the machine was upgraded with digitally controlled motor drives. In 1971, the U.S. Patent and Trademark Office granted a patent to Klaus Cappel for his invention and its use as a motion simulator. But Stewart platform evolved into a popular research topic of robotics only in the 1980s and in the 1990s. There has been a steady increase in the research of parallel manipulators and Stewart platforms.

5.3   COMPARISON OF SERIAL AND PARALLEL SYSTEMS

Serial machines are different from parallel machine tools in many ways. Few attempts have been made to compare the properties of the two systems. But the common performance criterion is the key to converge to a final decision. Tlusty et al. [2] have compared commercial SKM and PKM mainly for stiffness and acceleration capabilities. It was revealed that the performance of constant strut leg PKM is better mainly due to high stiffness than the more flexible variable strut machine. But the constant strut machine will have limitations in the workspace.

The serial machine's working volume can be a parallelepiped due to three Cartesian axes. Whereas the parallel machines give a variety of shapes of the working volume. It depends basically on the type of joints in their structure. The shape of the working volume can be spherical, cylindrical, or elliptical. Commercial parallel machines have a large footprint and give a small workspace as

---

* Polyhedra is a plural of polyhedron and is often defined as a geometric object with flat faces and straight edges.

TABLE 5.1

Basic Comparison between SKM and PKM

Property	SKM	PKM
Working volume/total size of machine	2	1
Error accumulation	1	3
Accuracy	3	2
Static stiffness	3	3
Axis acceleration	1	3
Cutting forces	1	3
Machining of ≤5 faces in single setup	1	2–3
Range of angular motion (reaching to 90°) dexterity	3	2

1, least suitable; 2, average; 3, most suitable.

compared to their size. However, the ratio of machine working volume to the overall machine size is better in serial systems. Serial machines exhibit a homogenous stiffness throughout its workspace, as minor change in the stiffness can harm the volumetric accuracy.

In PKM, the stiffness was an issue in the earlier versions, but now, the commercial PKM like the "Variax" has a theoretical stiffness value of $175\,N\,\mu m^{-1}$ compared with $35\,N\,\mu m^{-1}$ for a typical conventional serial machine tool [3]. However, such a high stiffness can be a potential benefit in cutting very hard materials [4].

Serial machines accumulate errors for every axis due to its inherent design of being a serial chain of elements. All the geometrical errors of each axis get multiplied to become a final volumetric error for the tool actual location. In contrary, parallel structures consist of simple elements. There is no bending moments involved as all closed loop kinematic chains share the load of the moving platform. In serial structure, the outer most axes bear the entire load, thus becomes vulnerable to more structural bending errors.

Although PKMs have less sources to produce errors but today serial machines are taking the lead for getting higher accuracies. A couple of examples on the accuracies attained from both the micro- and standard-size machines are reviewed in Ref. [5]. Moreover, errors can be compensated in serial and parallel machines using off-line and online state-of-the-art error compensation techniques. In a PKM, the movement and compensation in single axis requires a movement in all actuators, while in serial, only one actuator per axis is needed. Some properties of machine tools compared both in serial and parallel systems are shown in Table 5.1.

5.4 PRECISION DESIGN OF A PKM

The precision of a PKM will only be secured by the type of the concept chosen that generates precision motion. Previous concepts used in serial machine for precision bearings and precision actuators could be converted into precision joints that is key elements here to secure precise motion. PKMs are designed for high load-carrying capacity, good dynamic performance, and precise positioning; therefore, a systematic approach is required to form an integrated virtual environment for PKM design, analysis, validation, path planning, and remote control to be used in the early stage for conceptual design of PKM. Zhang et al. [6] have discussed the virtual integrated system specifically for parallel kinematic manipulators.

Motion is guaranteed in PKMs by a sequence of actuation provided by the actuators in which axial stiffnesses will contribute to the overall PKM stiffness. It is understood that the overall stiffness will also depend on the posture of the PKM, hence the importance of the concept chosen for the machine. Maintaining the stops for different postures is provided by the joints themselves which will contribute to the overall stiffness (Figure 5.1).

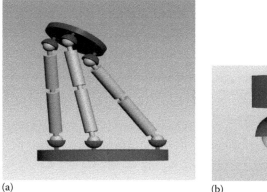
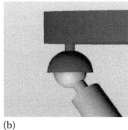

(a) (b)

FIGURE 5.1 (See color insert following page 174.) (a) PKM machine and (b) spherical joint.

5.5 WHY DO WE WANT A PKM?

PKM offers advantages as mentioned previously over a serial machine. It is required to secure better workspace with low singularities inside. To make a PKM design as an alternative for the machine tool application, certain specifications are to be met which make the design competitive enough to be comparable with its serial counterpart. There are certain objectives set for this purpose.

5.5.1 LOW COST

The parallel kinematic design already offers low structural cost due to the light weight and similar elements. If these machines are made in large quantities, the element manufacturing cost will be cheaper. But, for a one-off prototype, this benefit cannot be exploited fully because the number of components will be very few. The other costly item can be a multi-axis controller required for precise movement and actuation.

5.5.2 DEGREES OF FREEDOM

All six DOF are desired for the full movement of the manipulator. Three translations and rotations are required to be available in the large workspace. The main issue in PKM is the identification of translations and rotations available in a particular direction or portion in the workspace.

5.5.3 LARGE WORKSPACE VOLUME WITHOUT SINGULARITIES

In PKM, workspace volume can take different shapes depending upon the number of limbs, length of limbs, and type of joints, e.g., the shapes can be cylindrical, spherical, or elliptical. Singularities can be found within or at the boundary of the workspace. Singularities found at the boundaries are mostly due to the maximum stretch of the limbs and are called inverse kinematic singularity. In forward kinematic singularities, manipulator can have additional DOF and redundant motion can take place even if the actuators are locked [7,8].

The manipulator workspace must avoid both inverse and forward singularities and the limb interference as well. The selection of the type of joints will also play a decisive role in formulating the manipulator workspace and hence dexterity.

5.5.4 HIGH STIFFNESS

Due to the light weight structural components and joints, only PKM with constant length struts can offer high stiffness as compared to the SKMs. In milling application, if six variable length limbs in a configuration similar to Stewart platform are used, then it is required to be seen whether the available stiffness can reach to comparable level of serial machine stiffness in the PKM full workspace.

5.5.5 HIGH AGILITY

The parallel kinematic machine must be highly agile so that it can accelerate more than $3g$ which is a normal agility found in state-of-the-art machine tools. In fact, agility is not a drawback in PKM systems. Rather, they are better in agility performance than the serial machines where the axes are stacked on one another and the lower most axis has the lowest speed. However, in PKM system, high velocities are available only in some configurations depending upon the condition number of the Jacobian which will be defined and explained in Section 5.8.3. The place where the condition number value is large, high velocity of the end-effector is not possible.

5.5.6 REPEATABILITY IN MOVEMENT

The system must have a very precise positioning accuracy that can be specified in the submicron region. Positioning accuracy of the axis will ensure the fine repeatability in movement. A very tight tolerance level is to be met for the manufacturing of machine's mechanical components as well as a high performance controller is also required.

5.5.7 LOW INERTIA

The micro-PKM system will be light weight structures offering low noise and vibrations, low inertia, and higher natural frequencies than the serial machines. This objective is actually an inherent design characteristic of a PKM.

5.6 PKM CONFIGURATIONS AND CHARACTERISTIC ISSUES

Many configurations of PKM exist already and the number of possible combinations and concepts has increased tremendously. An overview of the PKM characteristics can be given using different types of joints and varying number of limbs. In Figure 5.2, three properties are discussed using different joints

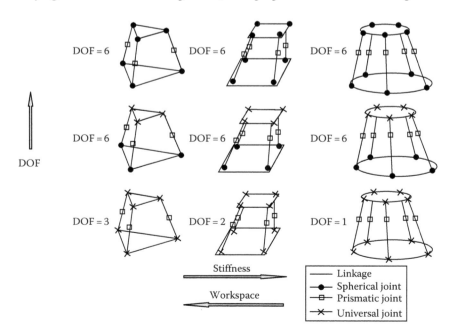

FIGURE 5.2 PKM characteristics.

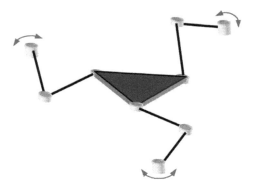

FIGURE 5.3   Planar parallel kinematic mechanism.

and number of limbs. In the first row, all the kinematic limbs are SPS (spherical–prismatic–spherical), and the DOF are calculated. In the lower rows, universal joints are being used, and the DOF of the manipulator have reduced considerably. The number of legs is varied from 3 to 5 and the corresponding stiffness of the system will increase with the addition of each leg. In contrary, workspace will shrink by adding more legs as each limb will put the constraint on the mobility of the mechanism. A trade-off is required between high stiffness and large dexterous workspace requirement.

Over all, PKM can be classified in to two groups on the basis of DOF. The mechanisms which give only two DOF are often called planar systems (Figure 5.3). Spatial mechanisms are the complex form of planar mechanisms with three to six DOF. DOF is dependent on the type of joints and number of limbs. Delta robot shown in Figure 5.4c is a light weight manipulator used in medical applications.

Some of the spatial mechanisms can be found with one axis extended to give a large workspace. This extended axis is like a serial outermost or the biggest axis. Manipulator can have the translational slides on the static platform. Most of these manipulators are used in the hanging configuration with vertical and horizontal guideways. Figure 5.5a is called hexaslide. The upper platform joints can translate in the fixed guideways to increase the manipulator workspace.

Classification can also be created in the PKM manipulator according to the limb configurations used. For example, Figure 5.6 shows the SPS kinematic limb with or without the inclination angle in the base. Figure 5.7 shows different manipulators using variety of joints at different places in the manipulator. SPS and PSS (prismatic–spherical–spherical) are the six DOF manipulators but different joint configurations give them different resulting characteristics in terms of workspace, dexterity, singularity, and stiffness, etc. Figure 5.8 shows the 3SPS system with a customized ball joint example for a specific application.

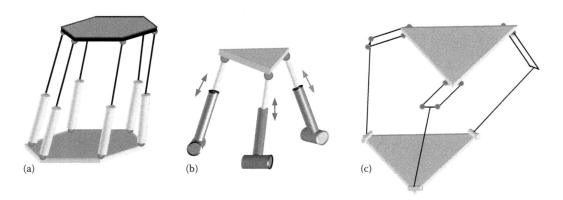

(a)   (b)   (c)

FIGURE 5.4   Examples of spatial mechanisms: (a) Stewart–Gough, (b) 3-RPS, and (c) delta.

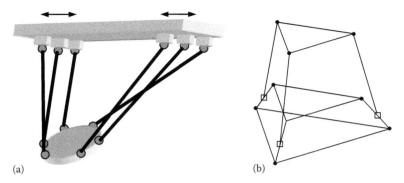

FIGURE 5.5 Spatial mechanisms with (a) linear horizontal and (b) vertical guideways.

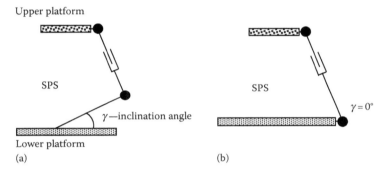

FIGURE 5.6 SPS configuration (a) with and (b) without inclination angle.

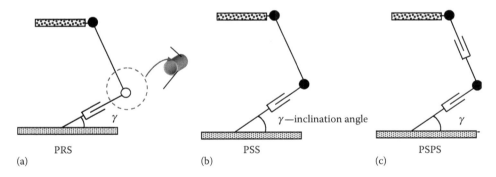

FIGURE 5.7 Different configurations for a variety of joints.

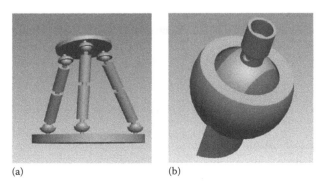

FIGURE 5.8 (a) 3SPS system structure with spherical joints and (b) customized ball joint example.

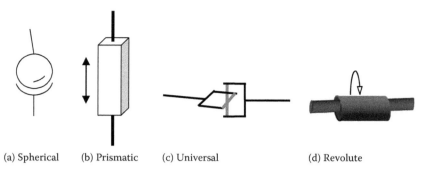

(a) Spherical   (b) Prismatic   (c) Universal   (d) Revolute

FIGURE 5.9   Different types of joints used in PKM: (a) spherical, (b) prismatic, (c) universal, and (d) revolute.

Some typical conventional types of joints being used in the PKM structures are shown in Figure 5.9. All of these joints are used in combination with other joints to form a closed loop kinematic chain. The combination of a number of closed loop chains forms a complete PKM system. There are four conventional type of joints used in PKM. They are spherical, prismatic, universal, and revolute joints. Every joint has its own rotational and translational capability.

Spherical joints offer three DOF, i.e., the three rotations around three Cartesian axes. Universal joints offer two DOF and the revolute joints give single DOF. Prismatic joints are different in nature and provide pure translations. Each of these joints causes a separate effect on the system's overall mobility. More recently, the research interest has grown up to make innovative joint types so that more complex and smart machine tool structures can be designed. Figure 5.8 shows one of such example in which an intelligent spherical joint can self-lock itself causing the manipulator to lock at any point of the workspace.

5.6.1   DOF CALCULATION

Gogu [9] has presented more than 30 different equations that are used in literature, to quickly calculate the DOF or mobility of the system, as there is no accepted method to calculate the DOF. So far, Grubler's equation is widely used in the literature for this purpose. Hunt [10] has presented the mobility equation with some modification of the Grubler's formula.

$$M = b(n - g - 1) + \sum_{i=1}^{p} f_i \qquad (5.1)$$

where
n is the number of limbs
g is the number of joints
f_i is the DOF of ith joint
b is the mobility number; $b = 3$ for planar mechanisms and $b = 6$ for spatial ones.

Tsai proposed to subtract the passive DOF as the internal DOF cannot be used to transmit motion or torque about an axis.

$$M = b(n - g - 1) + \sum_{i=1}^{p} f_i - f_p \qquad (5.2)$$

where f_p is the passive DOF about an axis.

Different examples can be carried out to verify the generality of the Tsai and the Hunt equations. Three cases are taken for the purpose, i.e., SPS (spherical–prismatic–spherical),

SPU (spherical–prismatic–universal), and SPR (spherical–prismatic–revolute) and the system mobility is calculated with the help of Tsai's equation.

For SPS, $M = 6(8-9-1)+6(3)+3(1)-3 = 6$

For SPU, $M = 6(8 - 9 - 1)+3(3)+3(1) + 3(2) - 3 = 3$

For SPR, $M = 6(8 - 9 - 1)+3(3)+3(1) + 3(1) - 3 = 0$

The formula for the SPR system is giving an illogical answer as SPR system can move in the z direction and it should have DOF equal to one. Zero DOF means that the system is locked and the end-effector cannot move at all. If we use the Hunt's equation, the SPS system will have nine DOF which is again an illogical result. SPU and SPR will have six and three DOF, respectively. It is found from the quick equations of mobility that the general rule of thumb for the calculation of PKM system mobility has not been found yet.

5.7 PRINCIPLE IN THE DESIGN OF PKM

A method is formulated based on the forward and inverse kinematics of the PKM manipulator. In forward kinematics, the input actuator values are given to the manipulator and the task of the manipulator is to find the corresponding workspace. In contrary, in inverse kinematics, workspace location is given as an input to the manipulator and it computes the actuator values to go to the particular point of the workspace. Manipulator Jacobians are formulated from the closed loop kinematic chains formed from the vectorial relationships. As a starting point, a general analytical modeling method for the PKM is first explained below.

5.7.1 KINEMATIC MODELING—GENERAL FORMULATION

Before modeling a specific PKM system, a general modeling method needs to be explained. The number of limbs can be ranged from three to six depending upon the stiffness requirement of the system. In this study, three limbs are proposed initially in the upcoming examples. Three limbs form an equilateral triangle and 120° apart from each other. Equation 5.3 refers to the common base platform designed in PKM systems as shown in Figure 5.10. The joints location with respect to the platform center can be given as follows:

$$\begin{bmatrix} a_{1x} \\ a_{1y} \\ a_{1z} \end{bmatrix} = \begin{bmatrix} r_p \\ 0 \\ 0 \end{bmatrix}, \begin{bmatrix} a_{2x} \\ a_{2y} \\ a_{2z} \end{bmatrix} = \begin{bmatrix} r_p c_\alpha \\ r_p s_\alpha \\ 0 \end{bmatrix} \quad \text{and} \quad \begin{bmatrix} a_{3x} \\ a_{3y} \\ a_{3z} \end{bmatrix} = \begin{bmatrix} r_p c_\beta \\ r_p s_\beta \\ 0 \end{bmatrix} \quad (5.3)$$

The moving platform is defined with a radius r_p. The rotation representation of the moving platform is based on the Euler method which gives the transformation rotation matrix of the upper

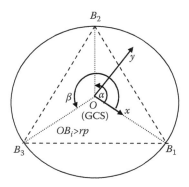

FIGURE 5.10 Lower platform—position of joints.

platform with respect to the global reference O. The rotation representation of the moving platform based on the Euler method of angular transformation is given by

1. Angle φ about the z-axis, followed by
2. Rotation by an angle ψ about the x-axis, followed by
3. Rotation by an angle θ about the y-axis

$$R = R_{Y(\theta)} R_{X(\psi)} R_{Z(\varphi)} \tag{5.4}$$

$$R = \begin{bmatrix} \cos\theta & 0 & \sin\theta \\ 0 & 1 & 0 \\ -\sin\theta & 0 & \cos\theta \end{bmatrix} \begin{bmatrix} 1 & 0 & 0 \\ 0 & \cos\psi & -\sin\psi \\ 0 & \sin\psi & \cos\psi \end{bmatrix} \begin{bmatrix} \cos\varphi & -\sin\varphi & 0 \\ \sin\varphi & \cos\varphi & 0 \\ 0 & 0 & 1 \end{bmatrix} \tag{5.5}$$

$$R = \begin{bmatrix} u_x & v_x & w_x \\ u_y & v_y & w_y \\ u_z & v_z & w_z \end{bmatrix} = \begin{bmatrix} c_\theta c_\varphi + s_\psi s_\theta s_\varphi & -c_\theta s_\varphi + s_\psi s_\theta c_\varphi & c_\psi s_\theta \\ c_\psi s_\varphi & c_\psi c_\varphi & -s_\psi \\ -s_\theta c_\varphi + s_\psi c_\theta s_\varphi & s_\theta s_\varphi + s_\psi c_\theta c_\varphi & c_\psi c_\theta \end{bmatrix} \tag{5.6}$$

Equation 5.6 gives the transformation rotation matrix of the upper platform with respect to the global reference O, where c stands for cos and s stands for sin. u_x, v_x, w_x, etc., are defined as the direction cosines of the rotation matrix. The translations can also be introduced with the Euler transformation. By applying the Euler transformation for the upper platform, new locations can be found for the moving manipulator in global coordinate system (GCS) of the base platform. If translations are introduced with the Euler transformation, then

$$[T]_L^U = [\text{Trans}][R_{ZXY}] \tag{5.7}$$

By applying the Euler transformation for the upper platform, new locations can be found for the moving manipulator in GCS of the base platform (Figure 5.11).

$$\{U_p\}_{GCS} = [T]_L^U \{U_p\}_{UCS} \tag{5.8}$$

For each upper joint A_i, global coordinates can be calculated using the following transformation:

$$\begin{bmatrix} A_{ix} \\ A_{iy} \\ A_{iz} \\ 1 \end{bmatrix}_{GCS} = \begin{bmatrix} u_x & v_x & w_x & x \\ u_y & v_y & w_y & y \\ u_z & v_z & w_z & z \\ 0 & 0 & 0 & 1 \end{bmatrix}_L^U \begin{bmatrix} A_{ix} \\ A_{iy} \\ A_{iz} \\ 1 \end{bmatrix}_{UCS} \tag{5.9}$$

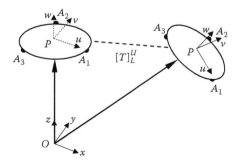

FIGURE 5.11 Euler transformation for each upper joint.

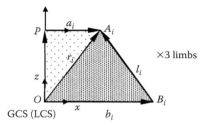

FIGURE 5.12 Forward (ΔOPA_i) and inverse kinematics (ΔOB_iA_i): transformation from lower (global) to upper coordinate system.

The three components of the upper joints in terms of global reference can be written as follows:

$$r_{ix} = x + u_x a_{ix} + v_x a_{iy} + w_x a_{iz}$$
$$r_{iy} = y + u_y a_{ix} + v_y a_{iy} + w_y a_{iz} \tag{5.10}$$
$$r_{iz} = z + u_z a_{ix} + v_z a_{iy} + w_z a_{iz}$$

where a_i is the upper platform vector as shown in Figure 5.12.

The type of the lower platform joints will determine the constraints to be put on the system. By using the constraints, the equations can be derived for the forward kinematic solution. The vector position of the joint with respect to the GCS is comprised of the lower platform radius and the limb length.

$$\vec{r_i} = \vec{b_i} + \vec{l_i} \tag{5.11}$$

The actuator can be placed either in b_i or l_i. The module of the actuator vector can be found using the square of length which may result in a quadratic equation. If l_i is taken as the actuator then the modulus length of ith leg can be written as

$$l_i = \pm\sqrt{r_i^2 + b_i^2 - 2r_ib_i} \tag{5.12}$$

Negative value can be discarded as it has no physical meaning. The manipulator's Jacobian provides the link between the actuators and the chosen points on the end-effector. From the Jacobian information, workspace, dexterity, and singularities can be analyzed. If the manipulator can move in all the six directions, then the end-effector location vector can be represented as

$$X = \begin{bmatrix} x & y & z & \psi & \theta & \varphi \end{bmatrix}^T \tag{5.13}$$

If the manipulator has three prismatic actuators, then the vector for actuators will be

$$q = \begin{bmatrix} b_1 & b_2 & b_3 \end{bmatrix}^T \tag{5.14}$$

If f is an n-dimensional implicit function of q and X, then

$$f(X,q) = 0 \tag{5.15}$$

Differentiating with respect to time, a relationship can be obtained between the input joint rates and the end-effector output velocity:

$$\frac{\partial f}{\partial X} \cdot \begin{bmatrix} \dot{x} \\ \dot{y} \\ \dot{z} \\ \dot{\psi} \\ \dot{\theta} \\ \dot{\phi} \end{bmatrix} = \frac{\partial f}{\partial q} \cdot \begin{bmatrix} \dot{b}_1 \\ \dot{b}_2 \\ \dot{b}_3 \end{bmatrix} \tag{5.16}$$

A two part Jacobian showing the relationship between the actuators q and the end-effector location X is shown in Equation 5.16 and can be summarized in the following form:

$$J_x \dot{X} = J_q \dot{q} \tag{5.17}$$

where

J_x is the Jacobian matrix of the end-effector location
J_q is the Jacobian of actuator
The overall Jacobian matrix J can be written as $J X = q$ where $J = J_q^{-1} J_x$.

The Jacobian formed here has a very important role to play. Inversion of Jacobian also depends upon the final shape of the matrix. To deal with these issues, two case-studies will be considered next.

5.7.2 CASE STUDY OF 3-PRS AND PSS SYSTEMS

PRS (prismatic–revolute–spherical) and PSS (prismatic–spherical–spherical) systems are analyzed for its workspace and dexterity calculation. A general architecture shown in Figure 5.13b for the PRS mechanism is adopted in this study. The PRS mechanism has been modeled using inverse and forward kinematics and an interesting discussion is made to model the kinematic constraints. The mechanism is analyzed with the inclination angle in the base platform. In both the base and the moving platforms, joints are arranged at the corners of equilateral triangle due to three number of limbs as shown in Figure 5.10. In the PSS system, revolute joint of the PRS is replaced by the spherical joint. For the base,

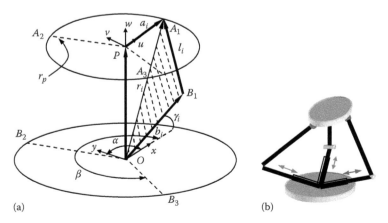

(a)   (b)

FIGURE 5.13 (a) Vector model of the inclined PRS manipulator and (b) general architecture. (Pond, G.T. and Carretero, J.A., Kinematic analysis and workspace determination of the inclined PRS parallel manipulator. *15th CISM-IFToMM Symposium on Robot Design, Dynamics and Control*, Saint-Hubert, Montreal, Quebec, Canada, 2004.)

x-axis is along OB_1 as shown in Figure 5.13a. OB_1 is making an angle α with OB_2 and β with OB_3, respectively. $B_1B_2B_3$ is an equilateral triangle; therefore, α and β are equal to 120° and 240°, respectively.

For moving platform, u-axis is along PA_1 and the radius of moving platform is r_p as shown in Figure 5.13.

$$PA_1 = PA_2 = PA_3 = r_p \qquad (5.18)$$

Moving platform coordinates with respect to point P are given in Equation 5.2. To show a_i coordinates globally, it must be multiplied by the rotation matrix to transform the vector into global coordinates as done before in Equation 5.9.

$$r_i = P_i + \left\{ U_p \right\}_{GCS} \cdot a_i \qquad (5.19)$$

The lower platform local coordinates with respect to point O are as follows. γ is the inclination angle of the base platform.

$$\begin{bmatrix} b_{1x} \\ b_{1y} \\ b_{1z} \end{bmatrix} = \begin{bmatrix} b_1 c_\gamma \\ 0 \\ b_1 s_\gamma \end{bmatrix}, \quad \begin{bmatrix} b_{2x} \\ b_{2y} \\ b_{2z} \end{bmatrix} = \begin{bmatrix} b_2 c_\gamma c_\alpha \\ b_2 c_\gamma s_\alpha \\ b_2 s_\gamma \end{bmatrix}, \quad \text{and} \quad \begin{bmatrix} b_{3x} \\ b_{3y} \\ b_{3z} \end{bmatrix} = \begin{bmatrix} b_3 c_\gamma c_\beta \\ b_3 c_\gamma s_\beta \\ b_3 s_\gamma \end{bmatrix} \qquad (5.20)$$

5.7.2.1 Direct Kinematic Solution

Using Equation 5.10 and Figure 5.13, following equations can be deduced from observation:

$$r_{1y} = y + u_y r_p = 0 \qquad (5.21)$$

$$r_{2y} = r_{2x} \tan(\alpha) \qquad (5.22)$$

$$r_{3y} = r_{3x} \tan(\beta) \qquad (5.23)$$

Equations 5.21 through 5.23 are considered constraints in the literature [8,12] for the revolute joints. These equations are formulated by putting the moving platform projection on the base platform. Another provision to model the joint constraints is provided in the formulation of Jacobian. The conventional Jacobian formulation is based on the closed loop vectorial relationship. The vector loop relationship comprises the upper and lower platform radius, actuated and unactuated limbs, etc. The rotation of the ith leg in the PRS system can be possible only in the plane generated by the actuator b_i and limb l_i. This vectorial relationship defines the plane of movement but do not consider the type of joints. If the revolute joints are replaced by the universal or spherical joints as in the case of PSS, then this modeling approach of single plane vectorial relationship cannot be used. The modeling method must cover the out of the plane movement. Therefore, it can be stated that the mechanical constraints have been modeled specifically in the PRS case for the revolute joints and the vectorial relationship based on a single plane creates a Jacobian which also include the constraint for the revolute joint in the kinematic chain. For universal or spherical joints, the modeling approach must show the out of the plane movement usually through a cross product.

By using Figure 5.14 and Equation 5.10 for r_{2x}, r_{2y}, r_{3x}, and r_{3y} and inserting them in Equations 5.22 and 5.23:

$$y + u_y r_p \cos(\alpha) + v_y r_p \sin(\alpha) = \tan(\alpha)\left\{ x + u_x r_p \cos(\alpha) + v_x r_p \sin(\alpha) \right\} \qquad (5.24)$$

$$y + u_y r_p \cos(\beta) + v_y r_p \sin(\beta) = \tan(\beta)\left\{ x + u_x r_p \cos(\beta) + v_x r_p \sin(\beta) \right\} \qquad (5.25)$$

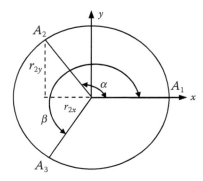

FIGURE 5.14 Upper platform projected to the GCS.

Equation for y can be deduced by using Equation 5.21.

$$y = -u_y r_p \tag{5.26}$$

By putting y from Equation 5.26 in Equation 5.24, and solving for "x":

$$x = -u_x r_p \cos(\alpha) - v_x r_p \sin(\alpha) + \frac{r_p}{\tan(\alpha)}\Big[u_y\big(\cos(\alpha)-1\big)+v_y \sin(\alpha)\Big] \tag{5.27}$$

By placing the expressions for both x and y in Equation 5.25, equation for ϕ can be expressed with the help of Equation 5.6 as follows:

$$\phi = \tan^{-1}\left(\frac{-R}{S}\right) \tag{5.28}$$

where

$$R = A\big(\cos(\theta)-\cos(\psi)\big)+B\sin(\theta)\sin(\psi) \quad \text{and} \quad S = A\sin(\theta)\sin(\psi)-B\cos(\theta)+C\cos(\psi)$$

with

$$A = \cos(\alpha)-\cos(\beta), \; B = \sin(\alpha)-\sin(\beta), \; \text{and} \quad C = \frac{(\cos(\beta)-1)}{\tan(\beta)} - \frac{(\cos(\alpha)-1)}{\tan(\alpha)}$$

Here, rotations around x- and y-axes, i.e., ψ and θ, respectively, are taken as the independent motion as well as z-axis translation. Whereas x, y, and φ (rotation around z) are calculated through the forward kinematic solution. Kinematic constraint equations are formed for these dependent motions in terms of independent variables ψ and θ.

5.7.2.2 Inverse Displacement Solution

The vector loop diagram comprising the actuated prismatic joint, constant length leg, and upper joint location can be written as (see Figure 5.13a)

$$\vec{r}_i = \vec{b}_i + \vec{l}_i \tag{5.29}$$

where b_i is the distance from the global reference O to the lower platform joint B_i, and l_i is the corresponding leg length. In the PRS system, b_i is considered as the actuator, and l_i a constant length, then

$$\vec{l_i} = \vec{r_i} - \vec{b_i} \tag{5.30}$$

and the module of the vector l_i is deduced by using Equation 5.30, length modulus of ith leg can be shown as

$$l_i^2 = \left(r_{ix} - b_{ix}\right)^2 + \left(r_{iy} - b_{iy}\right)^2 + \left(r_{iz} - b_{iz}\right)^2 \tag{5.31}$$

Substituting b_i from Equation 5.20 in Equation 5.31, a quadratic equation will be formed for b_i. The following equation gives the relationship between the fixed length l_i, actuator position b_i, and the position of upper joints as shown in Figure 5.13a.

$$b_1 = r_{1x} \cos(\gamma) + r_{1z} \sin(\gamma) \pm \sqrt{\left(r_{1x} \cos(\gamma) + r_{1z} \sin(\gamma)\right)^2 - \left(r_{1x}^2 + r_{1y}^2 + r_{1z}^2 - l_1^2\right)} \tag{5.32}$$

$$b_2 = r_{2x} \cos(\gamma)\cos(\alpha) + r_{2y} \cos(\gamma)\sin(\alpha) + r_{2z} \sin(\gamma)$$
$$\pm \sqrt{\left(r_{2x} \cos(\gamma)\cos(\alpha) + r_{2y} \cos(\gamma)\sin(\alpha) + r_{2z} \sin(\gamma)\right)^2 - (r_{2x}^2 + r_{2y}^2 + r_{2z}^2 - l_2^2)} \tag{5.33}$$

$$b_3 = r_{3x} \cos(\gamma)\cos(\beta) + r_{3y} \cos(\gamma)\sin(\beta) + r_{3z} \sin(\gamma)$$
$$\pm \sqrt{\left(r_{3x} \cos(\gamma)\cos(\beta) + r_{3y} \cos(\gamma)\sin(\beta) + r_{3z} \sin(\gamma)\right)^2 - (r_{3x}^2 + r_{3y}^2 + r_{3z}^2 - l_3^2)} \tag{5.34}$$

The three equations for b_i provide the solution for inverse kinematics where independent variables, i.e., Ψ, θ, and z are given as an input and the actuation values are found from the solution. Only positive values of b_i are taken further because negative solution is not possible physically. Positive solution means that the actuator will result in the inward leaning of the constant length leg l_i.

5.7.2.3 Conventional Jacobian Analysis

The manipulator's Jacobian provides the link between the actuators and the chosen points on the end-effector. Conventional Jacobian is derived from the closed loop vectorial relationship which comprises the upper and lower platform radius, actuated and unactuated limbs, etc. The manipulator has a motion in Cartesian space in all six dimensions and has three prismatic actuators as given in Equations 5.13 and 5.14.

Any point f_i on the moving platform with respect to the base frame origin O and involving the three spherical joints can be defined by using the parametric equation of plane (Figure 5.15). The method is introduced by Kim and Ryu [13] to formulate a dimensionally homogenous Jacobian. The method defines the vector of kinematic parameters of three points anywhere in the same plane which carries the moving platform spherical joints. If the coordinates of three random points T_j are expressed in the absolute coordinate frame of point O (Figure 5.15), then the coordinates of moving platform joints A_i can be expressed in terms of the three random T_j points as

$$OA_i = \begin{bmatrix} k_{i,1}x_1 + k_{i,2}x_2 + k_{i,3}x_3 \\ k_{i,1}y_1 + k_{i,2}y_2 + k_{i,3}y_3 \\ k_{i,1}z_1 + k_{i,2}z_2 + k_{i,3}z_3 \end{bmatrix}, \quad i = 1,2,3 \tag{5.35}$$

where $k_{i,j}$ is a dimensionless constant. Here i denotes three joints and j denotes the three preselected points making the plane. As an example, by expanding Equation 5.35 for one of the joints

$$OA_1 = \begin{bmatrix} k_{1,1}x_1 + k_{1,2}x_2 + k_{1,3}x_3 \\ k_{1,1}y_1 + k_{1,2}y_2 + k_{1,3}y_3 \\ k_{1,1}z_1 + k_{1,2}z_2 + k_{1,3}z_3 \end{bmatrix} \tag{5.36}$$

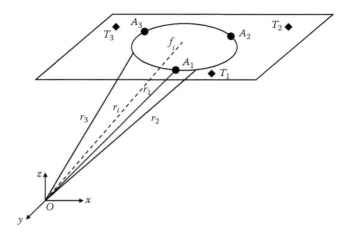

FIGURE 5.15 Representing the coordinates of platform joints in three T_j points with respect to global reference O.

Equation 5.36 gives the location of joint A_1 according to the three T_j points. But actually, only one T_j point is necessarily required to find the joint location of A_1. The other two T_j points are redundant to specify the location of A_1. Similarly, the other two joints A_2 and A_3 can easily be defined with their corresponding T_j point. Therefore, the following conditions can be put for Equation 5.36:

$$k_{i,j} = 1, \quad \text{when } i = j$$

$$k_{i,j} = 0, \quad \text{when } i \neq j$$

If the three T_j points correspond to the spherical joints A_1, A_2, and A_3, the parametric equation of plane for any point f_i on the moving platform will be given by

$$f_i = r_3 + k_{i,1}\left(r_1 - r_3\right) + k_{i,2}\left(r_2 - r_3\right) \tag{5.37}$$

that can be further simplified as

$$r_i = k_{i,1}r_1 + k_{i,2}r_2 + k_{i,3}r_3 \tag{5.38}$$

By expanding Equation 5.29 on right side and comparing with Equation 5.38

$$k_{i,1}r_1 + k_{i,2}r_2 + k_{i,3}r_3 = s_{bi}\left|b_i\right| + s_{li}\left|l_i\right| \tag{5.39}$$

where s_{li} is a unit vector along the limb l_i. Taking the time derivative of Equation 5.39 and projecting all elements onto vector s_{li}^T by dot multiplying it on both sides

$$k_{i,1}s_{li}^T \dot{r_1} + k_{i,2}s_{li}^T \dot{r_2} + k_{i,3}s_{li}^T \dot{r_3} = s_{li}^T s_{bi}\left|\dot{b_i}\right| \tag{5.40}$$

The two-form Jacobians given in Equation 5.17 can be compared with Equation 5.40. The elements J_x, J_q, $\dot{x}$, and $\dot{q}$ are given below where

$$J_x = \begin{bmatrix} k_{1,1}s_{l1}^T & k_{1,2}s_{l1}^T & k_{1,3}s_{l1}^T \\ k_{2,1}s_{l2}^T & k_{2,2}s_{l2}^T & k_{2,3}s_{l2}^T \\ k_{3,1}s_{l3}^T & k_{3,2}s_{l3}^T & k_{3,3}s_{l3}^T \end{bmatrix}_{3\times9}$$

(5.41)

$$\dot{X} = \begin{bmatrix} \dot{r_1} & \dot{r_2} & \dot{r_3} \end{bmatrix}^T_{9\times1}$$

(5.42)

$$J_q = \begin{bmatrix} s_{l1}^T s_{b1} & 0 & 0 \\ 0 & s_{l2}^T s_{b2} & 0 \\ 0 & 0 & s_{l3}^T s_{b3} \end{bmatrix}_{3\times3}$$

(5.43)

$$\dot{q} = \begin{bmatrix} \dot{b_1} & \dot{b_2} & \dot{b_3} \end{bmatrix}$$

(5.44)

and J will be found as the overall Jacobian and can be calculated by inverting J_q.

$$J = J_q^{-1} J_x$$

(5.45)

Expressions in J_x matrix will use the conditions specified for k_{ij} explained after Equation 5.36. The expanded expressions of J are given as

$$J = \begin{bmatrix} \dfrac{s_{l1x}}{F_1} & \dfrac{s_{l1y}}{F_1} & \dfrac{s_{l1z}}{F_1} & 0 & 0 & 0 & 0 & 0 & 0 \\ 0 & 0 & 0 & \dfrac{s_{l2x}}{F_2} & \dfrac{s_{l2y}}{F_2} & \dfrac{s_{l2z}}{F_2} & 0 & 0 & 0 \\ 0 & 0 & 0 & 0 & 0 & 0 & \dfrac{s_{l3x}}{F_3} & \dfrac{s_{l3y}}{F_3} & \dfrac{s_{l3z}}{F_3} \end{bmatrix}$$

(5.46)

where

$$F_1 = s_{b1x}s_{l1x} + s_{b1y}s_{l1y} + s_{b1z}s_{l1z}$$

$$F_2 = s_{b2x}s_{l2x} + s_{b2y}s_{l2y} + s_{b2z}s_{l2z}$$

$$F_3 = s_{b3x}s_{l3x} + s_{b3y}s_{l3y} + s_{b3z}s_{l3z}$$

The inversion of the Jacobian given in Equation 5.46 will be discussed later. For the PSS system, again the manipulator has a motion in Cartesian space in all six dimensions and has three prismatic actuators as shown in Equations 5.13 and 5.14 and in Figure 5.13. From the plane made by the ith limb as shown in Figure 5.13, the closed loop kinematic chain is given as

$$\overline{OP} + \overline{PA_i} = \overline{OB_i} + \overline{A_iB_i}$$

(5.47)

Differentiating Equation 5.47 with respect to time yields

$$v_p + w_A \times a_i = s_{bi}\left|\dot{b_i}\right| + l_i\left(w_i \times s_{li}\right)$$

(5.48)

where

a_i and s_{bi} denote the vector $\overline{PA_i}$ and $\overline{OB_i}$, respectively
w_i denotes the angular velocity of the ith limb

Now, by dot-multiplying both sides of Equation 5.48 by s_{li}^{T}

$$s_{li}^{\mathrm{T}} \cdot v_p + (a_i \times s_{li}) \cdot w_A = s_{li}^{\mathrm{T}} s_{bi} \dot{b}_i \tag{5.49}$$

Equation 5.49 can be written three times for the three limbs from which the two-form Jacobians can be deduced by using the relationship given in Equation 5.17, where

$$J_x = \begin{bmatrix} s_{l1}^{\mathrm{T}} & \left(a_1 \times s_{l1}\right)^{\mathrm{T}} \\ s_{l2}^{\mathrm{T}} & \left(a_2 \times s_{l2}\right)^{\mathrm{T}} \\ s_{l3}^{\mathrm{T}} & \left(a_3 \times s_{l3}\right)^{\mathrm{T}} \end{bmatrix}_{3 \times 6} \tag{5.50}$$

$$J_q = \begin{bmatrix} s_{l1}^{\mathrm{T}} s_{b1} & 0 & 0 \\ 0 & s_{l2}^{\mathrm{T}} s_{b2} & 0 \\ 0 & 0 & s_{l3}^{\mathrm{T}} s_{b3} \end{bmatrix}_{3 \times 3} \tag{5.51}$$

$$\dot{X} = \begin{bmatrix} \dot{x} & \dot{y} & \dot{z} & \dot{\psi} & \dot{\theta} & \dot{\varphi} \end{bmatrix}^{\mathrm{T}} \tag{5.52}$$

$$\dot{q} = \begin{bmatrix} \dot{b}_1 & \dot{b}_2 & \dot{b}_3 \end{bmatrix}^{\mathrm{T}} \tag{5.53}$$

An overall Jacobian can be found by inverting J_q and multiply it with J_x.

5.7.2.4 Jacobian with Screw Theory

Tsai [8] has explained the Jacobian formulation based on the theory of reciprocal screws. The Jacobian based on the screw theory leads to squared Jacobian. As the square Jacobian is easy to be solved further for determining the singularity conditions, workspace, and dexterity, it is preferred over the conventional rectangular Jacobian discussed above. Carretero et al. [11] have presented a Jacobian based on the screw theory for the inclined PRS manipulator as

$$J = \begin{bmatrix} \dfrac{\left(a_1 \times s_{l1}\right)^{\mathrm{T}}}{s_{l1} \cdot s_{b1}} & \dfrac{s_{l1}^{\mathrm{T}}}{s_{l1} \cdot s_{b1}} \\[2mm] \dfrac{\left(a_2 \times s_{l2}\right)^{\mathrm{T}}}{s_{l2} \cdot s_{b2}} & \dfrac{s_{l2}^{\mathrm{T}}}{s_{l2} \cdot s_{b2}} \\[2mm] \dfrac{\left(a_3 \times s_{l3}\right)^{\mathrm{T}}}{s_{l3} \cdot s_{b3}} & \dfrac{s_{l3}^{\mathrm{T}}}{s_{l3} \cdot s_{b3}} \\[2mm] \left(a_1 \times s_{2,1}\right)^{\mathrm{T}} & s_{2,1}^{\mathrm{T}} \\[1mm] \left(a_2 \times s_{2,2}\right)^{\mathrm{T}} & s_{2,2}^{\mathrm{T}} \\[1mm] \left(a_3 \times s_{2,3}\right)^{\mathrm{T}} & s_{2,3}^{\mathrm{T}} \end{bmatrix} \tag{5.54}$$

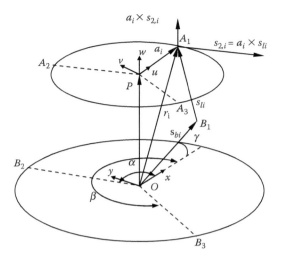

FIGURE 5.16　Vector representation of the cross products involved in screw theory Jacobian.

The first three rows in Equation 5.54 form the Jacobian of actuation (J_a) in which the reciprocal screw is found for all the screws when the actuators are locked. The other three rows form the Jacobian of constraints (J_c) in which reciprocal screw is determined for all the active and passive screws. The overall Jacobian is given by the combination of two Jacobians.

$$J = \begin{bmatrix} J_a \\ J_c \end{bmatrix} \tag{5.55}$$

This Jacobian is a 6×6 square Jacobian in Equation 5.54. The vectors a_i and s_{bi} are available readily from Equations 5.3 and 5.20 whereas s_{li} can be found using Equation 5.29. $s_{2,i}$ is a vector resulting from the cross product of a_i and s_{li} as shown in Figure 5.16. Using Equation 5.29, s_{li} can be found as

$$\begin{bmatrix} s_{lix} \\ s_{liy} \\ s_{liz} \end{bmatrix} = \begin{bmatrix} (r_{ix} - s_{bix})/l_i \\ (r_{iy} - s_{biy})/l_i \\ (r_{iz} - s_{biz})/l_i \end{bmatrix} \tag{5.56}$$

Equation 5.56 gives the relationship for s_{li}. This relationship will be used in further analysis in both the conventional Jacobian and the screw theory Jacobian.

The screw theory Jacobian formulation perfectly deals with the joint constraints. In screw theory, every joint is split into the individual twists in every direction such that the spherical joint is replaced by the three twists in the three Cartesian directions.

5.7.3　Case Study of 3-SPS Systems

The SPS mechanism is generally called Stewart–Gough platform in which six identical limbs join the moving platform with the base platform. Each limb is comprised of SPS chain. The six limbs of the manipulator will result in a 6×6 square Jacobian, that is easy to study further. Raghavan [14] has applied the continuation method and found that 6-SPS platform could have 40 solutions. However, it is possible that some of them are not feasible. In this example, SPS mechanism will consist of three legs. 3-SPS mechanism will lead to the rectangular Jacobian matrix that will

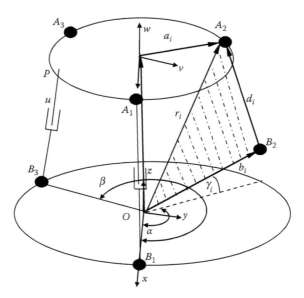

FIGURE 5.17 SPS mechanism architecture.

require further transformations to explore fully. Figure 5.17 shows the mechanism architecture. Point O on the lower platform specifies the global reference as shown in Figure 5.17. For the base, x-axis is along OB_1. OB_1 is making an angle α with OB_2 and β with OB_3, respectively. $B_1B_2B_3$ is an equilateral triangle; therefore, α and β are equal to $120°$ and $240°$, respectively. For moving platform, u-axis is along PA_1. The radius of moving platform is r_p and the joint coordinates with respect to point P are given in Equation 5.3. γ is the inclination angle of the base platform. For the base platform, a bigger radius is supposed, e.g., $2 \times r_p$ and the coordinates with respect to point O are as follows:

$$\begin{bmatrix} b_{1x} \\ b_{1y} \\ b_{1z} \end{bmatrix} = \begin{bmatrix} 2r_p c_\gamma \\ 0 \\ 2r_p s_\gamma \end{bmatrix}, \begin{bmatrix} b_{2x} \\ b_{2y} \\ b_{2z} \end{bmatrix} = \begin{bmatrix} 2r_p c_\gamma c_\alpha \\ 2r_p c_\gamma s_\alpha \\ 2r_p s_\gamma \end{bmatrix} \text{ and } \begin{bmatrix} b_{3x} \\ b_{3y} \\ b_{3z} \end{bmatrix} = \begin{bmatrix} 2r_p c_\gamma c_\beta \\ 2r_p c_\gamma s_\beta \\ 2r_p s_\gamma \end{bmatrix} \tag{5.57}$$

The rotation representation of the moving platform is the same as adopted previously. The direct kinematic solution for the SPS system will follow the same explanation as for the PRS and the PSS systems. This modeling method may effectively resolve the problem of unfeasible solutions.

5.7.3.1 Inverse Displacement Solution

From Figure 5.17, the closed loop vector diagram gives the equation in terms of vectors as already shown in Equation 5.47:

$$\overline{A_i B_i} = p + {}^B R_A {}^A a_i - b_i \tag{5.58}$$

The length of the ith limb is obtained by taking the dot product of the vector $\overline{A_i B_i}$ with itself and taking the square root after expanding.

$$d_i^2 = \left[p + {}^B R_A {}^A a_i - b_i \right]^T \left[p + {}^B R_A {}^A a_i - b_i \right] \quad \text{for } i = 1,2,3 \tag{5.59}$$

Expanding Equation 5.59 yields

$$d_i = \pm\sqrt{p^{\mathrm{T}}p + [{}^{A}a_i]^{\mathrm{T}}[{}^{A}a_i] + b_i^{\mathrm{T}}b_i + 2p^{\mathrm{T}}\left[{}^{B}R_A\,{}^{A}a_i\right] - 2p^{\mathrm{T}}b_i - 2\left[{}^{B}R_A\,{}^{A}a_i\right]^{\mathrm{T}}b_i} \qquad (5.60)$$

$$\text{for} \quad i = 1, 2, 3$$

where d_i denotes the length of the ith limb. Equation 5.60 shows the inverse kinematic solution. There are two possible solutions for limb lengths at each given location of the moving platform. The negative limb length is physically not feasible. The vectors, a_i, b_i and rotation matrix R are given in Equations 5.3, 5.57, and 5.6, respectively. Final equations extracted from Equation 5.60 are given below:

$$d_1^2 = P_x^2 + P_y^2 + P_z^2 + 2r_p(P_x u_x + P_y u_y + P_z u_z) + r_p^2\left(5 - 4\left(u_x c_\gamma + u_z s_\gamma\right)\right)$$
$$- 4r_p\left(P_x c_\gamma + P_z s_\gamma\right) \qquad (5.61)$$

$$d_2^2 = P_x^2 + P_y^2 + P_z^2 + 5r_p^2 + 2r_p\left[P_x\left(u_x c_\alpha + v_x s_\alpha\right) + P_y\left(c_\alpha u_y + s_\alpha v_y\right) + P_z\left(c_\alpha u_z + s_\alpha v_z\right)\right]$$
$$- 4r_p\left(c_\gamma c_\alpha P_x + s_\alpha P_y c_\gamma + s_\gamma P_z\right) - 4r_p^2 c_\gamma\left(c_\alpha^2 u_x + s_\alpha c_\alpha v_x + s_\alpha c_\alpha u_y + s_\alpha^2 v_y\right)$$
$$- 4r_p^2 s_\gamma\left(c_\alpha u_z + s_\alpha v_z\right) \qquad (5.62)$$

$$d_3^2 = P_x^2 + P_y^2 + P_z^2 + 5r_p^2 + 2r_p\left[P_x\left(u_x c_\beta + v_x s_\beta\right) + P_y\left(c_\beta u_y + s_\beta v_y\right) + P_z\left(c_\beta u_z + s_\beta v_z\right)\right]$$
$$- 4r_p\left(c_\gamma c_\beta P_x + s_\beta P_y c_\gamma + s_\gamma P_z\right) - 4r_p^2 c_\gamma\left(c_\beta^2 u_x + s_\beta c_\beta v_x + s_\beta c_\beta u_y + s_\beta^2 v_y\right)$$
$$- 4r_p^2 s_\gamma\left(c_\beta u_z + s_\beta v_z\right) \qquad (5.63)$$

As compared to the PSS manipulator, the SPS is formed by shifting the actuator position. The SPS limb is supposed to be equivalent to UPS (universal–prismatic–spherical). The spherical joint gives rotations around three Cartesian directions. In SPS, the spherical joints on both sides of the limb form SS pair. In the SS pair, the rotation around the z-axis can be considered as the SS limb rotation around the limb's own axis. This rotation is a passive movement and can be taken as redundant. If this redundant rotation is not taken into account, then the lower spherical joint is rotating only around x and y directions, thus making itself equivalent to a universal joint. There is a possibility that this redundant rotation in spherical joints may help in increasing the dexterous workspace at the boundaries where due to the UPS arrangement, manipulator may find it difficult to maneuver.

5.7.3.2 Conventional Jacobian Formulation

Manipulator has a motion in Cartesian space in all six dimensions as given in Equation 5.13. SPS manipulator has three prismatic actuators

$$\dot{q} = \begin{bmatrix} \dot{d_1} & \dot{d_2} & \dot{d_3} \end{bmatrix}^{\mathrm{T}} \qquad (5.64)$$

Differentiating Equation 5.47 with respect to time yields

$$v_p + w_A \times a_i = d_i w_i \times s_i + \dot{d_i}\, s_i \qquad (5.65)$$

where
    a_i and s_i denote the vector $\overline{PA_i}$ and a unit vector along $\overline{A_i B_i}$, respectively
    w_i denotes the angular velocity of the ith limb
    Now, by dot multiplying both sides of Equation 5.65 by s_i

$$s_i \cdot v_p + (a_i \times s_i) \cdot w_A = \dot{d}_i \tag{5.66}$$

Equation 5.66 can be written three times for the three limbs from which the two-form Jacobians can be deduced by using the relationship given in Equation 5.17, where

$$J_x = \begin{bmatrix} s_1^T & (a_1 \times s_1)^T \\ s_2^T & (a_2 \times s_2)^T \\ s_3^T & (a_3 \times s_3)^T \end{bmatrix}_{3 \times 6} \tag{5.67}$$

$$J_q = I_{3 \times 3} \tag{5.68}$$

The overall Jacobian in the expanded form can be found using Equation 5.45.

$$J = \begin{bmatrix} s_{1x} & s_{1y} & s_{1z} & (a_{1y}s_{1z} - a_{1z}s_{1y}) & (a_{1z}s_{1x} - a_{1x}s_{1z}) & (a_{1x}s_{1y} - a_{1y}s_{1x}) \\ s_{2x} & s_{2y} & s_{2z} & (a_{2y}s_{2z} - a_{2z}s_{2y}) & (a_{2z}s_{2x} - a_{2x}s_{2z}) & (a_{2x}s_{2y} - a_{2y}s_{2x}) \\ s_{3x} & s_{3y} & s_{3z} & (a_{3y}s_{3z} - a_{3z}s_{3y}) & (a_{3z}s_{3x} - a_{3x}s_{3z}) & (a_{3x}s_{3y} - a_{3y}s_{3x}) \end{bmatrix} \tag{5.69}$$

where s_i can be defined using the equation

$$s_i |d_i| = r_i - b_i \quad \text{or} \quad \begin{bmatrix} s_{ix} \\ s_{iy} \\ s_{iz} \end{bmatrix} = \begin{bmatrix} (r_{ix} - b_{ix})/d_i \\ (r_{iy} - b_{iy})/d_i \\ (r_{iz} - b_{iz})/d_i \end{bmatrix} \tag{5.70}$$

The inversion of the Jacobian given in Equation 5.69 is discussed later.

5.7.3.3 Screw Jacobian

The screw Jacobian for the 3-SPS system has problems to formulate. Tsai [8] has formulated the screw Jacobian for the 6-SPS system by considering the UPS as an equivalent kinematic chain to the SPS system, considering the limb rotation around its own axis as passive. But that Jacobian is only the Jacobian of actuators for the six numbers of limbs. The same Jacobian of actuators for the three limbs is given as

$$J_x = \begin{bmatrix} (a_1 \times s_{3,1})^T & s_{3,1}^T \\ (a_2 \times s_{3,2})^T & s_{3,2}^T \\ (a_3 \times s_{3,3})^T & s_{3,3}^T \end{bmatrix}_{3 \times 6} \tag{5.71}$$

The Jacobian of constraints may not be formulated because when the actuated joint screw is taken into consideration for the reciprocity, then it becomes a six system of screws for which reciprocal screw cannot be found. This is the limitation of the screw theory. In the SPS case, if only three limbs are taken and the Jacobian of actuator is formed, then the Jacobian formed will be rectangular as shown in Equation 5.71. The same is true in the case of PSS manipulator as after locking the prismatic actuator, it becomes the same SS limb and when unlocked it becomes six-screw system.

5.8   COMPUTATION

The Jacobian inversion will lead to the determination of the manipulator dexterous workspace, total workspace, and dexterity. Some issues in the computation of the workspace and the corresponding dexterity are discussed. The conventional Jacobians found in PRS is 3×9 and for PSS and SPS are 3×6, respectively. The matrix inversion of these conventional Jacobians is an issue. Conventional Jacobians as shown previously will usually give a rectangular (over or under determined) matrix unless the number of actuators are equal to the degrees of mobility. Rectangular Jacobian needs further transformations to allow inversion. Decomposition of these rectangular matrices may lead to square submatrices. This transformation process may lead to incorrect results if the inversion is not properly made. A pseudotechnique is suggested. Moreover, the screw theory method gives by default square Jacobians that are easy to invert by definition.

5.8.1   METHODS OF DECOMPOSITION

The matrix decomposition can be used to invert the Jacobian matrix. Actually, the decomposition of rectangular matrix may lead to ease the matrix inversion. There are a few methods available to decompose $m \times n$ matrices. For example, for the case where $m \geq n$, householder decomposition works to give two matrices. The two factors of J are the product of an orthonormal matrix Q and an upper triangular matrix R.

$$J = Q \times R \quad \text{for} \begin{cases} r(J) = n \\ m \geq n \end{cases} \tag{5.72}$$

The same decomposition method cannot help in all the three cases because the conventional Jacobian found are both under-determined problems. Also the rank in those cases is found equal to three which is less than the number of columns. As both cases are under-determined, pseudoinverse based on singular value decomposition is then a convenient method. The method factorizes the matrix J where $m < n$ and when the rank of the matrix J is not equal to the number of columns. The idea is to turn a single rectangular matrix into its factors of square matrices. The square matrices are then used to take their own inverses and get multiplied again in the reverse order. The whole workspace of the manipulator can be checked for the matrix inversion. The J matrix is decomposed based on the singular value decomposition into the following three components:

$$J = U \times S \times V^{\mathrm{T}} \quad \text{for} \begin{cases} r(J) \neq n \\ m < n \end{cases} \tag{5.73}$$

where V is $n \times n$ orthogonal matrix and consists of the eigenvectors of $J^{\mathrm{T}}J$. U is $m \times m$ orthogonal matrix and consists of the eigenvectors of JJ^{T}. S is the $m \times n$ diagonal matrix that consists of the square roots of the eigenvalues of JJ^{T} arranged in descending order.

5.8.2   METHODS OF INVERSION

As previously mentioned, matrix inversion is the next step after the decomposition. Pseudoinverse provides a partial solution for the inverse of the rectangular matrices. The Moore–Penrose generalized matrix inverse is a unique $n \times m$ pseudoinverse of a given $m \times n$ matrix and has some properties of the full inverse.

For the fully ranked system, the actual definition of the pseudoinverse is presented below:

$$\left(J^{-1}\right)_{\text{pseudo}} = \left(J^{\mathrm{T}}J\right)^{-1} J^{\mathrm{T}} \quad \text{for} \begin{cases} r = n \\ m > n \end{cases} \tag{5.74}$$

$$\left(J^{-1}\right)_{\text{pseudo}} = J^{\text{T}}\left(JJ^{\text{T}}\right)^{-1} \quad \text{for} \begin{cases} r = n \\ m < n \end{cases} \tag{5.75}$$

The interesting limitation in pseudoinversion is that if X matrix is the pseudoinverse of C rectangular matrix, then $X \times C$ will be an identity matrix but $C \times X$ will not be an identity matrix. It will act like an identity matrix on a portion of a space in the sense that $C \times X$ is symmetric. If both sides of a kinematic system is multiplied with J^{-1}

$$\{Y\} = J\{X\} \Rightarrow J^{-1}\{Y\} = [I]\{X\} \tag{5.76}$$

and by further multiplying Equation 5.76 with J written as

$$\underbrace{[J][J^{-1}]}_{\neq I}\{Y\} = [J]\{X\} \tag{5.77}$$

This means that the Jacobian will get inversed but cannot be inversed back further.

For the overall Jacobians found for the three systems, the pseudoinverse, based on the singular value decomposition, is used. Tucker [15] has explained that for the complete computation of the Moore–Penrose pseudoinverse, singular value decomposition is used. For Equation 5.73, the pseudoinverse matrix is written as

$$\left(J^{-1}\right)_{\text{pseudo}} = V \times (S^{-1})_{\text{pseudo}} \times U^{\text{T}} \quad \text{for} \ r(J) \neq n \tag{5.78}$$

With some issues around the inversion of such matrices, a few parameters are introduced to measure the manipulability within the workspace.

5.8.3 EFFECT OF CONDITION NUMBER

Condition number is a performance index of the manipulator to check its dexterity, velocity, and accuracy performance. The dexterity is the superior capability of the PKM to access workspace with the full orientations in 3D space. The condition number is defined as the square root of the ratio of the largest to the smallest eigenvalues of the Jacobian matrix.

$$k(J) = \sqrt{\frac{\lambda_{\text{max}}}{\lambda_{\text{min}}}} \tag{5.79}$$

The locations where the condition number reaches near to a value of 1, dexterity reaches to the maximum. A matrix is said to be well conditioned when k is small (1–3) and ill conditioned when k is large (3–10). A well-conditioned Jacobian matrix gives a more uniform set of possible velocities. Actually the minimum value of $k(J)$ is 1. But a zero value of $1/k(J)$ or a very high value of condition number can bring the manipulator to a singular location. The value of condition number increases near the workspace boundaries where the dexterity becomes very low.

Condition number also gives a performance measure of the positioning accuracy of the end-effector. A possible error amplification factor for the system expresses how a relative error in q gets multiplied and leads to a relative error in X [16].

$$\frac{\|\Delta X\|}{\|X\|} \leq \|J^{-1}\|\|J\|\frac{\|\Delta q\|}{\|q\|} \tag{5.80}$$

The inversion of the Jacobian may lead to an amplification of the motion given by the condition number $k(J)$.

$$k(J) = \left\| J^{-1} \right\| \cdot \left\| J \right\|$$ (5.81)

where ‖·‖ is two-norm of the matrix. The condition number can also be defined as the magnitudes of the major and minor semiaxes of the manipulability hyperellipsoid. Geometrically, the associated Jacobian matrix J describes a hyperellipsoid having lengths defined by its singular values, i.e., considered as an image of the maximum and minimum velocity amplification factor [16]. The condition number tells about how close the manipulability hyperellipsoid is to be a hypersphere. The hyperellipsoid will become more slender when the condition number is high according to the two part Jacobian relationship given in Equation 5.17. When the condition number rapidly changes, velocity of the manipulator changes rapidly due to the velocity amplification factor. Hence, condition number works as an observer in the manipulator workspace to check for its dexterity, accuracy, and velocity.

Equation 5.82 shows another measure of manipulator performance defined in Ref. [17] and is called kinematic manipulability. Rao [18] have used this measure for the workspace calculation. $J^T J$ or JJ^T forms square Jacobian from the original rectangular Jacobian. The squared matrix can get inversed and determinants can be taken easily in the space.

$$m(J) = \sqrt{\det(JJ^T)}$$ (5.82)

A searching program is made to evaluate the dexterity, dexterous workspace, and total workspace. To find out the total and dexterous workspace, the condition given in Equation 5.83 is required to be implemented in the searching program.

$$\text{cond}(J) \leq \text{cond}(J)_{\max}$$ (5.83)

The following conditions are met to formulate the workspaces in both the examples:

For dexterous workspace: usually cond(J) ≤ 1.5
For total workspace: usually cond(J) ≤ 5

Figure 5.18 presents the algorithm of the searching program. Workspace and dexterity are found using the value of maximum condition number to be searched in a defined direction throughout the whole space. The maximum condition number is given as the input and the program starts search from the defined reference point in a certain direction. When it reaches to a maximum value in that direction, the program stops. It then restarts searching from the reference point in another direction and in this way completes one full turn in one plane. All the planes are searched one by one to give a final workspace plot.

5.9   SOME RESULTS TO DISCUSS

Workspace and dexterity for the manipulators are found below with $r_p = 500$ mm, $h = 1000$ mm, and actuator expansion = 500 mm. The input ranges for both ψ and θ are kept from −1 to 1 rad. The total and dexterous workspaces for the SPS system are presented in Figure 5.19. Both the workspace and dexterity are calculated using conventional Jacobian. Cross section of the dexterity for SPS system shown in Figure 5.20 has three sides actually depicting the effect of three legs in the geometry. If high number of legs are added in the system, the corresponding corners in the final shape of the workspace or dexterity will emerge. The size of the dexterous workspace is found smaller proving that the dexterous workspace is the subset of the total workspace.

Dexterity is also calculated using forward and pseudoinverse approach of the conventional Jacobian. Similar dexterity is found from forward and pseudoinverse approaches of the conventional Jacobian. The results also show an increase trend in dexterity, dexterous, and total workspaces

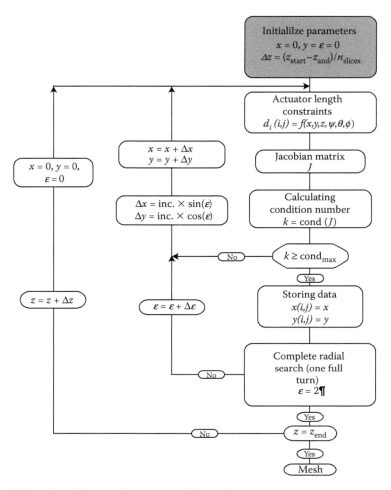

FIGURE 5.18 Schematic of the searching program for workspace.

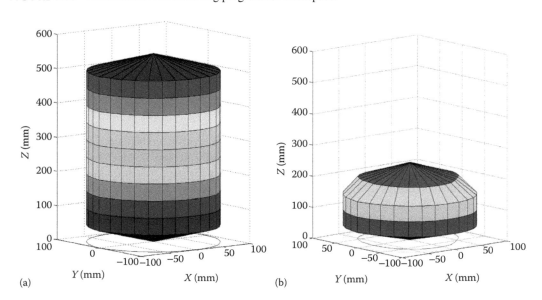

FIGURE 5.19 (See color insert following page 174.) (a) Total workspace (SPS), $\gamma = 0°$ and (b) dexterous workspace, $\gamma = 0°$.

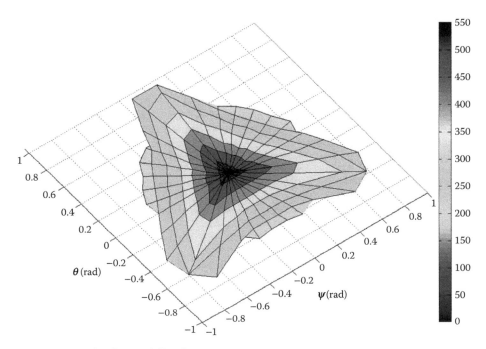

FIGURE 5.20 (See color insert following page 174.) Dexterity corresponding to the total w/s for SPS using forward and pseudoinverse approach of conventional Jacobian.

by increasing the inclination angle γ. The size and shape of the dexterous w/s and the total workspace is better in the SPS as compared to the PRS or PSS mechanisms. The information required by the designer is the workspace generated by the PKM and its corresponding dexterous workspace. As the dexterous workspace is the subset of the total workspace, this superior workspace will secure a better dexterity for the cutting tool, for example.

REFERENCES

1. I. Fassi, G. Legnani, and D. Tosi. Geometrical conditions for the design of partial or full isotropic hexapods. *Journal of Robotic Systems.* 22(10), 507–518, 2005.
2. J. Tlusty, J. Ziegert and S. Ridgeway. Fundamental comparison of the use of serial and parallel kinematics for machine tools. *Annals of the CIRP.* 48(1), 1999.
3. L. Kieckhafer, P. Sheldon and M. McGuire. A revolutionary new machining center. *European Production Engineering.* 19(1-12), 44–46, 1995.
4. M. Geldart, P. Webb, H. Larsson, M. Backstrom, N. Gindy and K. Rask. A direct comparison of the machining performance of a variax 5 axis parallel kinetic machining center with conventional 3 and 5 axis machine tools. *International Journal of Machine Tools & Manufacture.* 43, 1107–1116, 2003.
5. A. Khalid and S. Mekid. Design of precision desktop machine tools for meso-machining. *Proceedings of the International Conference of Intelligent Production Machines and Systems.* 165–170, 2006.
6. D. Zhang, L. Wang, and S.Y.T. Lang. Parallel kinematic machines: Design, analysis and simulation in an integrated virtual environment. *Transactions of the ASME, Journal of Mechanical Design.* 127, 580–588, 2005.
7. R.E. Stamper. A three degree of freedom parallel manipulator with only translational degrees of freedom, PhD thesis, University of Maryland, 1997.
8. L.W. Tsai. *Robot Analysis: The Mechanics of Serial and Parallel Manipulators*, John Wiley & Sons, New York, 1999.
9. G. Gogu. Mobility of mechanisms: A critical review. *Mechanism and Machine Theory.* 40, 1068–1097, 2005.

10. K.H. Hunt. *Kinematic Geometry of Mechanisms*, Oxford Science Publications, Oxford, UK, 1978.
11. G.T. Pond and J.A. Carretero. Kinematic analysis and workspace determination of the inclined PRS parallel manipulator. *15th CISM-IFToMM Symposium on Robot Design, Dynamics and Control*, Saint-Hubert, Montreal, Quebec, Canada, 2004.
12. J.A. Carretero, M.A. Nahon, R.P. Podhorodeski, and C.M. Gosselin. Kinematic analysis and optimization of a new three degree of freedom spatial parallel manipulator. *ASME Journal of Mechanical Design*. 122(1), 17–24, 2000.
13. S.-G. Kim and J. Ryu. New dimensionally homogeneous Jacobian matrix formulation by three end-effector points for optimal design of parallel manipulators. *IEEE Transactions on Robotics and Automation*. 19(4), 731–737, 2003.
14. M. Raghavan. The Stewart platform of general geometry has 40 configurations. *ASME Journal of Mechanical Design*. 115, 277–282, 1993.
15. M. Tucker and N.D. Perreira. Generalized inverses for robotic manipulators. *Mechanism and Machine Theory*. 22(6), 507–514, 1987.
16. J.P. Merlet. Jacobian, manipulability, condition number and accuracy of parallel robots. *ASME Journal of Mechanical Design*. 128, 199–206, 2006.
17. T. Yoshikawa. Manipulability of robotic mechanisms. *International Journal of Robotics Research*. 4(2), 3–9, 1985.
18. A.B. Koteswara Rao, P.V.M. Rao, and S.K. Saha. Workspace and dexterity analysis of hexaslide machine tools. *International Conference on Robotics & Automation*, Taipei, Taiwan, pp. 4104–4109, 2003.

6 Precision Control

Tan Kok Kiong, Andi Sudjana Putra,
and Sunan Huang

CONTENTS

… controlling motion parameters and compensating for errors

6.1 BACKGROUND AND MOTIVATION

Precision control is one of the core issues required by ultraprecision machines discussed previously. A well-chosen control strategy will enable a complete control of the mechanical system to achieve a better positioning and often compensate for errors either implicitly [1] or explicitly as it will be explained in the subsequent chapters. The field of high-precision actuators is now an interesting subject of research worth spending millions of dollars annually. Precision control discussed here refers to control of motion parameters of the order of micrometer or better.

Precision is a key to many successful automation processes. A process may be designed with sophisticated working principles and excellent control algorithms, but if the precision of the process is not up to the requirements, then the process will not be able to serve the purpose, resulting in unsuccessful operations.

Fundamental and applied works, technologies, and manufacturing processes are now moving toward product miniaturization, with the requirements of motion control down to the order of submicrometer level. This is driven by the emergence of current technologies, such as biotechnology and nanotechnology. In the early 1980s, semiconductors and biomedical industries started to demand for high-precision actuators to execute a more precise positioning and manufacturing throughout their processes. The requirements pertaining to the precision of motion vary substantially according to the applications of the devices. As such, high-precision actuators are now in high demand, and are expected to perform various types of actions from rotation to translation, high-torque capability, wide speed range, etc. The application areas of high-precision actuators are as diverse as aerospace, microelectronic, biomedical, and nanotechnology.

An example of the application of high-precision motion control is in the area of microfabrication. A microelectromechanical system (MEMS) is a synergistic integration of mechanical and electronic components on a common platform through the utilization of fabrication on micro- or submicrometer scales (often termed as microfabrication) [2]. A MEMS component consists of sensors, actuators, and control electronics in a compact and small structure. Examples of MEMS components are pressure sensors, flow sensors, inkjet dispenser, and micromotors, which now appear in many industrial devices. The development of MEMS started with a bulk pressure sensor in the beginning of 1980s to over one million micromirror arrays in a few millimeters chip today. The advantages often associated to MEMS are reduction in cost for mass production, reduction in size, reduction in power consumption, and improvement of functionalities and capabilities.

In microfabrication processes, a precision of $10\,\mu m$ or less is typical. Currently, structures such as transmission gears, friction drives, and motors can also be manufactured utilizing microfabrication, resulting in small and compact devices. Microfabrication is also required in the fabrication of integrated-circuits (IC). Standard processes in IC fabrication include, for example, thin film deposition, photolithography, and *dopant* introduction, all of which demands precision at an atomic level. Over the years, the density of the components contained in a single chip increases exponentially, which puts even greater demands on precision.

Precision control is closely related to servo control. The term *servo* originates from a Latin word *servus*, which means *servant* or *follower*. Along with this perspective, a servo control system can be defined as a system that is able to control some variables of interest to track user-specified objectives closely, in the form of reference signals. The history of servo control can be traced back to 1788, when James Watt invented the fly ball governor to regulate the speed of a steam engine. A boost to the development in this field came along with World War II, when servo control was used in diverse military applications, including the precise guidance and control of missiles, tracking of military targets, and development of navigation systems.

Today, servo control has become an integral part of almost every automation system or process, including, but not limited to, manufacturing systems, chemical and petrochemical processes, transportation, military systems, and biomedical systems. While this broad definition of servo control still holds, the expectations in terms of the tracking performance of servo control systems have risen significantly, in line with the ever tightening and stringent requirements associated with the products of today, and the processes to achieve them.

Generally, the objective of a servo control system is to force a controlled signal to follow or track a reference input signal, sometimes also called the set point, at high speed and accuracy, and to remain robust to keep the controlled signal on track, despite possible undesirable disturbances affecting the system.

In the domain of motion control applications, servo control will be concerned with the control of motion enablers (i.e., the actuators) to achieve desired motion profiles of the load to which they are attached, in terms of direct-motion variables such as proximity, position, velocity, and acceleration, or in terms of motion-induced variables such as force and torque. Disturbances to the achievement of this objective may arise in the form of load changes and the existence of motion impeding forces such as friction, backlash, and cogging forces.

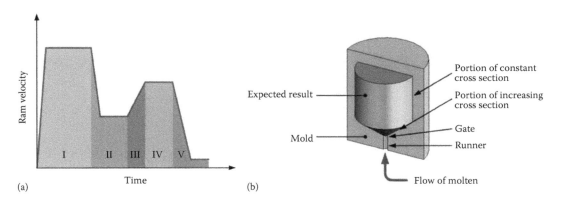

FIGURE 6.1   Injection-molding system: (a) optimal ram velocity profile and (b) outlook of injection molding.

The reference signal of a servo control system is the command input to the servo control system. It can be specified to optimize the quality of a product or the performance of a process. Two examples will be provided to reiterate this point.

Example 1: Injection-Molding Machine

The ram velocity profile of an injection-molding machine affects the precision and consistency at which the plastic components are produced. The optimal profile depends on factors such as the specific resin material properties and the ram geometry, and it is typically designed and generated from computer simulation software with a model of the whole system. Figure 6.1 shows an example of a ram velocity profile. With the desired profile as the reference signal, the servo control of the injection molding system will work to force the ram velocity to track this profile with minimum offset.

The velocity profile presented in Figure 6.1 is a typical ram velocity profile to fill a mold of increasing cross section initially, and having a constant cross section thereafter. The profile for this part can be divided into five sections. The profile in Section I is used for filling the runner, when velocity is very rapidly increased and then held constant. In Section II, the velocity is rapidly reduced to eliminate jetting at the gate. Section III corresponds to filling of the increasing cross section portion of the cavity, while Section IV is concerned with filling of the constant cross section portion of the cavity. Finally, the velocity is reduced in Section V to eliminate flushing and over-packing. All of the above objectives have to be achieved in short time available for mold filling.

Example 2: Flight Control

An experienced pilot is able to takeoff an airplane smoothly, causing minimal discomfort to the passengers and crew onboard. The velocity profile of the airplane along the runway determines how smooth the takeoff will be. Figure 6.2 shows a possible velocity profile for a smooth takeoff. Pilots will track this profile via a servo control system, either on the actual airplane or for trainee pilots via a flight simulator.

The performance of a servo control system is therefore evaluated by indicators measuring how closely the objective reference signal is tracked. A common indicator will be based on the magnitude of the root-mean-square tracking error over the profile. For point-to-point tracking (i.e., step change in the reference signal), classical performance indicators can be used, including the rise time, overshoot, and steady-state error.

6.2   FUNDAMENTALS OF MOTION CONTROL

The core of a servo drive is the control system. It orchestrates and links the different components to yield the highest level of performance achievable. Until the 1940s, the sensors and the actuators

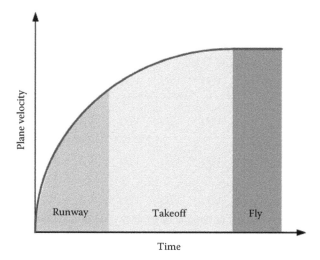

FIGURE 6.2 Velocity profile for plane taking-off.

were the stand-alone devices. There were neither signal transmission capabilities nor standards to allow the devices to communicate with each other. Even in the early years following the birth of the transistor and the rise of electronics, sensors were still pure measurement devices. The users will manually establish a measurement connection, take measurement, and record it into a log book or datasheet. The measurement data from the sensor ended in the sensor itself. Further and subsequent use of the data cannot be enabled by any function of the sensor.

Thus, control systems, until the 1950s, were still manual-based. The control operator will take a reading visually by looking at the display of the measurement instrument and manually adjust some actuator such as a valve to bring the control variable to a desired level. This form of control is also called open-loop control, wherein the operator has to provide the link between the sensor and the actuator, acting as the controller.

The control operator requires much experience and skills on open-loop control to achieve good performance, since the only form of feedback is via the operator himself. Such a form of control is seriously inadequate in applications with a huge number of control loops where it is simply ineffi-cient to have the operator to tune individual loop. It did not take long to realize that automatic and closed-loop control is a necessity in these applications.

The first innovation was to simply retrofit the sensor with a relay at the output, so that the relay can be used for on–off control of the system. This was the early form of closed-loop control. However, it was the advances in analog transmission standards from the 1940s, such as the 3–15 psi and 4–20 mA standards, which provided the key steps to realize truly closed-loop control systems. With different components of the control system adhering to the same transmission standards, it became possible to integrate all of them via a computer-based controller. Direct digital and supervisory con-trol systems arose in the 1960s as a result of the analog transmission standards developed, advancing the development of closed-loop control systems. The overall closed-loop control structure is shown in Figure 6.3.

6.2.1 SYSTEM MODELING AND PERFORMANCE ASSESSMENT

In general, the characteristics of a plant (e.g., the actuator in Figure 6.3) are obtained from its physical behavior, i.e., physical principles. A very common form of this behavior is described using d'Alembert principles as follows:

$$\ddot{y} + 2\zeta\omega_0\,\dot{y} + \omega_0^2 y = K_0 u \tag{6.1}$$

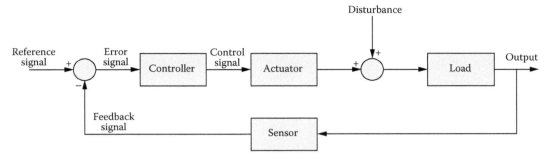

FIGURE 6.3 Overall structure of control system.

where

  y is the displacement

  $2\zeta\omega_0\dot{y}$ denotes its damping characteristics, in which ζ is the damping ratio

  ω_0 is the natural frequency

  ω_0^2 signifies the inertia characteristics

  K_0 is an input parameter

  u is the input signal

The order of the plant system is 2 (a second-order system), as suggested by the highest derivative in Equation 6.1.

This system can be expressed in a state-variable form as follows:

$$\begin{bmatrix} \dot{x}_1 \\ \dot{x}_2 \end{bmatrix} = \begin{bmatrix} 0 & 1 \\ -\omega_0^2 & -2\zeta\omega_0 \end{bmatrix} \begin{bmatrix} x_1 \\ x_2 \end{bmatrix} + \begin{bmatrix} 0 \\ K_0 \end{bmatrix} u$$

$$y = \begin{bmatrix} 1 & 0 \end{bmatrix} \begin{bmatrix} x_1 \\ x_2 \end{bmatrix}$$

(6.2)

where

$$\begin{bmatrix} x_1 \\ x_2 \end{bmatrix} = \begin{bmatrix} y \\ \dot{y} \end{bmatrix}$$

In general, any set of differential equations of any order can be transformed into state-variable form as follows:

$$\dot{x} = Ax + Bu$$ (6.3)

$$y = Cx + Du$$ (6.4)

where

  x is the state

  y is the output

  u is the input

  A is the system matrix

  B is the input matrix

  C is the output matrix

  D is the feed-through matrix

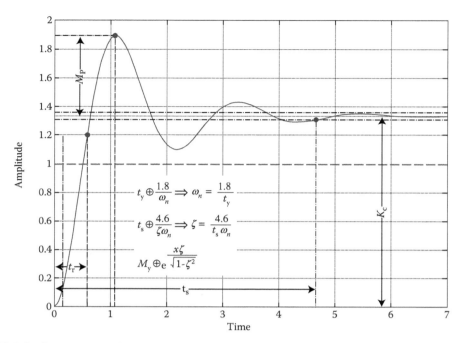

FIGURE 6.4   Step response of a second-order system.

Since this system of Equation 6.1 possesses the properties of a linear system,* the analysis is facilitated by the use of Laplace transform, transforming Equation 6.1 into

$$\left(s^2 + 2\zeta\,\omega_0 s + \omega_0^2\right)Y(s) = K_0 U(s) \tag{6.5}$$

From Equation 6.5, the transfer function of the system is

$$G(s) = \frac{Y(s)}{U(s)} = \frac{K_0}{s^2 + 2\zeta\,\omega_0 s + \omega_0^2} \tag{6.6}$$

Transfer function offers important information pertaining to the behavior of the system. It can be seen from Equation 6.6 that the transfer function relates the input to the output, i.e., the output of the system for a unit impulse input.

The behavior of the system upon the introduction of step input is presented in Figure 6.4, where K_c is the gain. The characteristics of the system, for example, the overshoot (M_p), rise time (t_r), and settling time (t_s), depend on the parameters of the system itself.

The role of motion control is to force those characteristics to meet the required specifications so that the system can deliver the required output satisfactorily, for example, with a certain overshoot or settling time. This can be explained by referring to Figure 6.4. Suppose the transfer function of the actuator is $G_{act}(s)$ and the combined sensor and control is $G_{ctr}(s)$, the transfer function of the system can be expressed as

$$\frac{Y(s)}{R(s)} = \frac{G_{act}(s)}{1 + G_{ctr}(s)}$$

showing that the output of the system is affected by $G_{ctr}(s)$.

---

* The term linear here refers to mathematically linear and not physically linear motion.

6.2.2 LINEAR DYNAMICS

The dynamics of the actuator can be viewed comprising of two components: a dominantly linear model and an uncertain and nonlinear remnant which nonetheless must be considered in the design of the control system if high-precision motion control is to be efficiently realized. The linear dynamics of actuators depend on the physical behavior of the actuators themselves.

For illustration purposes, in order to describe the fundamentals of motion control, especially on the design and performance assessment of the controller, the linear motor will be used as an example. The same technique can be extended to other actuators according to their respective specifications.

In the dominant linear model, the mechanical and electrical dynamics of a linear motor can be expressed as follows:

$$M\ddot{x} + D\dot{x} + F_{\text{load}} = F_{\text{m}} \tag{6.7}$$

$$K_{\text{e}}\dot{x} + L_{\text{a}}\frac{\mathrm{d}I_{\text{a}}}{\mathrm{d}t} + R_{\text{a}}I_{\text{a}} = u \tag{6.8}$$

$$F_{\text{m}} = K_{t}I_{\text{a}} \tag{6.9}$$

where x, M, D, F_{m}, and F_{load} denote the mechanical parameters: position, inertia, viscosity constant, generated force, and load force, respectively, while u, I_{a}, R_{a}, and L_{a} denote the electrical parameters: input dc voltage, armature current, armature resistance, and armature inductance, respectively; and K_t denotes an electrical–mechanical energy conversion constant.

Since the electrical time constant is typically much smaller than the mechanical time constant, the delay due to electrical transient response may be ignored, giving the following simplified model:

$$\ddot{x} = -\frac{K_1}{M}\dot{x} + \frac{K_2}{M}u - \frac{1}{M}F_{\text{load}} \tag{6.10}$$

where

$$K_1 = \frac{K_{\text{e}}K_t + R_{\text{a}}D}{R_{\text{a}}}, \quad K_2 = \frac{K_t}{R_{\text{a}}} \tag{6.11}$$

The dominant linear model has not included extraneous nonlinear effects which may be present in the physical structure.

A rotational dc motor can generally be seen as a circuitry of resistance R, inductance L, and electromotive voltage V_{emf}, connected to an inertial load J as presented in Figure 6.5.
In this model, the input of the motor is V_{app}, while the output is the rotational speed ω. Therefore, the problem of interest is what V_{app} is to be applied if certain ω is desired.

The torque τ of the motor shaft is

$$\tau = K_{\text{m}}i \tag{6.12}$$

where
  τ is the torque
  K_{m} is the armature constant of the motor
  i is the current

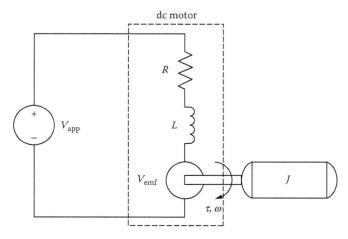

FIGURE 6.5   dc motor model.

The back emf voltage, V_{emf}, of the motor is

$$V_{emf} = K_b \omega \tag{6.13}$$

where
  V_{emf} is the back emf voltage
  K_b is the emf constant
  ω is the speed

The load and the motor are related by the following equation:

$$J \frac{d\omega}{dt} = \tau \tag{6.14}$$

where J is the inertia of the load.
  The electrical part of the motor can be described by

$$V_{app} - V_{emf} = L \frac{di}{dt} + Ri \tag{6.15}$$

Problem 6.1

A rotational dc motor has the following characteristics $R = 2.5\,\Omega$, $L = 0.9\,H$, $K_m = 8.5$, $K_b = 8.5$, and $J = 9.2\,kg \cdot m^2$.
  Construct the model of the motor and assess its transient performance (overshoot, rise time, and settling time).

Solution 6.1

The governing equations of dc motor can be rearranged to yield the following differential equations:

$$J \frac{d\omega}{dt} = \tau$$

$$= K_m i \tag{6.16}$$

$$\frac{d\omega}{dt} = \frac{K_m}{J} i$$

$$V_{app} - V_{emf} = L\frac{di}{dt} + Ri$$

$$L\frac{di}{dt} = -Ri - V_{emf} + V_{app}$$

$$= -Ri - K_b\omega + V_{app}$$

$$\frac{di}{dt} = -\frac{R}{L}i - \frac{K_b}{L}\omega + \frac{1}{L}V_{app}$$

(6.17)

Let $x_1 = \omega$ and $x_2 = i$. The system of dc motor can then be expressed as

$$\begin{bmatrix} \dot{x}_1 \\ \dot{x}_2 \end{bmatrix} = \begin{bmatrix} 0 & \dfrac{K_m}{J} \\ -\dfrac{K_b}{L} & -\dfrac{R}{L} \end{bmatrix}\begin{bmatrix} x_1 \\ x_2 \end{bmatrix} + \begin{bmatrix} 0 \\ \dfrac{1}{L} \end{bmatrix}V_{app}$$

$$y = \begin{bmatrix} 0 & 1 \end{bmatrix}\begin{bmatrix} x_1 \\ x_2 \end{bmatrix}$$

(6.18)

Inputting the above parameters, the following can be obtained:

$$\begin{bmatrix} \dot{x}_1 \\ \dot{x}_2 \end{bmatrix} = \begin{bmatrix} 0 & 0.9239 \\ -9.4444 & -2.7778 \end{bmatrix}\begin{bmatrix} x_1 \\ x_2 \end{bmatrix} + \begin{bmatrix} 0 \\ 1.1111 \end{bmatrix}V_{app}$$

$$y = \begin{bmatrix} 0 & 1 \end{bmatrix}\begin{bmatrix} x_1 \\ x_2 \end{bmatrix}$$

(6.19)

or equivalent to

$$G(s) = \frac{Y(s)}{U(s)} = \frac{1.0265}{s^2 + 2.77778s + 8.7257}$$

(6.20)

The step response of this system can now be determined via computer simulation; the result of which is presented in Figure 6.6.

The transient responses of the dc motor are (refer to Figure 6.4)

1. Overshoot $M_p = 9.37\% = 0.0937$
2. Rise time $t_r = 0.55\,\text{s}$
3. Settling time $t_s = 2.01\,\text{s}$

Problem 6.2

Continuing the example in Problem 1 improves the transient performance of the system. For example, try to get the overshoot 4% lower, the rise time 0.05 s faster, and the settling time 0.5 s faster than the original performance.

Solution 6.2

Having known the original performance of the motor, the improvement of its transient performance can be achieved, with the following desired performance:

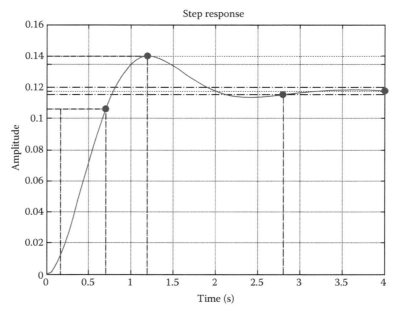

FIGURE 6.6 Step response of the dc motor.

1. Overshoot $M_p < 5\% = 0.05$
2. Rise time $t_r < 0.50\,\mathrm{s}$
3. Settling time $t_s = 1.5\,\mathrm{s}$

From the transfer function of the dc motor, one can relate to Equation 6.6 to determine the following characteristics $K_0 = 1.1111$, $\omega_0 = 3.0731$, and $\zeta_0 = 0.4520$. The load of the motor is not to be changed, so J is left unchanged. Also, the motor is not to be changed, so K_m and K_b remain unchanged as well. Therefore, only R and L are to be changed.

To make $t_r < 0.50$ (e.g., 0.40), the natural frequency $\omega_0 = (1.8/0.40) = 4.5$, so that L should be (note that $-(K_b/L) = -\omega_0^2 = -20.25$) 0.4198 H.

To make $M_p < 0.05$ (e.g., 0.045), the damping ratio $\zeta = 0.71$ and therefore $-2\zeta\omega_0 = -2(0.71)(4.5) = -6.39$, so that R should be (note that $-(R/L) = -6.39$) 2.6825 Ω.

With the new value of L and R, the system is reconstructed and the new step response is presented in Figure 6.7, where the transient characteristics are $M_p = 3.2\% = 0.032$, $t_r = 0.52\,\mathrm{s}$, and $t_s = 1.35\,\mathrm{s}$.

The new value of L and R can be implemented by adding a series resistor of 0.1825Ω and a parallel inductor of 0.7868H. Clearly, the performance of the dc motor can be improved with simple modifications.

6.2.3 Nonlinear Dynamics

The predominant nonlinear effects underlying a motor system are the various friction components (coulomb, viscous, and stiction), force ripples (detent and reluctance forces), and hysteresis arising from imperfections in the underlying components. A reduction of these effects, either through proper physical design or via the control system, is of paramount importance if high-speed and high-precision motion control is to be achieved. Compensation via proper physical design usually introduces mechanical complexity and extra manufacturing costs. On the other hand, control algorithms also have the advantage of preserving the maximum force (and hence acceleration) achievable. Thus, the control algorithm is preferable to compensate for the uncertainties.

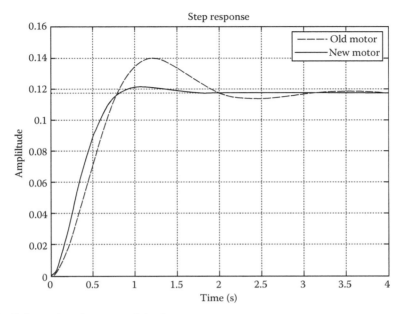

FIGURE 6.7 Enhanced performance of the dc motor.

6.2.3.1 Force Ripple

Undesirable force ripples, inevitably present in servo control systems, have long attracted attention from researchers, and a great deal of efforts has been devoted to overcome the difficulties associated with the nonlinear rippling effects. The two primary sources of force ripple are the cogging (or detent) force and the reluctance force. The cogging force arises as a result of the mutual attraction between the magnets and iron-cores translator. Cogging manifests itself by the tendency of the translator to align in a number of preferred positions, regardless of excitation states. This force exists even in the absence of any winding current and it exhibits a periodic relationship with respect to the position of the translator relative to the magnets. There are two potential causes of the periodic cogging force in motion control, resulting from the slotting and the finite length of iron-core translator. The reluctance force is due to the variation of the self-inductance of the windings with respect to the relative position between the translator and the magnets. Thus, the reluctance force also has a periodic relationship with the translator-magnet position.

A first-order model for the force ripple can be described as a periodic sinusoidal type signal:

$$F_{\text{ripple}}(x) = A(x)\sin(\omega x + \phi) \tag{6.21}$$

Higher harmonics of the ripple may be included in higher-order models.

6.2.3.2 Friction

Friction is present in nearly all moving mechanisms and it is one major obstacle to achieving precise motion control. Several characteristic properties of friction have been observed, which can be broken down into two categories: static and dynamic. The static characteristics of friction, including the stiction friction, the kinetic force, the viscous force, and the Stribeck effect, are functions of steady-state velocity. The dynamic phenomena include presliding displacement, varying breakaway force, and frictional lag. Many empirical friction models have been developed which attempt to capture specific components of observed friction behavior, but generally it is acknowledged that a

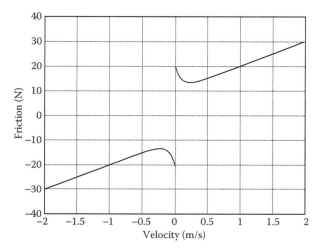

FIGURE 6.8 Tustin friction model.

precise and accurate friction model is difficult to be obtained in an explicit form, especially for the dynamical component. For many purposes, however, the Tustin model has proven to be useful and it has been validated adequately in many successful applications. The Tustin model may be written as

$$F_{\text{friction}} = \left[F_c + (F_s - F_c) e^{-|\dot{x}/\dot{x}_s|^\delta} + F_v |\dot{x}| \text{sgn}(\dot{x}) \right] \tag{6.22}$$

where
  F_s denotes static friction
  F_c denotes the minimum value of coulomb friction
  $\dot{x}_s$ and F_v are lubricant and load parameters
  δ is an additional empirical parameter

Figure 6.8 graphically illustrates this friction model. The effects of friction can be greatly reduced using high-quality bearings such as aerostatic or magnetic bearings.

6.2.3.3 Hysteresis

Hysteresis is an input/output nonlinearity with effects of nonlocal memory, i.e., the output of the system depends not only on the instantaneous input, but also on the history of its operation. Figure 6.9 shows the real-time open-loop response of a piezoelectric actuator. Typically, the magnitude of this hysteresis can constitute about 10%–15% of the motion range [3]. Hysteresis generally impedes high-precision motion and hysteresis analysis is thus a key step toward the realization of high-performance actuation systems.

As elaborated in Ref. [4], hysteresis is a friction-like phenomenon. In Ref. [5], a spring coupled to a pure coulomb friction element is used to model a piezoelectric drive with hysteresis. In Ref. [6], the Maxwell slip model is used to represent hysteresis. In Ref. [7], a more complex dynamical friction model from a frictional force is presented. The model contains a state variable representing the average deflection of elastic bristles which are a visualization of the topography of the contacting surfaces. The resulting model shows most of the known friction behavior like hysteresis, friction lag, varying break-away force, and stick-slip motion. It is comprehensive enough to capture dynamical hysteresis effects as follows:

$$F = \sigma_0 z + \sigma_1 \dot{z} + \sigma_2 \dot{x} \tag{6.23}$$

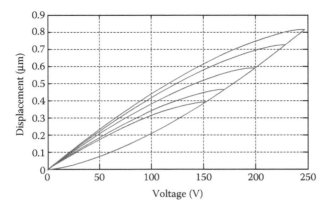

FIGURE 6.9 Hysteresis phenomenon.

with

$$\dot{z} = \dot{x} - \frac{|\dot{x}|}{h(\dot{x})} z$$

$$h(\dot{x}) = \frac{F_c + (F_s - F_c) e^{-(\dot{x}/\dot{x}_s)^2}}{\sigma_0}$$

where F_c, F_s, $\dot{x}_s$, σ_0, σ_1, and σ_2 are positive constants which are typically unknown.

6.2.3.4 Incorporating Nonlinear Dynamics

The three most common nonlinearities have been described above. Other types of nonlinearities, such as saturation and dead-zone, may also appear dominantly in certain actuators (not discussed in this chapter), but similar techniques to incorporate them in the system can be applied as well.

Having defined the important nonlinearities of the system, the dynamics of the actuator may now be described by

$$\ddot{x} = -\frac{K_1}{M}\dot{x} + \frac{K_2}{M}u - \frac{1}{M}\left(F_{\text{load}} + F_{\text{ripple}} + F_{\text{friction}} + F_{\text{hysteresis}}\right) \tag{6.24}$$

A common nonlinear function $F_1^*(x, \dot{x})$ may be used to represent the nonlinear dynamical effects due to force ripple, friction, and other unaccounted dynamics, collectively. The system of Equation 6.24 can thus be alternatively described by

$$\ddot{x} = -\frac{K_1}{M}\dot{x} + \frac{K_2}{M}u - \frac{1}{M}F_{\text{load}} + F_1^*(x, \dot{x}) \tag{6.25}$$

Let

$$\frac{K_2}{M}f(x, \dot{x}) = -\frac{1}{M}F_{\text{load}} + F_1^*(x, \dot{x}) \tag{6.26}$$

It follows that

$$\ddot{x} = -\frac{K_1}{M}\dot{x} + \frac{K_2}{M}u + \frac{K_2}{M}f(x, \dot{x}) \tag{6.27}$$

with the tracking error e of first-order defined as $e = x_d - x$, where x_d is the desired value.

Equation 6.27 may be expressed as

$$\ddot{e} = -\frac{K_1}{M}\dot{e} - \frac{K_2}{M}u - \frac{K_2}{M}f(x,\dot{x}) + \frac{K_2}{M}\left(\frac{M}{K_2}\ddot{x}_d + \frac{K_1}{K_2}\dot{x}_d\right) \tag{6.28}$$

Since

$$\frac{\mathrm{d}}{\mathrm{d}t}\int_0^t e(t)\mathrm{d}t = e \tag{6.29}$$

the system state variables are assigned as $x_1 = \int_0^t e(t)\mathrm{d}t = e$, $x_2 = e$, and $x_3 = \dot{e}$. Denoting $x = [x_1 \ x_2 \ x_3]^T$, Equation 6.28 can then be put into the equivalent state space form:

$$\dot{x} = Ax + Bu + Bf(x,\dot{x}) + B\left(-\frac{M}{K_2}\ddot{x}_d - \frac{K_1}{K_2}\dot{x}_d\right) \tag{6.30}$$

$$A = \begin{bmatrix} 0 & 1 & 0 \\ 0 & 0 & 1 \\ 0 & 0 & -K_1/M \end{bmatrix}, \quad B = \begin{bmatrix} 0 \\ 0 \\ -K_2/M \end{bmatrix} \tag{6.31}$$

6.3 CONTROL DESIGN STRATEGIES

The complexity of motion control poses challenges to develop systematic design procedures to meet the control performance requirements. In the face of such challenges, it is not possible to expect one particular design to apply equally well to all systems, especially when nonlinearities occur significantly. A control engineer needs a set of analysis and design tools that cover a wide range of situations. When facing a particular application, the engineer will need to utilize the tools which are most appropriate for the problem in hand. In Section 6.3.1, several of such tools will be covered, namely proportional-integral-derivative (PID) feedback control, feedforward control, ripple compensation, and radial-basis-function (RBF) compensation. When several of these techniques are used in a system, the overall control signal is the addition of each control signal.

6.3.1 PID FEEDBACK CONTROL

Feedback control is a very common concept nowadays, referring to the process of measuring the output of a system and using that information to force the output to eventually follow a desired value. The term *control* refers to the action of forcing the output to the desired value, while *feedback* refers to the use of past output information to influence the current output. The structure of feedback control is as presented in Figure 6.3.

There are many schemes of feedback control; a very common and widely used scheme is the PID control, which uses the combination of constant gain, integrated, and differentiated error signal (the difference between the desired value and the actual output) to generate control signal, as presented in Figure 6.10.

In spite of the advances in mathematical control theory over the last 50 years, industrial servo control loops are still essentially based on the three-term PID controller. It is estimated that 90% of the control loops in the industry implement PID controller. The main reason is due to the widespread field acceptance of this simple controller which has been effective and reliable in most situations when adequately tuned. More complex and advanced controllers have fared less favorably under

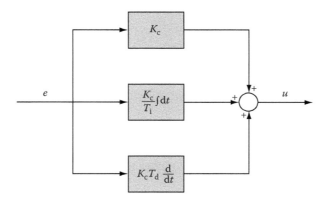

FIGURE 6.10   Structure of PID controller.

practical conditions, due to the higher costs associated with implementation and the higher demands in control tuning. It is very difficult for operators unfamiliar with advanced control to adjust the control parameters. Given these uncertainties, there is little surprise that PID controllers continue to be manufactured by the hundred thousand yearly and still increasing. In the composite control system, PID is used as the feedback control term. While the simplicity in a PID structure is appealing, it is also often proclaimed as the reason for poor control performance whenever it occurs. In this design, advanced optimum control theory is applied to tune PID control gains. The nominal portion of the system (without uncertainty) is given by

$$\dot{x}(t) = Ax(t) + Bu(t) \tag{6.32}$$

with

$$u(t) = u_{\mathrm{PID}}(t) = Kx = kx_1 + k_{d1}x_2 + k_{d2}x_3 \tag{6.33}$$

where x_1, x_2, and x_3 are defined in Equation 6.30. This is a PID control structure which utilizes a full-state feedback. Table 6.1 presents the effect of tuning PID parameters to the output of the system.

Problem 6.3

Suppose a plant can be described in the following transfer function:

$$G(s) = \frac{e^{-s}}{2s+1}.$$

Design a PID controller to improve the performance of the system.

Solution 6.3

Based on Figure 6.10 and Table 6.1, the PID parameters can be designed as follows $K_c = 2, T_i = 2$, and $T_d = 0.3$. Figure 6.11 shows the improvement achieved with PID controller, where the steady-state error of the system is completely eliminated.

   The optimal PID control parameters are obtained using the LQR (linear quadratic regulator) technique that is well known in modern optimal and robust control theory and it has been widely used in many applications. It has a very nice robustness property, i.e., if the process is of single-input and single-output, then the control system has at least a phase margin of 60° and a gain margin of infinity. Under mild assumptions, the resultant closed-loop system is always stable. This attractive property appeals to the practitioners. Thus, the LQR theory has received considerable attention since 1950s.

TABLE 6.1
Effect of Increasing PID Parameters

Parameters	Rise Time	Overshoot	Settling Time	Steady-State Error
K_c	↓	↑	×	↓
$1/T_i$	↓	↑	↑	O
T_d	×	↓	↓	×

The PID control using the LQR technique is given by

$$u_{\text{PID}} = -\left(r_0 + 1\right)B^{\text{T}}Px\left(t\right) \tag{6.34}$$

where P is the positive definite solution of the Riccati equation:

$$A^{\text{T}}P + PA - PBB^{\text{T}}P + Q = 0 \tag{6.35}$$

and $Q = H^{\text{T}}H$ where H relates to the states weighting parameters in the usual manner. Note that r_0 is independent of P and it is introduced to weigh the relative importance between control effort and control errors. Note for this feedback control, the only parameters required are the parameters of the second-order model and a user-specified error weight r_0. A full theoretical analysis of the stability properties of this PID feedback control scheme can be found in Ref. [8].

Where other state variables are available (e.g., velocity, acceleration, etc.), a full-state feedback controller may also be used for the feedback control component. Adaptive and robust control has also been investigated in a previous study as an alternative to the PID feedback control, where the feedback control signal is adaptively refined based on parameter estimates of the nonlinear system model, using prevailing input and output signals. The achievable performance is highly dependent on the adequacy of the model and the initial parameter estimates. Furthermore, full adaptive control schemes can greatly drain the computational resources available.

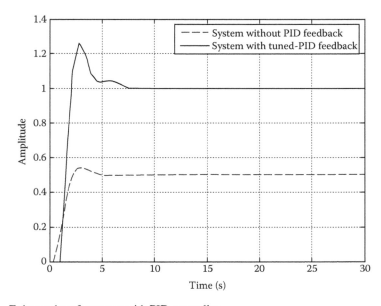

FIGURE 6.11 Enhanced performance with PID controller.

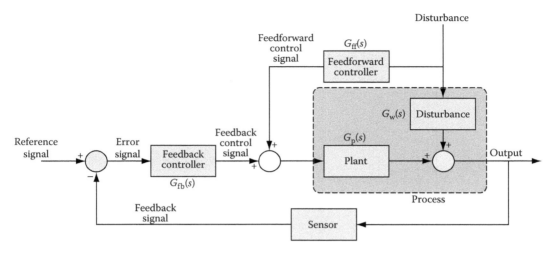

FIGURE 6.12   Feedforward–feedback control structure.

6.3.2   FEEDFORWARD CONTROL

Feedforward control is used to reject a measurable disturbance before it affects the output of the process. Feedforward control is typically used in tandem with a feedback control; in this configuration, feedforward is used solely to reject the major disturbance, while the feedback is used to track the reference value, as well as rejecting other possible disturbances. The feedforward plus feedback configuration is also referred to as a two degree of freedom (2 DOF) control. To achieve maximum advantage and outweigh the cost of implementation, it is therefore important that the disturbance to be rejected is major, and that a reasonably good model of the disturbance is available. Figure 6.12 presents the classical feedforward–feedback control structure.

In the simplest term, based on Figure 6.12, the design of feedforward control is started by defining the transfer function of the output with respect of the disturbance as follows:

$$H(s) = \frac{G_w(s) - G_p(s)G_{ff}(s)}{1 + G_{fb}(s)G_p(s)} \tag{6.36}$$

Therefore, to eliminate the disturbance, one needs to make $G_w(s) - G_p(s)G_{ff}(s) = 0$. For a system where the deadtime and relative order of $G_w(s)$ are greater than those of $G_p(s)$, and where $G_p(s)$ has no right-half plane zeros, the following equation will hold:

$$G_{ff}(s) = G_p^{-1}(s)G_w(s) \tag{6.37}$$

Problem 4

For the system in Problem 3, design a feedforward control to reject the disturbance that can be modeled as $G_w(s) = (e^{-0.25s} / 6s + 1)$, where the disturbance is to occur at 10 s with an amplitude of 1.

Solution 4

Based on Equation 6.37, the feedforward controller can be designed to be $G_{ff}(s) = (2s + 1/6s + 1)$. Figure 6.13 shows the improvement achieved with feedforward controller.

The design of the feedforward control component is straightforward. From Equation 6.31, the term

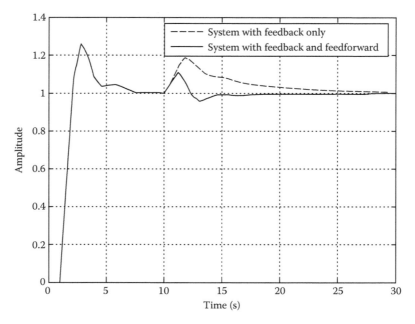

FIGURE 6.13   Enhanced performance with feedforward.

$$B\left(-\frac{M}{K_2}\ddot{x}_d - \frac{K_1}{K_2}\dot{x}_d\right)$$

may be neutralized using a feedforward control term in the control signal. The feedforward control is thus designed as

$$u_{FF}(t) = \frac{M}{K_2}\ddot{x}_d + \frac{K_1}{K_2}\dot{x}_d \tag{6.38}$$

Clearly, the reference position trajectory must be continuous and twice differentiable; otherwise a precompensator to filter the reference signal will be necessary. The only parameters required for the design of the feedforward control are the parameters of the second-order linear model.

Additional feedforward terms may be included for direct compensation of the nonlinear effects, if the appropriate models are available. For example, if a good signal model of the ripple force is available like in Equation 6.21, then an additional static term in the feedforward control signal, $u_{FF}(x)$ $= -F_{ripple}(x_d)$, can effectively compensate the ripple force. In fact, in the proposed overall strategy, an adaptive feedforward control (AFC) component for ripple compensation has been included.

In the same way, a static friction feedforward precompensator can be installed if a friction model is available. In Ref. [9], an efficient way of friction modeling using relay feedback is proposed where a simple friction model (incorporating coulomb and viscous friction components) can be obtained automatically. This model can be used to construct an additional feedforward control signal, based only on the reference trajectories. In addition, if the motion control task is essentially repetitive, an iteratively refined additional feedforward signal can further reduce any control-induced tracking error. A possible scheme based on iterative learning control (ILC) can be found in Ref. [10]. The basic idea in ILC is to exploit the repetitive nature of the tasks as experience gained to compensate for the poor or incomplete knowledge of the system model and the disturbances. Essentially, the ILC

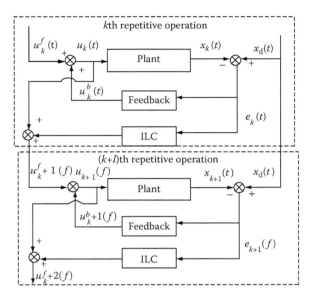

FIGURE 6.14 Iterative learning control.

structure includes a feedforward control component which refines the feedforward signal to enhance the performance of the next cycle based on previous cycles. A block diagram of the ILC scheme is depicted in Figure 6.14. As all others feedforward control schemes, the performance is critically dependent on the accuracy of the model parameters.

6.3.3 RIPPLE COMPENSATION

From motion control viewpoints, force and torque ripples are highly undesirable. A control strategy is available to deal with this issue, using a dither signal as a "Trojan horse" to cancel the effects of force ripples. The construction of dither signal requires knowledge of the characteristics of force ripples which can be obtained from simple step experiments. For greater robustness, real-time feedback of motion variables can be used to adaptively refine the dither signal characteristics.

A common approach to compensate the force and torque ripples is via AFC. It is assumed that the force ripple can be equivalently viewed as a response to a virtual input described in the form of a periodic sinusoidal signal:

$$u_{\text{ripple}} = A(x)\sin(\omega x + \phi) = A_1(x)\sin(\omega x) + A_2(x)\cos(\omega x) \tag{6.39}$$

The dither signal is thus designed correspondingly to eradicate this virtual force as

$$u_{\text{AFC}} = a_1(x(t))\sin(\omega x) + a_2(x(t))\cos(\omega x) \tag{6.40}$$

Perfect cancellation will be achieved when

$$a_1^*(x) = -A_1(x),\ a_2^*(x) = -A_2(x) \tag{6.41}$$

Feedforward compensation schemes are well known to be sensitive to modeling errors which inevitably result in significant remnant ripples. An adaptive approach can be adopted so that a_1 and a_2 will be continuously adapted based on desired trajectories and prevailing tracking errors.

Let

$$a = \begin{bmatrix} a_1(x) \\ a_2(x) \end{bmatrix}, \quad \theta = \begin{bmatrix} \sin(\omega x) \\ \cos(\omega x) \end{bmatrix}, \quad a^* = \begin{bmatrix} -A_1(x) \\ -A_2(x) \end{bmatrix} \tag{6.42}$$

The system output due to AFC is then given by

$$x_a = P[a - a]^{\mathrm{T}}\theta \tag{6.43}$$

where P denotes the system. Equation 6.43 falls within the standard framework of adaptive control theory. Possible update laws for the adaptive parameters will therefore be

$$\dot{a}_1(x(t)) = -ge\sin(\omega x) \tag{6.44}$$

$$\dot{a}_2(x(t)) = -ge\cos(\omega x) \tag{6.45}$$

where $g > 0$ is an arbitrary adaptation gain.

Differentiating Equations 6.44 and 6.45 with respect to time, it follows that

$$\dot{a}_1(t) = -ge\dot{x}_\mathrm{d}\sin(\omega x) \tag{6.46}$$

$$\dot{a}_2(t) = -ge\dot{x}_\mathrm{d}\cos(\omega x) \tag{6.47}$$

In other words, the adaptive update laws (Equations 6.46 and 6.47) can be applied as an adjustment mechanism such that $a_1(t)$ and $a_2(t)$ in Equation 6.40 converge to their true values.

6.3.4 RBF COMPENSATION

The 2 DOF control may suffice for many practical control requirements. However, if further performance enhancement is necessary, a third control component may be enabled. An RBF is applied to model the nonlinear remnant, and this is subsequently used to linearize the closed-loop system by neutralizing the nonlinear portion of the system. RBF, the basis of many neural networks, is often utilized for modeling nonlinear functions that are not explicitly defined. For the RBF, the hidden units within the neural network provide a set of basis functions as these units are expanded into the higher-dimensional hidden-unit space.

Since $f(\dot{x}, x)$ is a nonlinear smooth function (unknown), it can be represented as

$$f(\dot{x}, x) = \sum_{i=0}^{m} w_i \phi_i(\dot{x}, x) + \varepsilon = \sum_{i=0}^{m} \phi_i(\dot{x}, x) w_i + \varepsilon \tag{6.48}$$

where ε is the error, with $|\varepsilon| \leq \varepsilon_M$, where $\phi_i(\dot{x}, x)$ is the RBF given by

$$\phi_i(\dot{x}, x) = \frac{\exp\left(-\left(\|x_{\mathrm{vect}} - c_i\|^2\right)\big/2\sigma_i^2\right)}{\sum\limits_{j=0}^{m}\exp\left(-\left(\|x_{\mathrm{vect}} - c_j\|^2\right)\big/2\sigma_j^2\right)} \tag{6.49}$$

where $x_{vect} = [\dot{x}\ x]^T$.

The following assumptions are made

1. Ideal weights are bounded by known positive values so that $|w_i| \leq w_M$, $i = 1,2,\ldots, m$.
2. There exists an $\varepsilon_M > 0$ such that $|\varepsilon(x, \dot{x})| \leq \varepsilon_M$, $\forall\, x_{vect} \in \Im$ on a compact region $\Im \subset R^2$.

Let the RBF functional estimates for $f(\dot{x}, x)$ be given by

$$\hat{f}(\dot{x},x) = \sum_{i=0}^{m} \phi_i(\dot{x},x)\hat{w}_i \tag{6.50}$$

where $\hat{w}_i$ are estimates of the ideal RBF weights which are provided by the following weight-tuning algorithm:

$$\dot{\hat{w}}_i = r_i x^T PB\phi_i - r_2\hat{w}_i \tag{6.51}$$

where $r_1, r_2 > 0$, and P is the solution of Equation 6.35. Therefore, the RBF adaptive control component is given by

$$u_{RBF} = -\hat{f}(\dot{x},x) \tag{6.52}$$

6.3.5 INTERNAL MODEL CONTROL

To compensate dynamically for the effect of unmodeled mechanical behaviors on the quality of the straight motion, an internal model control (IMC) scheme (whose global approach of the notion of precision of the model avoids focusing on any specific phenomena or mechanical behavior effects) was added on top of the PID controller [11,12]. It consists of introducing into the control a variable characterizing the difference between the output of the process and modeling. Its value is then converted into an image of the extra torque required to compensate unmodeled loads and is added to the motor's command calculated by the PID controller. The control scheme is presented in Figure 6.15.

In Figure 6.15, block M represents the model of the process and M^{-1} its reverse, which exists and is stable [16]. Moreover, the calculus of the modeling error is achieved at the acceleration level of the control in order to avoid the presence of integral terms in the model M. It means that the modeling error is considered as the difference between the slide-table's translation acceleration predicted by

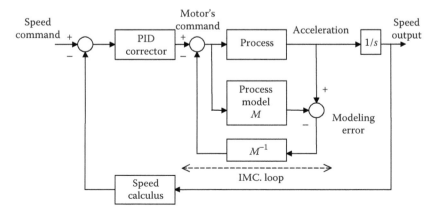

FIGURE 6.15 IMC applied to the PID speed control.

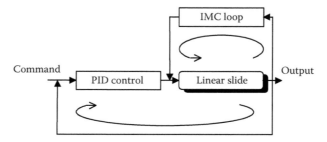

FIGURE 6.16   Separate action of the IMC loop and PID servo control.

the model and the one measured. The real acceleration is obtained by a double numerical derivation of the information provided by the linear incremental encoder. A low-pass numerical filter can be used to reduce the amount of high-frequency noise introduced by the derivations and focus the action of the IMC loop on very low-frequency disturbances. Indeed, if the modeling error calculated at the sample time n contains a lot of numerical noise, when used to modify the value of the motor's command at the time $n+1$, it acts as a powerful noise generator which goes against the initial purpose of the IMC. The order of the filter was limited to reduce dephasing. The very low cutoff frequency of the filter does not affect the natural frequency of the control as the IMC loop acts independently from the PID control, which determines the dynamic of the servo system (Figure 6.16)

6.4   CASE STUDY: DESIGN OF PIEZOELECTRIC ACTUATOR

The following case study discusses the design of an intracytoplasmic sperm injection (ICSI) installation. This case study involves the design of the actuation system, sensing system, and control system, constructing typical complete system. As is the usual practice in design, the application will have to be studied first.

ICSI is a laboratory procedure developed to help infertile couples undergoing in vitro fertilization (fertilization on the dish) due to male factor infertility [13]. It was first introduced by Palermo et al. [14] and has since been used throughout the world [15]. It involves the injection of a single sperm directly into the cytoplasm of an oocyte (egg cell) using a glass needle (injector), thereby increasing the likelihood of fertilization when there are abnormalities in the number, quality, or function of the sperm.

An oocyte, as presented in Figure 6.17, comprises three parts: zona pellucida (cell membrane), oolemma (vitelline membrane), and cytoplasm (vitelline). Zona pellucida is a thick, transparent,

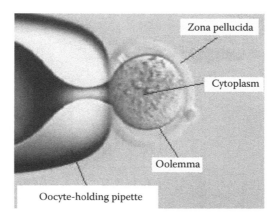

FIGURE 6.17   Structure of an oocyte.

elastic membrane that is usually difficult to pierce at a low speed. The oolemma, on the other hand, is a highly elastic membrane covering the cytoplasm, which contains the living substance and genetic information of the oocyte. In the ICSI procedure, the injector will first puncture zona pellucida, and then will advance to break the oolemma. Once the oolemma has been punctured, the sperm inside the needle can be released into the deep area of the cytoplasm.

The difficulties of this procedure arise from puncturing the zona pellucida and oolemma. With conventional (manual) process, deformation and vibration of the oocyte may occur, increasing the internal pressure of the oocyte, inducing emission of cytoplasm upon penetration, and eventually contributing to the oocyte death. To reduce the vibration and deformation of the oocyte, mercury has been used in ICSI experiment [16], but it can lead to serious health consequences, especially in the subsequent fetal development [17]. Furthermore, the high elasticity of oolemma does not allow the injector to readily penetrate the oocyte. The reported way of puncturing the oolemma is to apply negative pressure, thus drawing along a certain amount of cytoplasm, inducing further damage to the oocyte. Therefore, the conventional (manual) procedure depends heavily on the skill of the operator/embryologist [18].

The design of ICSI installation to be presented shortly is intended to address the above-mentioned issues of oocyte penetration to facilitate sperm injection. The requirement of this motion system is to execute a highly precise piercing motion through a soft, elastic, movable ball membrane of oocyte with a diameter of about $100\,\mu m$ and a needle with a diameter of $10\,\mu m$, without causing damage to the oocyte.

The actuation system to be applied is a linear-reciprocating motion, since this motion will penetrate the zona pellucida and oolemma in a gradual stepwise manner, rather than a lump motion that is more destructive to the oocyte.

6.4.1 Actuators: Piezoelectric Stack Actuator

In selecting the actuator, one has to consider the requirements of the application. In this case study, the resolution of the motion is of the order of micrometer, and it is therefore reasonably safe to design the actuator to perform motion down to submicrometer resolution. To meet this requirement, piezoelectric actuator is chosen. Furthermore, piezoelectric actuator also possesses other characteristics suitable for biomedical application, as discussed in Chapter 7.

To realize the linear-reciprocating motion with high precision, the piezoelectric actuator is designed in a stack construction, as presented in Figure 6.18. This construction has the advantage of generating larger displacement (due to its serial mechanical construction) with smaller electrical excitation (due to its parallel electrical connection). One end of the actuator is fixed, while the other end is in motion. The injector is attached to the motion end of the actuator.

The stack design of piezoelectric actuator is available off-the-shelf from many piezoelectric manufacturers, which come with different specifications in terms of

- Maximum displacement
- Excitation voltage
- Resolution of motion

The mathematical model of the actuator is now to be determined such that it represents the performance of the actuator and facilitate the controller design. One has to make a compromise between a simple modeling for noncomplicated controller design and accurate modeling for satisfactory result. While the order of the actuator may be high, it is generally acceptable to assume the actuator as a second-order system. The nonlinearity of the actuator is then incorporated in one term of the model accordingly.

In this case study, the following model will be used [19]

$$m\ddot{x} = -K_{\mathrm{f}}\dot{x} - K_{\mathrm{g}}x + K_{\mathrm{e}}\left(u(t) - F\right) \tag{6.53}$$

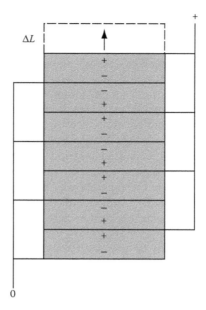

FIGURE 6.18 Construction of linear actuator.

where

$u(t)$ is the time-varying motor terminal voltage
$x(t)$ is the position
K_f is the damping coefficient produced by the motor
K_g is the mechanical stiffness
K_e is the input control coefficient
m is the effective mass
F is the system nonlinear disturbance

6.4.2 SENSOR: LVDT

A sensor is required in the actuation system to monitor the motion of the actuator. The signal obtained by the sensor can be displayed to the users/operators and, more importantly, feedback to the controller (to be designed in Section 6.4.3) so that the actuator can be controlled to follow certain trajectory.

The resolution of the sensor should be finer than the actuator; otherwise the sensor is unable to sense the motion of the actuator. Since in this case study the motion of the actuator is in submicrometer resolution, and the position is the parameter of interest, linear variable displacement transformer (LVDT) is used as the sensor. As can be found with many sensor manufacturers, LVDT works in typically nanometer-level resolution. The LVDT is to be attached to the piezoelectric actuator at the motion end of the actuator. The sensing signal of the LVDT is first amplified to attain a readable level. To increase the sensitivity of the sensor, it can be incorporated in a Wheatstone bridge circuit. More description about LVDT sensor, and other similar sensors, can be found in this chapter.

6.4.3 CONTROLLER DESIGN: ADAPTIVE CONTROLLER

Having designed the actuator and sensor, the system can now be configured in a closed-loop manner, as depicted in Figure 6.3. The piezoelectric stack actuator is the actuator in the system, while the LVDT is the sensor. The oocyte is the load, however, in this case study it is not considered to be

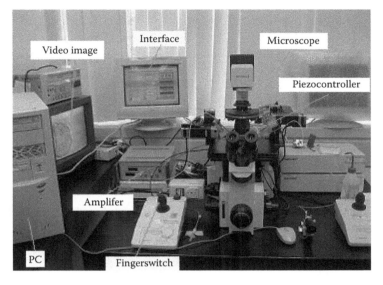

FIGURE 6.19 ICSI installation.

a part of the control system. It is assumed that there is no disturbance to the system. The controller is realized via computer (PC), where the control algorithm is to be designed. The entire system is shown in Figure 6.19.

To complete the system, the controller is now designed based on the actuator in Equation 6.53. Due to the nonlinearity of the piezoelectric actuator, an adaptive controller is to be implemented.

LEMMA 6.1

The internal state z is bounded.
The nonlinear function F can also be written as

$$F = (\sigma_1 + \sigma_2)\dot{x} + \sigma_0 z - \sigma_1 \frac{|\dot{x}|}{h(\dot{x})} z = (\sigma_1 + \sigma_2)\dot{x} + F_d(z, \dot{x}) \tag{6.54}$$

The first part $(\sigma_1 + \sigma_2)\dot{x}$ is a simple function of the velocity. The second part

$$\left(\sigma_0 - \frac{\sigma_1 |\dot{x}|}{h(\dot{x})} \right) z$$

is scaled by z due to the dynamical perturbations in hysteresis. Since z and $h(\dot{x})$ are bounded

$$\left| F_d(z, \dot{x}) \right| = \left\| \left(\sigma_0 - \sigma_1 \frac{|\dot{x}|}{h(\dot{x})} \right) z \right\| \leq k_1 + k_2 |\dot{x}| \tag{6.55}$$

where k_1 and k_2 are constants.

Consider the system of Equation 6.53. The tracking control problem is stated as follows: Find a control mechanism for every bounded smooth output reference $x_d(t)$ with bounded time derivatives so that the controlled output $x(t)$ converges to the reference $x_d(t)$ as closely as possible.

Assumption 6.1

The desired trajectories $x_d(t)$, $\dot{x}_d(t)$, and $\ddot{x}_d(t)$ are continuous and available, and bounded.

Define the position tracking error $e(t)$ and the corresponding filtered tracking error $s(t)$ as follows:

$$e(t) = x_d(t) - x(t) \tag{6.56}$$

and error

$$s = K_I \int_0^t e(\tau)\, d\tau + K_P e + \dot{e} \tag{6.57}$$

where K_I, $K_P > 0$ are chosen such that the polynomial $s^2 + K_P s + K_I$ is Hurwitz. Differentiating $s(t)$ and using Equation 6.53, the dynamics in terms of $s(t)$ is

$$\frac{m}{K_e}\dot{s} = \frac{m}{K_e}\left(K_I e + K_P \dot{e} + \ddot{x}_d\right) + \frac{K_f}{K_e}\dot{x} + \frac{K_g}{K_e}x - (u - F)$$

$$= \frac{m}{K_e}\left(K_I e + K_P \dot{e} + \ddot{x}_d\right) + \left(\frac{K_f}{K_e} + \sigma_1 + \sigma_2\right)\dot{x} + \frac{K_g}{K_e}x - u + F_d(z, \dot{x}) \tag{6.58}$$

By using straightforward exact model knowledge, the control input is

$$u = K_v s + a_m\left(K_I e + K_P \dot{e} + \ddot{x}_d\right) + a_{k\sigma}\dot{x} + a_{ge}x + F_d \tag{6.59}$$

where $K_v > 0$ is a constant, $a_m = (m/K_e)$, $a_{k\sigma} = (K_f/K_e) + \sigma_1 + \sigma_2$, and $a_{ge} = (K_g/K_e)$. Substituting the control input given by Equation 6.59 into the open-loop expression of Equation 6.53, the closed-loop filtered tracking error system is obtained, i.e., $(m/K_e)\dot{s} = -K_v s$. Since $K_v > 0$, the resulting system is asymptotically stable. Unfortunately, the hysteresis is unknown a priori in practice. In addition, it is also difficult to obtain the precise values of m, K_e, K_f, and K_g. Motivated by this observation, an adaptive control technique, by replacing m/K_e, $(K_f/K_e) + \sigma_1 + \sigma_2$, K_g/K_e, and F_d with the estimates $\hat{a}_m$, $\hat{a}_{k\sigma}$, $\hat{a}_{ge}$, and $\hat{k}_1\,\mathrm{sgn}(s) + \hat{k}_2|\dot{x}|\mathrm{sgn}(r)$, respectively, is designed as follows:

$$u = K_v s + \hat{a}_m\left(K_I e + K_P \dot{e} + \ddot{x}_d\right) + \hat{a}_{k\sigma}\dot{x} + \hat{a}_{ge}x + \hat{k}_1\,\mathrm{sgn}(s) + \hat{k}_2|\dot{x}|\mathrm{sgn}(s) \tag{6.60}$$

Substituting the control Equation 6.60 into Equation 6.58, one has

$$\frac{m}{K_e}\dot{s} = -K_v r + \tilde{a}_m\left(K_I e + K_P \dot{e} + \ddot{x}_d\right) + \tilde{a}_{k\sigma}\dot{x} + \tilde{a}_{ge}x - \hat{k}_1\,\mathrm{sgn}(s) - \hat{k}_2|\dot{x}|\mathrm{sgn}(s) + F_d \tag{6.61}$$

where $\tilde{a}_m = (m/K_e) - \hat{a}_m$, $\tilde{a}_{k\sigma} = (K_f/K_e) + \sigma_1 + \sigma_2 - \hat{a}_{k\sigma}$, and $\tilde{a}_{ge} = (K_g/K_e) - \hat{a}_{ge}$. Taking a positive definite function

$$V(t) = \frac{1}{2}\frac{m}{K_e}s^2 + \frac{1}{2\gamma_1}\tilde{a}_m^2 + \frac{1}{2\gamma_2}\tilde{a}_{k\sigma}^2 + \frac{1}{2\gamma_3}\tilde{a}_{ge}^2 + \frac{1}{2\gamma_4}\tilde{k}_1^2 + \frac{1}{2\gamma_5}\tilde{k}_2^2 \tag{6.62}$$

where $\tilde{k}_1 = k_1 - \hat{k}_1$ and $\tilde{k}_2 = k_2 - \hat{k}_2$. Its time derivative becomes

$$\dot{V} = \frac{m}{K_e}\dot{s}s + \frac{1}{\gamma_1}\tilde{a}_m\dot{\tilde{a}}_m + \frac{1}{\gamma_2}\tilde{a}_{k\sigma}\dot{\tilde{a}}_{k\sigma} + \frac{1}{\gamma_3}\tilde{a}_{ge}\dot{\tilde{a}}_{ge} + \frac{1}{\gamma_4}\tilde{k}_1\dot{\tilde{k}}_1 + \frac{1}{\gamma_5}\tilde{k}_2\dot{\tilde{k}}_2$$

$$= -K_v s^2 + \left[\tilde{a}_m\left(K_1 e + K_p\dot{e} + \ddot{x}_d\right) + \tilde{a}_{k\sigma}\dot{x} + \tilde{a}_{ge}x\right]s$$

$$+ \left[-\hat{k}_1\,\mathrm{sgn}\,(r) - \hat{k}_2\,|\dot{x}|\,\mathrm{sgn}\,(r) + F_d\right]s \tag{6.63}$$

$$+ \frac{1}{\gamma_1}\tilde{a}_m\dot{\tilde{a}}_m + \frac{1}{\gamma_2}\tilde{a}_{k\sigma}\dot{\tilde{a}}_{k\sigma} + \frac{1}{\gamma_3}\tilde{a}_{ge}\dot{\tilde{a}}_{ge} + \frac{1}{\gamma_4}\tilde{k}_1\dot{\tilde{k}}_1 + \frac{1}{\gamma_5}\tilde{k}_2\dot{\tilde{k}}_2$$

By using the inequality (Equation 6.55), it is shown that

$$\dot{V} \le -K_v s^2 + \left[\tilde{a}_m\left(K_1 e + K_p\dot{e} + \ddot{x}_d\right) + \tilde{a}_{k\sigma}\dot{x} + \tilde{a}_{ge}x\right]s - \hat{k}_1\,|s| - \hat{k}_2\,|\dot{x}||s| + k_1\,|s| + k_2\,|\dot{x}||s|$$

$$+ \frac{1}{\gamma_1}\tilde{a}_m\dot{\tilde{a}}_m + \frac{1}{\gamma_2}\tilde{a}_{k\sigma}\dot{\tilde{a}}_{k\sigma} + \frac{1}{\gamma_3}\tilde{a}_{ge}\dot{\tilde{a}}_{ge} + \frac{1}{\gamma_4}\tilde{k}_1\dot{\tilde{k}}_1 + \frac{1}{\gamma_5}\tilde{k}_2\dot{\tilde{k}}_2 \tag{6.64}$$

$$= -K_v s^2 + \left[\tilde{a}_m\left(K_1 e + K_p\dot{e} + \ddot{x}_d\right) + \tilde{a}_{k\sigma}\dot{x} + \tilde{a}_{ge}x\right]s + \left[\tilde{k}_1\,|r| + \tilde{k}_2\,|\dot{x}||s|\right]$$

$$+ \frac{1}{\gamma_1}\tilde{a}_m\dot{\tilde{a}}_m + \frac{1}{\gamma_2}\tilde{a}_{k\sigma}\dot{\tilde{a}}_{k\sigma} + \frac{1}{\gamma_3}\tilde{a}_{ge}\dot{\tilde{a}}_{ge} + \frac{1}{\gamma_4}\tilde{k}_1\dot{\tilde{k}}_1 + \frac{1}{\gamma_5}\tilde{k}_2\dot{\tilde{k}}_2$$

Choosing the update laws

$$\dot{\hat{a}}_m = \gamma_1\left[\left(K_1 e + K_p\dot{e} + \ddot{x}_d\right)s - \gamma_{11}\hat{a}_m\right] \tag{6.65}$$

$$\dot{\hat{a}}_{k\sigma} = \gamma_2\left[\dot{x}s - \gamma_{21}\hat{a}_{k\sigma}\right] \tag{6.66}$$

$$\dot{\hat{a}}_{g\sigma} = \gamma_3\left[xs - \gamma_{31}\hat{a}_{ge}\right] \tag{6.67}$$

$$\dot{\hat{k}}_1 = \gamma_4\left[|s| - \gamma_{41}\hat{k}_1\right] \tag{6.68}$$

$$\dot{\hat{k}}_2 = \gamma_5\left[|\dot{x}||s| - \gamma_{51}\hat{k}_2\right] \tag{6.69}$$

where $\gamma_1, \gamma_{11}, \gamma_2, \gamma_{21}, \gamma_3, \gamma_{31}, \gamma_4, \gamma_{41}, \gamma_5, \gamma_{51} > 0$, it follows that

$$\dot{V} \le -K_v s^2 + \gamma_{11}\tilde{a}_m\hat{a}_m + \gamma_{21}\tilde{a}_{k\sigma}\hat{a}_{k\sigma} + \gamma_{31}\tilde{a}_{ge}\hat{a}_{ge} + \gamma_{41}\tilde{k}_1\hat{k}_1 + \gamma_{51}\tilde{k}_2\hat{k}_2 \tag{6.70}$$

By replacing the estimate variable, it follows that

$$\tilde{a}_m\hat{a}_m = \tilde{a}_m\left(a_m - \tilde{a}_m\right) = -\tilde{a}_m^2 + \tilde{a}_m a_m \tag{6.71}$$

$$\tilde{a}_{k\sigma}\hat{a}_{k\sigma} = -\tilde{a}_{k\sigma}^2 + \tilde{a}_{k\sigma}a_{k\sigma} \tag{6.72}$$

$$\tilde{a}_{ge}\hat{a}_{ge} = -\tilde{a}_{ge}^2 + \tilde{a}_{ge}a_{ge} \tag{6.73}$$

$$\tilde{k}_1\hat{k}_1 = -\tilde{k}_1^2 + \tilde{k}_1 k_1 \tag{6.74}$$

$$\tilde{k}_2\hat{k}_2 = -\tilde{k}_2^2 + \tilde{k}_2 k_2 \tag{6.75}$$

Substituting the Equations 6.71 through 6.75 into Equation 6.70 yields

$$\dot{V} \le -K_v \left\{ \begin{array}{l} \left[s^2 + \dfrac{\gamma_{11}}{K_v}\left(\tilde{a}_m - \dfrac{1}{2}a_m\right)^2 + \dfrac{\gamma_{21}}{K_v}\left(\tilde{a}_{k\sigma} - \dfrac{1}{2}a_{k\sigma}\right)^2 + \dfrac{\gamma_{31}}{K_v}\left(\tilde{a}_{ge} - \dfrac{1}{2}a_{ge}\right)^2 \right. \\[4mm] + \dfrac{\gamma_{41}}{K_v}\left(\tilde{k}_1 - \dfrac{1}{2}k_1\right)^2 + \dfrac{\gamma_{51}}{K_v}\left(\tilde{k}_2 - \dfrac{1}{2}k_2\right)^2 - \dfrac{\gamma_{11}}{4K_v}a_m^2 - \dfrac{\gamma_{21}}{4K_v}a_{k\sigma}^2 \\[4mm] \left. - \dfrac{\gamma_{31}}{4K_v}a_{ge}^2 - \dfrac{\gamma_{41}}{4K_v}k_1^2 - \dfrac{\gamma_{51}}{4K_v}k_2^2 \right] \end{array} \right\} \tag{6.76}$$

which is guaranteed negative as long as either

$$|s| > \sqrt{\dfrac{\gamma_{11}}{4K_v}a_m^2 - \dfrac{\gamma_{21}}{4K_v}a_{k\sigma}^2 - \dfrac{\gamma_{31}}{4K_v}a_{ge}^2 - \dfrac{\gamma_{41}}{4K_v}k_1^2 - \dfrac{\gamma_{51}}{4K_v}k_2^2} = B_s \tag{6.77}$$

or

$$|\tilde{a}_m| > \sqrt{\dfrac{1}{4}a_m^2 - \dfrac{\gamma_{21}}{4\gamma_{11}}a_{k\sigma}^2 - \dfrac{\gamma_{31}}{4\gamma_{11}}a_{ge}^2 - \dfrac{\gamma_{41}}{4\gamma_{11}}k_1^2 - \dfrac{\gamma_{51}}{4\gamma_{11}}k_2^2} + \dfrac{1}{2}a_m = B_{am} \tag{6.78}$$

$$|\tilde{a}_{k\sigma}| > \sqrt{\dfrac{\gamma_{11}}{4\gamma_{21}}a_m^2 - \dfrac{1}{4}a_{k\sigma}^2 - \dfrac{\gamma_{31}}{4\gamma_{21}}a_{ge}^2 - \dfrac{\gamma_{41}}{4\gamma_{21}}k_1^2 - \dfrac{\gamma_{51}}{4\gamma_{21}}k_2^2} + \dfrac{1}{2}a_{k\sigma} = B_{ak\sigma} \tag{6.79}$$

$$|\tilde{a}_{ge}| > \sqrt{\dfrac{\gamma_{11}}{4\gamma_{31}}a_m^2 - \dfrac{\gamma_{21}}{4\gamma_{31}}a_{k\sigma}^2 - \dfrac{1}{4}a_{ge}^2 - \dfrac{\gamma_{41}}{4\gamma_{31}}k_1^2 - \dfrac{\gamma_{51}}{4\gamma_{31}}k_2^2} + \dfrac{1}{2}a_{ge} = B_{age} \tag{6.80}$$

$$|\tilde{k}_1| > \sqrt{\dfrac{\gamma_{11}}{4\gamma_{41}}a_m^2 - \dfrac{\gamma_{21}}{4\gamma_{41}}a_{k\sigma}^2 - \dfrac{\gamma_{31}}{4\gamma_{41}}a_{ge}^2 - \dfrac{1}{4}k_1^2 - \dfrac{\gamma_{51}}{4\gamma_{41}}k_2^2} + \dfrac{1}{2}k_1 = B_{k1} \tag{6.81}$$

$$|\tilde{k}_2| > \sqrt{\dfrac{\gamma_{11}}{4\gamma_{51}}a_m^2 - \dfrac{\gamma_{21}}{4\gamma_{51}}a_{k\sigma}^2 - \dfrac{\gamma_{31}}{4\gamma_{51}}a_{ge}^2 - \dfrac{\gamma_{41}}{4\gamma_{51}}k_1^2 - \dfrac{1}{4}k_2^2} + \dfrac{1}{2}k_2 = B_{k2} \tag{6.82}$$

Thus, $\dot{V}$ is negative outside the compact set

$$M_\delta = \left\{ \left(s, \tilde{a}_m, \tilde{a}_{k\sigma}, \tilde{a}_{ge}, \tilde{k}_1, \tilde{k}_2\right) \, \middle\| \, \left[s, \tilde{a}_m, \tilde{a}_{k\sigma}, \tilde{a}_{ge}, \tilde{k}_1, \tilde{k}_2 \right] \middle\| \le \delta \right\} \tag{6.83}$$

where $\delta = \max(B_s, B_{am}, B_{ak\sigma}, B_{age}, B_{k1}, B_{k2})$. Thus, s, $\tilde{a}_m$, $\tilde{a}_{k\sigma}$, $\tilde{a}_{ge}$, $\tilde{k}_1$, and $\tilde{k}_2$ are uniformly and ultimately bounded. The value of $|s|$ can be made small by increasing K_v. Therefore, the following theorem can be established.

THEOREM 1:   *Asymptotic convergence of robust adaptive controller*

Consider the plant of Equation 6.53 and the control objective of tracking the desired trajectories: x_d, $\dot{x}_d$, and $\ddot{x}_d$. The control law given by Equation 6.60 with Equations 6.65 through 6.69 ensures that the system states and parameters are uniformly bounded.

Remark 1

There is a sliding mode used in the controller. It is well known that a large gain will induce chattering in the presence of fast dynamics. To solve this problem, one has to use very small adaptation

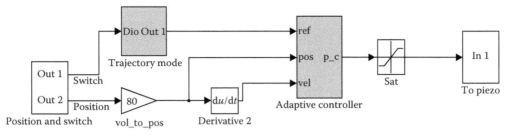

FIGURE 6.20   Software design for SIMULINK® model.

gains. Furthermore, σ-modification term is introduced into the adaptive law to prevent the parameters $\hat{a}_m$, $\hat{a}_{k\sigma}$, $\hat{a}_{ge}$, $\hat{k}_1$, and $\hat{k}_2$ increasing to infinity.

6.4.4   RESULTS

The system as designed in this case study has been implemented in the ICSI installation. The performance of the proposed adaptive control scheme has been tested using real-time experiments in the installation.

The actuator used in the installation is a stack piezoelectric actuator manufactured by Physik Instrument (PI), which has a travel length of 80 μm and it is equipped with an LVDT sensor with an effective resolution of 5 nm. The dSPACE control development and rapid prototyping platform is used as an interface between the actuation and sensing components and the computer to allow a closed-loop control of the system. MATLAB®/SIMULINK® can be used from within a dSPACE environment. The entire software design for SIMULINK is shown in Figure 6.20. The position and switch signals are feedback to the control system by the analog and digital input of the dSPACE card, respectively. The trajectory mode is forwarded to the adaptive controller as a reference signal and then the controller output is sent out to the piezoactuator. The control system is implemented by running Real-Time Windows of MATLAB.

To build the model of the actuator, the experimental data from the piezoelectric actuator was first collected. This procedure, known as system identification, includes sending input signal to the actuator and recording the output signal from the actuator through the sensor. A chirp signal was used since it allows using a limited amount of data with specified frequencies being put through the system. The sample rate was chosen as 2000 Hz, while the frequency of the chirp signal was selected as from 0 to 200 Hz. The input–output signals are shown in Figures 6.21 and 6.22. The dominant linear model was obtained to be

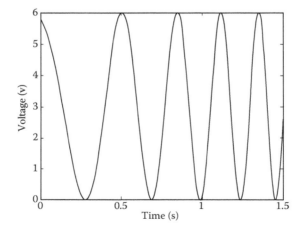

FIGURE 6.21   Input signal for identification.

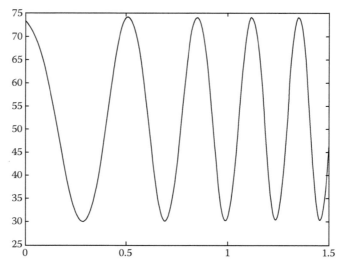

FIGURE 6.22 Output signal for identification.

$$\ddot{x} = -1081.6\dot{x} - 5.9785 \times 10^5 x + 4.2931 \times 10^6 u \tag{6.84}$$

This model presents the linear characteristic of the piezoelectric actuator without the nonlinearity, such as hysteresis.

The adaptive controller as presented in Equation 6.60 was applied to control the piezoelectric actuator. The parameters of the controller were selected as

$$K_v = 0.00001, \quad K_I = 400000, \quad K_P = 100 \tag{6.85}$$

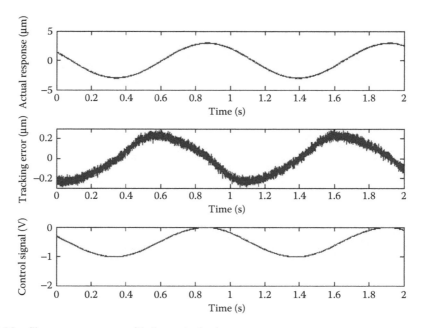

FIGURE 6.23 Sine wave response with the control scheme.

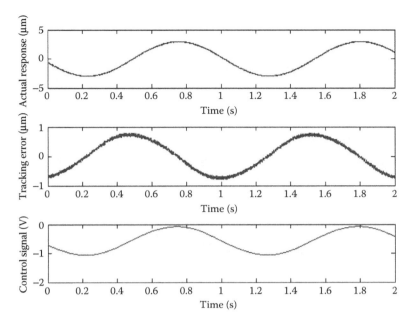

FIGURE 6.24   Sine wave response with PID control.

The initial values for $\hat{a}_m$, $\hat{a}_{k\sigma}$, and $\hat{a}_{ge}$ were chosen based on the identified model of Equation 6.84; the values were $\hat{a}_m(0) = 2.3293 \times 10^{-7}$, $\hat{a}_{k\sigma}(0) = 2.5194 \times 10^{-4}$, and $\hat{a}_{ge}(0) = 0.1184$. The initial values for $\hat{k}_1$ and $\hat{k}_2$ were chosen as 10^{-7} and 10^{-8}, respectively. Due to the inherent unmodeled high-frequency dynamics in the mechanical structure and other components in the system, which should not be excited, small adaptation factors were used $\gamma_1 = \gamma_2 = \gamma_3 = 10^{-22}$, $\gamma_4 = \gamma_5 = 10^{-20}$, and $\gamma_{11} = \gamma_{21} = \gamma_{31} = \gamma_{41} = \gamma_{51} = 0.0001$.

The reference signal for tracking was the sinusoidal trajectory $A\sin(\omega t)$, where $A = 3\,\mu m$ and $\omega = 6$ rad/s. Figure 6.23 shows the performance of the system with the adaptive control. It can be observed that the actual response to the sinusoidal trajectories is good. The tracking error is about $0.3\,\mu m$.

PID control could also have been used to control the system. With proper tuning, the result is shown in Figure 6.24 and the tracking error is about $1.0\,\mu m$. This shows that the adaptive controller can achieve better tracking performance than that of PID control, since the adaptive controller takes the nonlinearity of the actuator into consideration.

REFERENCES

1. S. Mekid and O. Olejniczak, High precision linear slide: Control and measurements, *International Journal of Machine Tools and Manufacture*, 40(7), 1051–1064, 2007.
2. F. Chen, H. Xie, and G. K. Fedder, A MEMS-based monolithic electrostatic microactuator for ultra-low magnetic disk head fly height control, *IEEE Transactions on Magnetics*, 37(4), 1915–1918, 2001.
3. T. S. Low and W. Guo, Modeling of a three-layer piezoelectric bimorph beam with hysteresis, *Journal of Microelectromechanical Systems*, 4(4), 230–237, 1995.
4. G. Robert, D. Damjanovic, and N. Setter, Separation of Nonlinear and Friction-like Contributions to the Piezoelectric Hysteresis, *12th IEEE International Symposium on Application of Ferroelectrics*, 2, 699–702, 2000.
5. M. Goldfarb and N. Celanovic, Modeling piezoelectric stack actuators for control of micromanipulation, *IEEE Control Systems*, 17, 69–79, 1997.
6. G. H. Choi, J. H. Oh, and G. S. Choi, Repetitive tracking control of a coarse-fine actuator, *Proceedings of the 1999 IEEE/ASME International Conference on Advanced Intelligent Mechatronics*, Atlanta, GA, pp. 335–340, 1999.

7. C. Canudas-De-Wit, H. Olsson, K. Astrom, and P. Lischinsky, A new model for control of systems with friction, *IEEE Transaction on Automatic Control*, 40(3), 419–425, 1995.

8. K. K. Tan, S. N. Huang, T. H. Lee, S. J. Chin, and S. Y. Lim, Adaptive robust motion control for precise trajectory tracking applications, *ISA Transactions*, 40, 17–29, 2001.

9. K. K. Tan, T. H. Lee, S. N. Huang, and X. Jiang, Friction modeling and adaptive compensation using a relay feedback approach, *IEEE Transaction on Industrial Electronics*, 48, 169–176, 2001.

10. K. K. Tan, H. F. Dou, Y. Q. Chen, and T. H. Lee, High precision linear motor control via relay tuning and iterative learning based on zero-phase filtering, *IEEE Transaction on Control Systems Technology*, 9, 244–253, 2001.

11. C. Foulard, S. Gentil, and J. P. Sandraz, Commande et regulation par calculateur numeriques: de la theorie aux applications, 5e edition, Eyrolles, Paris, pp. 174–184, 1987.

12. D. Lahmar, L. Loron, and M. Bonis, Asservissement en tres basse vitesse d'un moteur autosynchrone, *Conference Canadienne sur l'Automatisation Industrielle*, Montreal, 1992.

13. American Society for Reproductive Medicine. *Patient's Fact Sheet: Intra-Cytoplasmic Sperm Injection (ICSI)*, ASRM. 1, 2001.

14. G. Palermo, H. Joris, M. P. Derde, and A. C. van Steirteghem, Pregnancies after intra-cytoplasmic injection of single spermatozoon into an oocyte, *Lancet*, 340, 17–18, 1992.

15. K. Yanigada, H. Katayose, H. Yazawa, Y. Kimura, K. Konnai, and A. Sato, The usefulness of a piezo-micromanipulator in intra-cytoplasmic sperm injection in humans, *Human Reproduction*, 14(2), 448–453, 1998.

16. K. Ediz and N. Olgac, Effect of mercury column on the microdynamics of the piezo-driven pipettes, *Journal of Biomechanical Engineering*, 127, 531–535, 2005.

17. ATSDR, Toxicological profile of mercury. Agency for Toxic Substances and Disease Registry, Atlanta, GA., 1999.

18. K. K. Tan, S. C. Ng, and Y. Xie, Optimal intra-cytoplasmic sperm injection with a piezo micro-manipulation, *Proceedings of the 4th World Congress on Intelligent Control and Automation*, pp. 1120–1123, 2002.

19. H. Wulp, *Piezo-Driven Stages for Nanopositioning with Extreme Stability: Theoretical Aspects and Practical Design Considerations*, Delft University of Technology, p. 21, 1997.

7 Actuators, Transmission, and Sensors

Tan Kok Kiong and Samir Mekid

CONTENTS

If you make an error and you sense it but you don't correct it, you will make another error.

7.1 INTRODUCTION

Actuators, transmissions, and sensors are devices necessary to realize a control system design. If the system being designed is considered as a separate entity from its surroundings, actuators, transmissions, and sensors can be thought of as the devices that will interface and communicate with the outside world. Therefore, selection of actuators, transmissions, and sensors is also dependent on the operating conditions, apart from the application requirements.

Actuators are dedicated to generate desired movements by moving an object to certain position accurately, with an acceptable resolution. Besides the position requirement, actuators are also required to deliver sufficient force (for linear actuator) or torque (for rotational actuator). In general, good actuators are associated with the following criteria:

1. High stiffness in the active axis
2. Linearity or coaxiality
3. No backlash
4. Low disturbances
5. Availability of reverse motion
6. Fast time response to active control

Transmission can be considered as an integral part of actuators. The main purpose of transmission is to transmit motion from the prime mover (main actuator) to other system components or to the working objects. Transmission is usually employed when there is a need to manipulate the nature of the motion, for example direction reversals, speed reduction, change of motion axis, etc.

Sensors are used to obtain the information from the system and its operating and surrounding conditions. These parameters, such as position, speed, and acceleration, are sensed by sensors and then converted into electrical signals, which can then be processed for controlling purpose, or for display to the operators.

This chapter discusses the various types and characteristics of actuators, transmissions, and sensors.

7.2 ELECTRIC ACTUATORS AND ELECTRIC DRIVES

Electric drives form an important and integral part of almost all industrial and automation processes. In addition, they also find many applications in our day-to-day lives, improving the quality of living. Modern trains are now powered by electrical drives, while cars employ electric drive components to enable basic and advanced functions. Household appliances also employ electric drives to fulfill a wide range of functions, from simple function such as in electric shaver to more complicated functions necessary in an automatic washing machine.

Today, electric drives have evolved in a significant way from what they were some 50 years ago. They are no longer constructions of windings and magnet that rotate when activated by electrical power; they are now equipped with controllers and converters that allow them to perform very precise functions almost independently. Indeed, in the last 20 years, many developments related to electric drives have taken place, enhancing electric drives capabilities to take on the challenges posed by emerging new industries.

An electric drive is a system which performs a motoring function or a generating function, by converting electrical energy into mechanical energy in the former case, and converting mechanical energy to electrical energy in the latter case. It is, therefore, an intermediate stage in the conversion of energy. Figure 7.1 represents a general configuration of electric drives.

The electric motor is the core component of an electric drive as it is the motor that converts electrical energy into mechanical energy. One of the first electrical motor was invented in 1821 by Michael Faraday, which is a direct current (dc) motor in nature. Ever since, the control of electric drives has been of great interest to researchers and engineers. Faraday's invention was then followed by the invention of modern dc motor by Zenobe Gramme in 1873. Alternating current (ac) motors were later invented in 1889–1890.

The operation of a modern electric drive is mainly controlled via power electronics. One of the first inventions of power electronics is the mercury arc rectifier in 1920, followed by dc–ac

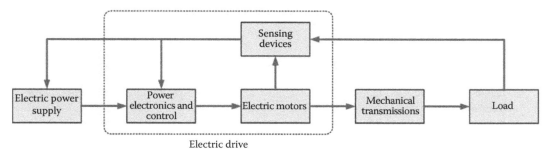

FIGURE 7.1 Configuration of an electric drive.

converter in 1930. The invention of semiconductors, and later on transistors in the late 1940s by William Schokley, John Bardeen, and Walter Brattain, further enhanced the development of power electronics. In the subsequent years, many more power electronic devices were invented, such as the thyristor and the power transistor. This contributes to the development of control circuitry of electric drives. In 1980s, digital control technology began its implementation in electric drives, riding on the advances brought forth by the application of microprocessors. This improves the performance of the drives in terms of accuracy, response speed, and cost. It also allows the implementation of modern complex control schemes such as self-learning and intelligent control approaches.

Digital communication has further revolutionized the way a full electric drive application is developed and enabled efficient control strategies.

The applications of electric drives as motion enablers are wide ranging from low-power applications in instrumentation, computer peripherals, and robots, to high-power applications in ship propulsion and industrial mills. A classification according to performance and power requirements is depicted in Figure 7.2.

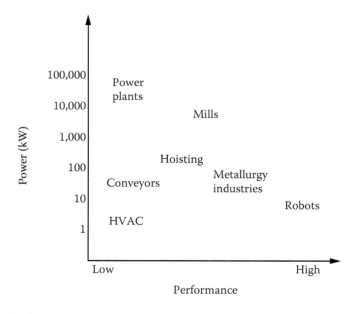

FIGURE 7.2 Application range of electric drive.

From the perspectives of a motion system, the benefits offered by an electric drive can be mainly summarized as follows:

1. It provides a wide range of power and speed, from mW to MW, and up to hundred thousand rpm.
2. It is configurable with many transmissions to suit specific application requirements with reduced noise.
3. It can operate under a wide range of environmental conditions and different load demand.
4. It can be controlled via simple to complicated controllers.
5. It covers all quadrants/modes of operations, forward–backward motion, and motoring–braking operations.
6. It has a high energy conversion ratio and fast response.
7. It requires simple maintenance.

However, an electric drive also has constraints which may limit its use in certain applications:

1. It requires continuous electric energy input, and is thus not feasible for mobile applications.
2. It has a lower power-to-weight ratio due to magnetic saturation and cooling problems.

The principles of electric actuators, or motors, are based on the field of electromagnetism, which serves as a bridge between electric excitation and magnetic motion. The development of electric motors followed the discovery of electromagnetism by renowned scientists, including Hans Christian Oersted, Michael Faraday, Hendrik Antoon Lorentz, and Heinrich Lenz.

Oersted first demonstrated the relationship between magnetism and electricity. When electric charge/current flows through a wire/conductor, a magnetic field is developed around the wire. Therefore, an electric current flowing through a winding of wire (known as coil) will create a magnet, more specifically an electromagnet since it becomes a magnet only via electric excitation.

Faraday showed that the inverse phenomenon is true as well. Current can be generated through magnetism. When a conductor is moved across a magnetic field, an electromotive force (emf) is induced across the conductor and a current will flow if there is a closed circuit around the conductor. Lenz subsequently demonstrated that the direction of induced current will be such that its magnetic action tends to oppose/resist the motion through which it is produced.

Lorentz showed yet another permutation of these variables. If an electric current is transmitted along the conductor lying across a magnetic field, there will be a force generated on the wire, causing it to move along a direction perpendicular to the field, known as the Lorentz force.

With these physical principles, the operational principle behind an electric motor can now be explained, as depicted in Figure 7.3.

This machine consists of a generated magnetic field and an armature winding. With the flow of current as indicated in Figure 7.3, the armature winding generates a magnetic field which is perpendicular to itself. This type of magnetic field, which is generated by electric current, is known as magnetomotive force (mmf). The interaction between field poles and armature poles, in the form of attraction and repulsion of opposite and similar poles, causes the motor to rotate. Commutator and brushes are used to maintain the direction of the current so as to force the motor to rotate continuously in one direction.

In an actual motor, the armature is fixed (and hence referred to as stator), whereas the field winding is allowed to rotate (and hence referred to as rotor) for ease of construction. Also, there can be numerous armature loops in the motor to provide a higher output power.

7.2.1 Stepper Motors

A stepper motor is a device that converts electrical impulses into mechanical movement. Each electrical pulse input causes a rotation of the motor over a certain discrete angle. A stepper motor can therefore

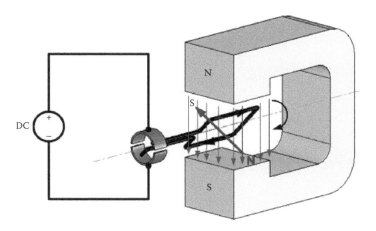

FIGURE 7.3 (See color insert following page 174.) Functional diagram of an electric motor.

be considered as a digital motor, as opposed to a continuous motor. This motor allows precise control of load with regard to speed and position, with a typical error of less than 5% per angle. A stepper motor is commonly operated with a dc voltage, although operation with an ac voltage is also possible.

The main advantages of a stepper motor are as follows:

1. It is compatible with digital systems, eliminating the necessity of digital-to-analog converters, and easy implementation of digital closed-loop control.
2. A wide range of step angles (from 1.8° to 90°) and torque (from a few Nm to tens of Nm) is available.
3. It allows bidirectional motion control and low-speed operation.
4. It has a low moment of inertia and high torque at low speed, reducing the acceleration or deceleration torque as well as the starting current.

The disadvantages of stepper motors are as follows:

1. It has a low efficiency.
2. It is very sensitive to misalignment of load and drive.

The operational diagram of a stepper motor is depicted in Figure 7.4.

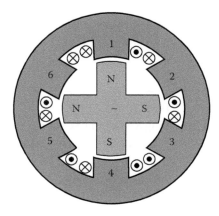

FIGURE 7.4 (See color insert following page 174.) Functional diagram of a stepper motor.

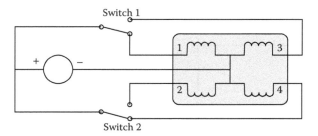

FIGURE 7.5   Full-step switching configuration.

The operating principle of a stepper motor is based on attraction of opposite magnetic poles and repulsion of similar magnetic poles. In Figure 7.4, the rotor (the inner part) is a permanent magnet structure with poles as indicated. The stator is made of windings that will become an electromagnet when activated. With activation as indicated, poles 1, 2, and 3 will be north-polarized, while poles 4, 5, and 6 will become south-polarized. The rotor will thus rotate one step counterclockwise as pole N repels pole 1 and is attracted to pole 5, while pole S repels pole 4 and is attracted to pole 2.

In practice, the number of teeth of the stator and rotor determines the incremental angular rotation of the motor.

A stepper motor can be operated in either full-stepping mode or half-stepping mode. In full-stepping mode, as presented in Figure 7.5 (showing a four-pole configuration), each switching action causes the rotor to advance one-fourth of a tooth. A full tooth is therefore achieved in four steps. The switching sequence under the full-stepping mode is presented in Table 7.1.

In half-stepping mode, as presented in Figure 7.6, each switching action causes the rotor to advance one-eighth of a tooth. A full tooth is therefore achieved over eight steps, resulting in a finer step as compared to half-switching mode. The switching sequence under the half-stepping mode is presented in Table 7.2.

Applications of stepper motors can be found in the textile industry, IC fabrication, and robotics, where positioning of parts or tools will be an integral part of the requirements of such applications.

The common types of stepper motors are variable-reluctance motors, permanent-magnet (PM) motors, and hybrid motors.

7.2.2 DC Motor

DC motors have been implemented widely in the field of adjustable speed drive because of their simple control requirement. The simplicity of the control comes from the fact that the armature mmf and field mmf are independent of each other, so that controlling one mmf while keeping the other constant can be easily accomplished.

TABLE 7.1
Full-Stepping Switching Sequence

Step	Switch 1	Switch 2
1	1	2
2	1	4
3	3	4
4	3	2

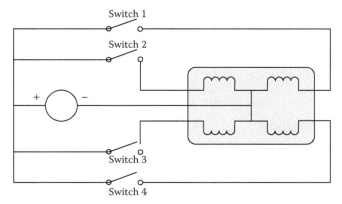

FIGURE 7.6   Half-step switching configuration.

TABLE 7.2
Half-Stepping Switching Sequence

Step	Switch 1	Switch 2	Switch 3	Switch 4
1	On	Off	On	Off
2	On	Off	Off	Off
3	On	Off	Off	On
4	Off	Off	Off	On
5	Off	On	Off	On
6	Off	On	Off	Off
7	Off	On	On	Off
8	Off	Off	On	Off

In the simplest definition, a dc motor is a motor that is driven by a dc electric supply. A dc motor's speed is relatively easy to be controlled, which is either via an applied armature voltage or an applied field voltage. Thus, dc motors are commonly used where variable speed and strong torque are required. Examples of applications of dc motors can be found commonly in the steel mill, printing press, crane, and hoists.

The main parts of a dc motor are the armature winding and field winding. The armature winding is the rotating component of the motor which is known as the rotor. This is attached to the shaft of the motor, which will in turn produce the rotating torque of the motor. The field winding is static and is connected to the frame of the motor. Field winding generates the primary magnetic field inside the motor.

The ends of the armature winding are connected to the commutator. The commutator is formed by insulated copper bars and acts as a switch which forces current to flow through the armature in the same direction. To improve commutation, commutating poles can be installed between regular field poles. Brushes are placed in contact with the commutator as the means to transmit the current in and out of the dc motors.

The construction of dc motors as explained above is a conventional construction known as brush dc motors. There is also another type of dc motors known as brushless dc motor, which, as the name suggests, do not have brushes. They are able to dispense with the brushes by using permanent magnets as rotors.

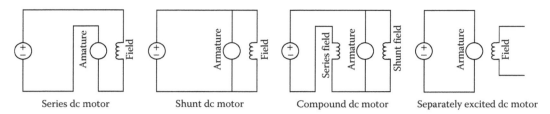

FIGURE 7.7 Types of dc motors.

There are four common types of dc motors, as presented in Figure 7.7:

1. Series motor
2. Shunt motor
3. Compound motor (combination of shunt and series)
4. Separately excited motor

The magnetic fields of the armature and the field of dc motors interact with each other, causing the motor to rotate. The four dc motors mentioned vary mainly in the way the field is generated. In series motors, the field is generated by the full electrical current from the source. In shunt motors, the field is generated by a fraction of the electrical current. Compound motors combine the characteristics of series and shunt motors. Separately excited motors use a separate field to generate the magnetic field. This can be from an external dc supply or, usually, from a permanent magnet. Each type of dc motors offers different characteristics as follows.

In a series motor (Figure 7.8), the torque varies as the square of the armature current. Thus, a series motor is suitable for applications that require a large torque generation with a small incremental current. On the other hand, the speed varies greatly and thus this motor is not suitable for applications where the load may change drastically. The application of a series motor is, for example, in a crane operation.

In a shunt motor (Figure 7.9), the speed is nearly constant within its operation range. The torque varies linearly with respect to the armature current, and hence the load. Therefore, this motor is suitable for applications that require continuous operations at a constant speed.

The characteristics of series motors and shunt motors are combined in compound motors (Figure 7.10). The speed changes with the load, but not as sharply as in a series motor. The applications

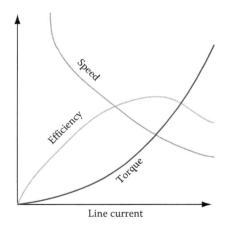

FIGURE 7.8 Characteristics of a series dc motor.

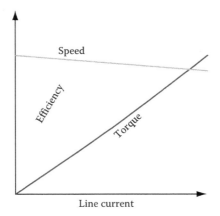

FIGURE 7.9   Characteristics of shunt dc motor.

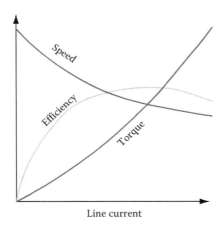

FIGURE 7.10   Characteristics of a compound dc motor.

of compound motors include elevators, conveyors, hoists, pumps, and presses. This motor is capable of starting heavy load like series motors, with safe operations at low torque like shunt motors.

PM motors, which are typical constructions of separately excited motors, only have one winding, that is the armature winding. The magnetic field is generated using a permanent magnet, hence its name. The brushless dc motor is a type of PM motors.

Brushless motors have the following advantages compared to the other types of dc motors:

1. It is smaller in size for a given power rating.
2. The field strength is not affected by the armature current since they are separate in nature.
3. There is a linear relationship between speed and torque, making it easier to control.
4. It produces a high torque at low speed and self-braking mechanism.
5. It requires low maintenance.

The disadvantages of brushless motors include

1. It is relatively easier to experience overheating.
2. The magnetic field is distorted at extreme low and extreme high temperatures.

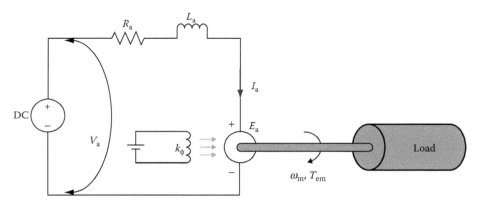

FIGURE 7.11   DC motor equivalent circuit.

DC motors can be represented as an electric circuit model as shown in Figure 7.11.

The performance characteristics of the dc motor are determined by the properties of the motor and the electrical variables applied to the motor. The properties of the motor include the resistance (R_a), the induction (L_a), and the field constant (k_φ). The speed of the motor (ω_m) is determined largely by its emf (which largely depends on the voltage), while the torque (T_{em}) is determined by its current. In short, the relationship among these parameters can be expressed as follows, referring to Figure 7.11

$$E_a = k_\varphi \omega_m \qquad (7.1)$$

$$V_a = E_a + I_a R_a \qquad (7.2)$$

$$T_{em} = k_\varphi I_a \qquad (7.3)$$

Furthermore, the relationship between speed and torque, and also other parameters, is

$$\omega_m = \frac{V_a}{k_\varphi} - \frac{R_a}{k_\varphi^2} T_{em} \qquad (7.4)$$

Control of dc motors mainly constitutes the control of its speed and torque. While control of torque can be accomplished by solely controlling the current, the control of speed can be achieved through a variety of approaches:

1. Armature voltage control, V_a
   This method is preferred for speed control below the base speed. Base speed is the maximum speed a motor can run at, at which the rated torque can still be obtained.
2. Field flux control, k_φ
   This method is preferred for speed control above the base speed. If the speed builds up above the base speed, the insulation can fail and this will in turn burn the motor.
3. Armature resistance control, R_a
   This method is more complicated than the other two mentioned above and therefore is seldom used.

In general, the control of a dc motor consists of controlling the armature voltage and/or the magnetic field. The relationships between these two parameters and the speed are as follows:

1. Increasing armature voltage increases motor speed, and vice versa.
2. Increasing magnetic field decreases motor speed, and vice versa.

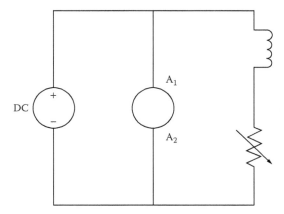

FIGURE 7.12   Shunt-field control.

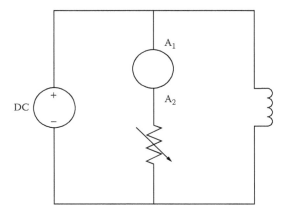

FIGURE 7.13   Armature-resistance control.

The approaches of controlling the speed of the motors are

1. Shunt-field control
   This approach is presented in Figure 7.12. The magnetic field intensity is adjusted by a rheostat installed in series to the field winding. The advantage of this approach is the high-efficiency achievable.
2. Armature-resistance control
   This approach is presented in Figure 7.13. The armature voltage is adjusted by a rheostat installed in series to the armature winding. Increasing the rheostat resistance will subsequently reduce the armature voltage and then reduces the speed of the motor. This approach has a poor efficiency, since a large amount of energy is dissipated as heat through the rheostat.
3. Other special types of control
   Other types of dc motor control include voltage–voltage control, multivoltage control, and Ward–Leonard control, not explained in this chapter.

7.2.3   AC Motor

An ac motor is obtained by replacing the dc power supply in Figure 7.3 with an ac power supply. AC motors are the economical workhorse of industries today, mainly because of their ruggedness, reliability, and low cost. Despite these merits, they are more complex than dc motors. AC motors, as

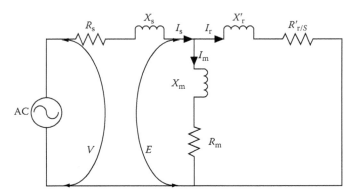

FIGURE 7.14 AC motor equivalent circuit.

the name implies, are driven by ac power supply, which requires both frequency and magnitude control. AC motors are mainly used in applications requiring a constant or slightly varying speed.

AC motors can be represented electrically as a circuitry of stator impedances, rotor impedances, and air gap impedances, each of which represents a component in its structure. Figure 7.14 represents the equivalent circuit of an ac motor for one phase. For a three phase motor, the same circuit is constructed according to the respective construction of the motor (a delta or a wye). In Figure 7.14, the parameters are according to the stator side of the motor; all other parameters are expressed with respect to the stator. The air gap impedances are included to take into account the magnetization effect that occurs at the air gap.

In principle, there are two types of ac motors:

1. Synchronous motors
2. Asynchronous motors (induction motors), including the squirrel-cage motors and wound-rotor motors

Between these two types, induction motors are by far more widely used in the industry.

A synchronous motor is a motor which moves in tandem with the phase of the ac that drives it. A synchronous motor consists of two main elements:

1. Armature winding
2. Field winding

In practice, most synchronous motors will have the armature winding static (and hence is called stator) and its field winding rotating (and hence is called rotor). While the armature winding is connected to an ac supply, the field winding is connected to a dc supply. This construction is typically implemented by installing a rectifier in the motor, so that the motor only requires one type of power supply, which is ac supply.

The rotor is attached to the motor shaft, so that the resulting rotation of the rotor can be delivered to the load through this shaft.

The operational principles behind a synchronous motor are as follows. The ac supplied to the armature winding, which is sinusoidal in nature, produces a rotating magnetic field in the motor which rotates in synchronization with the frequency of the supply current. The dc supplied to the field winding produces another nonrotating magnetic field, having fixed pole pairs. The interaction between the rotating and nonrotating magnetic field will thus cause the rotor to rotate at the speed synchronous to the supply frequency. This is essentially how the synchronous motor gets its name.

A synchronous motor is not a self-starting machine, which means it needs another mechanism to start the rotation from a stationary condition before it can rotate by itself at the synchronous speed. This is because, in a stationary mode, while the armature rotating poles sweep across the field poles, they tend to pull the field poles alternately back and forth, resulting in no motion. Therefore, an additional starting mechanism is required to bring it to its operating speed.

A squirrel cage winding (also called amortisseur winding) is placed in the rotor to bring the rotor into its synchronous speed from stationary. After achieving the synchronous speed, the rotor is energized with the necessary dc voltage and the motor can now operate independently.

The speed of a synchronous motor, as explained earlier, is determined by the frequency of the supply current. It is also determined by the number of poles of the motor. Mathematically, the speed of a synchronous motor is given by

$$N_s = \frac{120 \times f}{p} \tag{7.5}$$

where
  N_s is the synchronous speed in rpm
  f is the frequency of the electricity supply in Hz
  p is the number of motor's pole

For synchronous motors with a requirement to operate at more than one speed, the manipulation of poles grouping is an approach that can be implemented. In this approach, the poles are grouped to effectively reduce the number of poles, which in turn increase the speed. The options of speed variation, however, are not flexible.

When a synchronous motor is pulling a constant load, the current that excites armature winding and field winding is varied in a certain fashion. The adjustment of current depends on the power factor of the operating motor, as presented in Figure 7.15. At a given load, there is a unique field current that will give unity power factor, that is maximum power factor, other than which the power factor is reduced. As presented in Figure 7.15, unity power factor corresponds to minimum armature current. However, when the motor is required to drive a variable load, the field current is usually kept constant at maximum load.

Motor field can be excited by

1. A separate exciter set, driven by an induction motor
2. A constant dc voltage supply

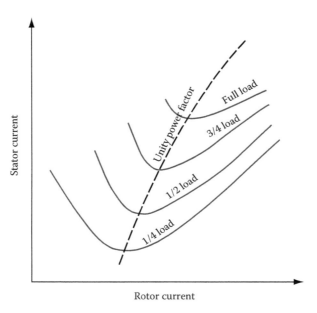

FIGURE 7.15   Characteristics of a synchronous motor.

There are several torques associated with the operation of a synchronous motor:

1. Starting torque (pull-in torque)
   The starting torque of a synchronous motor is the torque developed by the motor when the full-rated voltage is applied to the armature winding. A synchronous motor has a low starting torque of about 10% of the full torque.
2. Running torque
   Running torque is the torque developed by the motor during its full-rated operation. As the speed of the motor is largely fixed at its synchronous speed, the running torque is very much determined by the power of the motor.
3. Pull-out torque
   Pull-out torque is the maximum torque that the motor will develop without being pulled out of step; that is out of synchronization with the rotating field winding.

The performance of a synchronous motor is measured in terms of its power factor. Power factor is the ratio of actual power to apparent power of the motor. A synchronous motor operates at leading power factor, as opposed to an induction motor. Therefore, the application of synchronous motors in a system can improve the overall power factor.

Synchronous motors are typically used for the following purposes:

1. Power factor correction
   The application of synchronous motor as power factor correction is mainly due to its leading power factor. In a large installation with many electrical devices, which typically have lagging power factors, the overall power factor can be very poor and reduces its efficiency. The leading power factor of a synchronous motor improves the overall power factor of the installation. The power factor of a synchronous motor can be varied by adjusting the current of the field windings. This is a very desirable characteristic of a synchronous motor, as the power correction of power factor can be made adjustable. When a synchronous motor is employed exclusively for the purpose of correcting power factor, it is often called synchronous capacitor.

2. Voltage regulation
   At the end of a long transmission line, the voltage can vary greatly, especially with large inductive load. Connection or disconnection of inductive loads may cause the voltage to rise or drop significantly. This may be overcome by installing a synchronous motor with a voltage regulator to control its field winding voltage. The action of voltage regulator is such that, when the voltage drops, the field of the motor is strengthened so that its power factor rises and the line voltage is maintained. On the other hand, when the voltage rises, the field of the motor is weakened so that its power factor drops and again the line voltage is maintained.

3. Constant speed load
   As a synchronous motor runs at a constant speed, it is very suitable for applications that require a constant speed, for example in paper mills, centrifugal compressors, and dc generators.

Squirrel-cage motors are the workhorses among the motors used in the industry, having an advantage which is due to its simple construction and operation, ruggedness, and ease of manufacture.

A squirrel-cage motor consists of

1. Field winding as its stator
   The stator is constructed by a laminated steel core with slots in which coils are located. The coils are grouped and connected to form a polar area, producing a rotating magnetic field.
2. Squirrel cage bars (hence its name) as its rotor
   The rotor is constructed by laminated steel, while the windings consist of short-circuited conductor bars.

The gap between the stator and rotor is called air gap and is designed to be as small as possible to obtain the best power factor of the motor.

The operating principle of a squirrel-cage motor can be explained as follows. The field winding is energized by ac voltage, and this produces a rotating magnetic field inside the motor. As the magnetic field rotates, it cuts the squirrel cage bars and sets up voltage according to Faraday's principle. This voltage causes electrical current to flow through the squirrel-cage bars and develop another magnetic field, now on the rotor. The poles in the rotor interact with the rotation poles in the stator in an alternate attraction and repulsion, causing the rotor to rotate. However, the rotor does not rotate as fast as the rotating magnetic field, for, if it was so, the conductor would be stationary rather than cutting across the field. Therefore, the rotation of a squirrel-cage motor is always less than its correspondent synchronous motor (hence, always less than associated synchronous speed).

The difference between its operating speed and synchronous speed is known as slip, commonly expressed as the percentage of synchronous speed as follows:

$$s = \frac{N_s - N_\omega}{N_s} \times 100 \tag{7.6}$$

where s is the slip and N_ω is the operating speed. Therefore, the speed of the motor is formulated as follows:

$$N_\omega = (1 - s) \times N_s \tag{7.7}$$

Slip is dependent on the load of the motor. When the load increases, the slip becomes larger.

As in synchronous motors, squirrel-cage motors also require a certain amount of starting torque. This can range from 5% to 50% of the full rated-torque. However, squirrel-cage motors do not require additional equipment for starting.

A typical relationship between the torque and the speed of a squirrel-cage motor is presented in Figure 7.16. To achieve satisfactory operations, certain control schemes must be implemented to the motors. The control schemes must be able to allow the motors to perform the following functions:

1. Start and stop the motor
2. Speed regulation
3. Speed reversal
4. Motor protection

Squirrel-cage motors are applied in many diverse areas of applications in industry, due to its ruggedness and simplicity.

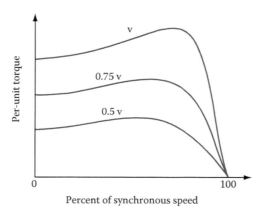

FIGURE 7.16 Torque-speed characteristic of a squirrel-cage motor.

A wound-rotor motor consists of

1. Field winding as its stator
   The stator is constructed exactly as one used in squirrel-cage motor.
2. Armature winding as its rotor

The rotor consists of coils of wire connected in regular succession, having the same number of poles as the stator. The ends of the windings are connected to slip rings.

   The operational principles of wound-rotor motors can be explained as follows. As in a squirrel-cage motor, the stator produces a rotating magnetic field in the motor. This magnetic field cuts through the rotor winding. The current induces in the rotor are carried to an external resistance through slip rings. Subsequently, rotor current produces another magnetic field, now in the rotor. The interaction between these two magnetic fields causes rotation of motor, in the same manner as in a squirrel-cage motor.

   As explained above, the induced current of the rotor is transmitted to an external resistance. The variation of this resistance can therefore be used to control the operations of the motor, since the intensity of the induced current directly affects the resulting magnetic field. The controller used in a wound-rotor motor is used for the following functions:

1. To start and stop the motor in a satisfactory manner
2. To regulate the speed of the motor by variation of the external resistance

Wound-rotor motors are implemented in applications that do not require exact speed regulation, operating intermittently in an acceptable lowefficiency, and requiring starting with a heavy load. Examples of such applications include hoists, cranes, elevators, pumps, and compressors.

7.2.4 LINEAR MOTOR

While rotary motors comprise almost 98% of motors applications, linear motors are also available with their unique merits. Linear motors are very much like rotary motors, having two windings that act as armature winding and field winding. The design of a linear motor can be thought of by cutting and unrolling a rotary motor. The result is a flat linear motor that produces linear forces. The construction of a linear motor is presented in Figure 7.17.

   The advantages of linear motors are as follows:

1. It provides a high precision and accuracy in a high-speed implementation as compared to its rotary counterpart.
2. It does not require coupling mechanism, as the output motion has already been a linear motion.

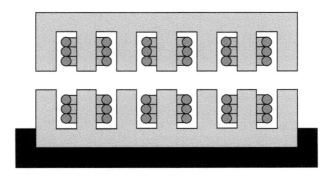

FIGURE 7.17   Linear motor.

3. It is able to develop a much higher magnetic flux without significant heat (hence low thermal losses), as it incorporates earth permanent magnet.
4. It offers high force density.

The applications of linear motors are mainly in industries or processes that require high accuracy and precision, such as semiconductor processes, precision metrology, and miniature systems.

There are several designs of liner motor as follows:

1. Force-platen

   This motor consists of a moving platen and a stationary platen. The moving platen consists of electromagnetic coils (with winding and iron core). Permanent magnets are placed on the stationary platen oriented at a right angle to the thrust axis, but slightly skewed in the vertical plane to reduce the thrust ripple. Force-platen motors feature a low-height profile and a wide range of available size. The applications include automobile and machine tools applications where high continuous and peak forces are required.

2. U-shaped

   This motor consists of a U-shaped motor armature and a permanent magnetic field generated by the track. The armature is a planar winding epoxy bonded to a plastic blade, which projects between double rows of magnet. The magnetic field works in conjunction with the electromagnetic field in the blade to produce linear motion. U-shaped motors are used in high-precision operations that require smooth motion, since these motors give a zero detent force. Furthermore, U-shaped motors are cost-effective and provide a long travel length. The drawback, however, includes resonance in high acceleration operation, high thermal losses, and inefficiency in magnetic utilization.

3. Tubular

   Tubular motors consist of a stationary thrust rod and a moving thrust block. The thrust rod is a permanent magnet, while the thrust block is an electromagnetic winding. Tubular motors feature a high force generation and high energy efficiency. The drawbacks include limited travel distance, tall height, and limited size and force range.

The modern linear motors give superior performance over many conventional motion drives. The direct energy conversion feature greatly reduces the contact-type nonlinearities and disturbances in linear motors. However, the inherent ability associated with mechanical coupling to suppress the effects of structural uncertainties and external disturbance is consequently lost. Due to its working principle, the presence of uncertainties and disturbances is a prominent factor limiting the performance of motion control system.

Force ripple is predominantly present in permanent magnet linear motors (PMLM), resulting in an undesirable performance. It can be minimized or even eliminated by an alternative design of the motor structure or spatial layout of the magnetic materials such as skewing the magnet, optimizing the disposition and width of the magnets, etc. These mechanisms often increase the complexity of the motor structure. PMLM, with a slot-less configuration, is a popular alternative since the cogging force component due to the presence of slots is totally eliminated. Nevertheless, the motor may still exhibit significant cogging force owing to the finite length of the iron-core translator. Finite element analysis confirms that the force produced on either end of the translator is sinusoidal and unidirectional. Since the translator has two edges (leading and trailing edges), it is possible to optimize the magnet length so that the two sinusoidal force waveform of each edge cancel out each other. However, this would again contribute some degree of complexity to the mechanical structure. A more practical approach to eliminate cogging force would be to adopt a sleeve-less or an iron-less design in the core of the windings. However, this approach results in a highly inefficient energy conversion process with a high leakage of magnetic flux due to the absence of material reduction in the core. As a result, the thrust force generated is largely reduced (typically by 30% or more). This solution is not acceptable for applications where high acceleration is necessary. In addition, iron-core motors, which produce high thrust force, are ideal for accelerating and moving large masses while maintaining stiffness during the machining and processing operations.

7.3 SOLID-STATE ACTUATORS AND PIEZOELECTRIC ACTUATORS

Solid-state actuators are typically used for achieving motion of the order of nanometer. They include mechanisms of special transducer materials such as magnetostrictive, electrostrictive, and piezoelectric materials.

Magnetostrictive materials have an ability to deform, that is to extend and to contract, in the presence of magnetic field. These materials are often referred to as rare-earth materials; typically consisting of transition metals. Recently, magnetostrictive materials have been used as solid-state speakers, vibration tables, sensors, and actuators. The most appealing characteristic of this material is less hysteresis as compared to other high-precision transducers, which greatly relax actuation control. However, magnetostrictive materials are more sensitive to heat, and this makes them less preferable to, for example, electrostrictive materials.

Electrostrictive materials have an ability to deform in the presence of electric field. Electrostrictive materials are nonpoled ceramics, made from lead–magnesium–niobate. Despite successful development of electrostrictive actuators and their low hysteresis, their quadratic, nonlinear relationship between voltage and deformation makes this type of actuators less popular than pizoelectric actuators. This type of actuators has been applied in a lifting mechanism with high precision [1].

Piezoelectric materials have an ability to deform in the presence of electric field. Piezoelectric materials are poled ceramics, made from lead–zirconium–titanate. The difference between piezoelectric and electrostrictive materials lies in the deformation process. In electrostrictive materials, an electric field separates the positive and negative charged ions, expanding the materials. In piezoelectric materials, the electric charge excites the atoms of the materials according to the poling direction.

Piezoelectric actuators have become increasingly popular in many precise positioning applications as these actuators are able to provide very precise positioning (in nanometer order) and generate high force (up to a few thousand Newton).

Piezoelectric actuators have shown a high potential in applications that require manipulation within the range of submicrometer. Piezoelectric actuators have been applied in many areas, such as MEMS (microelectromechanical systems), bioengineering, and nanotechnology.

The experiment conducted by French brothers Curie in 1880 is widely considered as the birth of piezoelectricity. In their experiment, Curies discovered that certain crystals generated an electric charge when they were subjected to a mechanical strain; this phenomenon was called piezoelectric effect. Furthermore, they also discovered that the same crystals generated strain when they were subjected to an electric charge; this phenomenon was called inverse piezoelectric effect. It took several decades before this remarkable phenomenon was used commercially in an ultrasonic submarine detectors developed during World War I. During World War II in 1940s, scientists discovered that certain ceramics could be made piezoelectric by subjecting them in an electric field during their transformation phase. This boosted the implementation of piezoelectricity since the production cost could then be reduced. The birth of nanotechnology in 1950s put the ultimate drive of its development and application.

Piezoelectric materials are currently used both as sensors (associated to its piezoelectric effect) and as actuators (associated to its inverse piezoelectric effect). They have the following beneficial characteristics that make them suitable for numerous high-precision applications:

1. High resolution, unaffected by stiction and friction
2. Not affected by magnetic field
3. Low power consumption, since power is only required during motion and not during holding phase
4. Clean operation compatibility
5. Fast response time

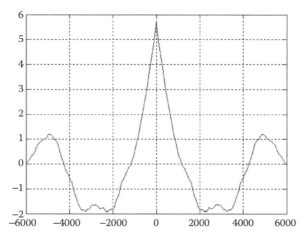

FIGURE 7.18 Autocorrelation of displacement.

The analysis of piezoelectric effect is conducted based on the linear theory of piezoelectricity, whereby the equations of linear elasticity are coupled to the equation of electric charge by means of piezoelectric constants.

The constitutive equations of piezoelectricity (as explained in Ref. [2]) are as follows:

$$T_{ij} = c_{ijkl}^{E} S_{kl} - e_{kij} E_k \qquad (7.8)$$

$$D_i = e_{ikl} S_{kl} + \varepsilon_{ij}^{S} E_k \qquad (7.9)$$

where
  T is the stress vector
  S is the strain vector
  E is the electric field vector
  D is the electric displacement vector
  c is the elastic stiffness constant matrix
  e is the piezoelectric constant matrix
  ε is the permittivity constant matrix
  superscripts "E" and "S" denote constant electric field and constant strain, respectively

In a piezoelectric material, stress/force, strain/displacement, and electric charge/voltage couple to each other, as shown in Equations 7.8 and 7.9. If an electric potential is applied across the electrodes of a piezoelectric material, there will be stress and strain induced in the material. In turn, this stress will lead to the build-up of an opposing electric charge which will affect the original stress and strain induced in the material.

The evidence of this iterated cause–effect phenomenon is now elaborated. An interesting finding is obtained when the displacement is measured with a cylinder in a relaxed condition. The autocorrelation of the output signal, that is the displacement (presented in Figure 7.18), shows that the small displacement variation contains of a significant random portion and also a systematic component that can be attributed to the coupling of the various parameters in a piezoelectric material as explained above.

7.4 TRANSMISSIONS (MECHANICAL ACTUATORS)

The choice of transmission system is critical to the performance of motion system. Not only must it be suitable to the actuators, it should also meet the requirements of the application.

In this section, some of the most prevalent transmissions for high-performance motion control will be discussed:

1. Friction drive
2. Lead screw
3. Ball screw
4. Flexures

7.4.1 MECHANICAL ACTUATORS

Actuators could also be mechanical-based systems dedicated to generate linear or rotation movement by moving an object to an accurate position. The mechanical component, for example the drive mechanism of the actuator, is a very important subsystem that has to comply with a number of criteria cited below. The electric component is required to deliver sufficient force or torque and secure a recommended resolution. As the required movement has to be very accurate, these actuators will be designed or chosen according to criteria described as follows:

1. High stiffness in the active axis
2. Linearity or coaxiality
3. No backlash
4. Low disturbances
5. Availability of reverse motion
6. Fast time response to active control

The choice of a drive mechanism is critical to motion system performance. There are many available technology solutions, but the most prevalent for high-performance motion control are as follows.

1. Friction drive actuators
2. Lead screw/ball screw with rotary servo motor and encoder
3. Lead screw/ball screw with open loop microstepper
4. Linear servo motor
5. Piezo ceramic linear motor (friction drive)
6. Combined actuator with master and slave roles

In ball screw-driven systems, the location of the feedback is extremely important.

7.4.2 FRICTION DRIVES DESIGN CONCEPTS

Depending on the specifications related to the moving slide table to be controlled, high stiffness and zero slippage are required together with high mechanical resolution. The material at the contact interface is much important in this case. The friction drive general configuration (Figure 7.19) is

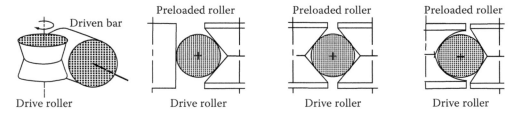

FIGURE 7.19 Rollers configurations.

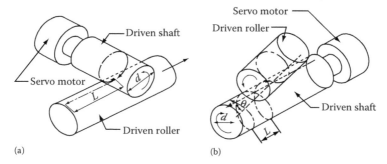

FIGURE 7.20 (a) Conventional friction drive and (b) angular friction drive.

composed of a round bar that is kept centered between twin-V rollers; the V geometry, allowing contact points and a double tangential stiffness that is estimated using Mindlin formulation [3]. Figure 7.19 suggests three possible combinations of the interface between the drive shaft and the driven roller for improved tangential stiffness and slippage avoidance.

One of the easiest ways to increase the resolution is to consider a small angle between the drive shaft and the driven roller. The resolution of the friction drive could be estimated by $L = \pi D$ (Figure 7.20a) while in Figure 7.20b, $L = \pi D \cdot \tan(\theta)$.

From the latter principle, Muzimuto [4] has suggested a twist roller that can achieve Angstrom resolution. Figure 7.21 shows the twist roller friction drive comprising three rollers to increase the stiffness. The three rollers are slightly inclined with respect to the driving shaft, which is linked to a motor. The flange is assembled to the moving carriage.

7.4.2.1 Stiffness

The contact stiffness at the contact interface has three components K_x, K_y, and K_z along x, y, and z axis, respectively (Figure 7.22).

Considering a standard friction drive, for any fine positioning with an axial stiffness K_T, any disturbance force F_δ will introduce an error δ_K in the direction of the motion and is estimated as follows:

$$\delta_K = \frac{F_\delta}{K_T} \tag{7.10}$$

Hence, a larger stiffness induces a lower error.

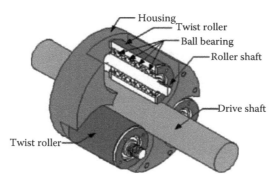

FIGURE 7.21 (See color insert following page 174.) Twist friction drive. (From Mizumoto, H., Yabuya, M., Shimizu, T., and Kami, Y., *Prec. Eng.*, 17, 57, 1995.)

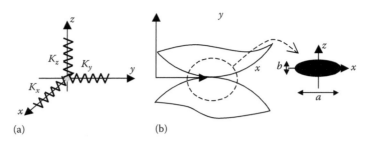

FIGURE 7.22   Stiffness of the contact interface.

The actuation stiffness is mainly related to the tangential stiffness at the contact interface. Mindlin [5] has suggested an equation to estimate the stiffness depending on the interface shape.

$$K_T = \frac{4Ea}{(2-\eta)(1+\eta)\Phi}\left(1 - \frac{P_x}{\mu_g P_z}\right)^{\frac{1}{3}}$$ (7.11)

where
  E is the Young modulus
  a is the dimension of the contact point
  η is Poisson's coefficient
  μ_g is the friction coefficient
  Φ is a function of the contact geometry
  P_x and P_z are tangential and normal contact pressure, respectively

The preload applied to both rollers preloaded against the moving shaft is the main parameter to be adjusted to obtain the required contact stiffness and to prevent slippage. This preload could be measured by using a compressive force sensor. The stiffness for three different interfaces, shown in Figure 7.19, has been computed for different tangential forces and the results are shown in Table 7.3.

TABLE 7.3
Stiffness Behavior of Various Friction Drives

Type	Tangential Stiffness Behavior	Friction Drive Concepts
V-C	(graph)	(diagram)

(a)

Graph details: Tangential stiffness (N/m) vs Preload (N). Legend: $F_{tan} = 1\,N$, $F_{tan} = 4\,N$, $F_{tan} = 8\,N$. $R_g = R_t = 17.5$ mm, $= 75°$, $\mu_g = 0.1$

TABLE 7.3 (continued)
Stiffness Behavior of Various Friction Drives

Type	Tangential Stiffness Behavior	Friction Drive Concepts

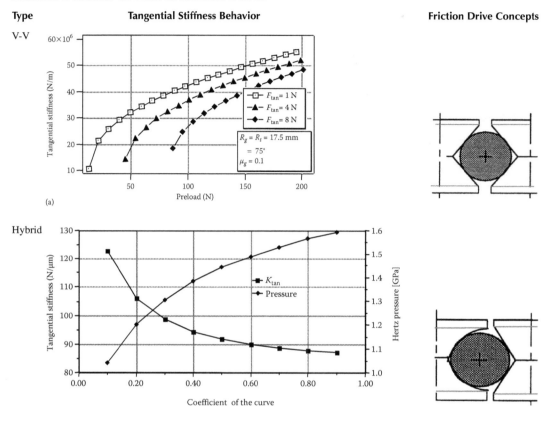

(a)

The tangential stiffness is very high (e.g., 120 N/μm) in case of a hybrid combination, that is a curved and a linear interface in V form with four contact points. It is not certain that this configuration, albeit high stiffness, will deliver a high resolution as the contact interface area is larger than in the case of two V rollers. A high resolution is usually obtained at low speed as stick-slip phenomenon may appear often at high speed depending on the applied preload. The driven bar could be provided with high hardness (60–70 HRC), straightness (4–15 μm/m), and smooth finish (0.25–0.5 μm Ra). These characteristics should be careful to comply with the specifications.

The birth of the slippage starts from the inner surface of the contact zone (Figure 7.23). If the contact zone is circular, then the slippage evolves in a concentric surface with respect to the contact zone. The shear stress at a distance ρ from the centre is estimated [3] as

$$
\left[
\begin{array}{l}
\tau_x(\rho) = \dfrac{3}{2}\mu_g \dfrac{P_z}{\pi a^2}\left(1 - \dfrac{\rho^2}{a^2}\right)^{1/2} \quad \text{if } a_o < \rho < a; \\[3mm]
\tau_x(\rho) = \dfrac{3}{2}\mu_g \dfrac{P_z}{\pi a^2}\left(1 - \dfrac{\rho^2}{a^2}\right)^{1/2} - \dfrac{3}{2}\mu_g \dfrac{P_z}{\pi a^2}\left(1 - \dfrac{\rho^2}{a^2}\right)^{1/2}\dfrac{a_o}{a}, \quad \text{if } \rho < a_o
\end{array}
\right.
$$

(7.12)

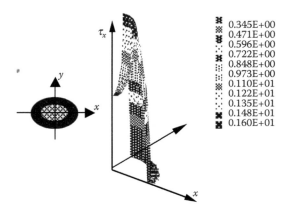

	0.345E+00
	0.471E+00
	0.596E+00
	0.722E+00
	0.848E+00
	0.973E+00
	0.110E+01
	0.122E+01
	0.135E+01
	0.148E+01
	0.160E+01

FIGURE 7.23 Distribution of the tangential shear stress at the start of slippage.

where
   μ_g is the friction coefficient
   P_z is the normal force
   a is the radius of the contact zone
   a_o is the initial diameter of slippage area

The reaction force could be calculated using Equation 7.13 as follows:

$$T_x = \int_0^a \int_0^{2\pi} \tau_x \, \rho \, d\rho \, d\theta = \mu_g P_z \left(1 - \frac{a_o^3}{a^3} \right) \tag{7.13}$$

Preferably, the dimensioning of the rollers has to meet two criteria:

1. Hertz contact stress admissible:

$$\frac{P_z}{\pi \, ab} < P_{\text{HertzMax}} \tag{7.14}$$

where
   P_z is the preload
   a, b are the dimensions of the contact surface

2. Maximum displacement precision avoiding geometric errors on the roller (radius R):

$$R = \frac{\Delta P}{\Delta \Phi} \tag{7.15}$$

where
   $\Delta \Phi$ is the encoder resolution
   ΔP is the required resolution

7.4.2.2 Case Study: Friction Drive Assembly

The friction drive may be coupled to the base of the translating system with reinforced flexures to minimize the effects of forced geometric congruence while the traction bar is linked to the carriage at one end via monolithic flexure (Figure 7.24).

7.4.2.3 Control Scheme for Better Positioning

The contact between two solid materials will always lead to nonlinearities observed in at the prerolling in fine positioning. From the point of view of contact mechanics, a model for prerolling

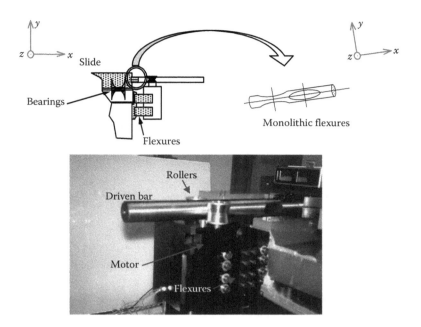

FIGURE 7.24 (See color insert following page 174.) Configuration of the friction drive.

has been written [5] to take into account the prerolling and the rolling resistance. Other effects may not be possible to model as most of the time these are not known. Therefore, to dynamically compensate for unmodeled phenomena, an internal model control (IMC) scheme is used. Its global approach of precision of the model avoids focusing on any specific phenomena or mechanical behavior effects [6]. This scheme was added on top of the PID controller (Figure 7.25).

Even at very low speed, rolling resistance induces microdynamics especially when it is submitted to periodical variation of loads (i.e., machining), hence for example the nonsuitability for ultrahigh-precision machining. However, the nonlinear behavior takes place at the contact interface and may cause loss of contact at high rolling speed [7]. The rolling resistance could be modeled to predict its evolution as shown in Figure 7.26b. A model was proposed depending on the type of materials at the interface [5]. The model could be taken into account by the controller. At low speeds, one of the most relevant problems in positioning control is the exhibition of the nonlinear behavior at prerolling stage as shown in Figure 7.26a.

The PID controller does not compensate for the phenomena and delays the positioning by Δt. Such an influence is addressed using an IMC controller instead of PID controller to compensate for temporal delay when it starts rolling.

7.4.2.4 Positioning Measurement with a Friction Drive

The friction drive may be coupled to the base with reinforced flexures to minimize the effects of forced geometric congruence while the traction bar is linked to the carriage at one end via monolithic

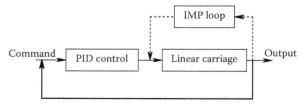

FIGURE 7.25 Representation of separate action of the IMC loop and PID servo control.

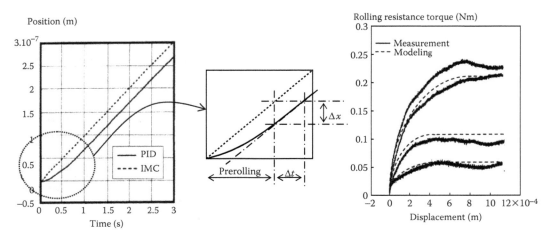

FIGURE 7.26 Error compensation in rolling resistance.

flexure (Figure 7.24). A principle to be respected is that the driving force of the actuator should operate through the axis of reaction. Hence, the slide will not be subject to parasitic forces or Abbé errors unless and an offset was considered.

The measurement of the straightness over the whole stroke was performed using a laser interferometer. A variation of less than one micrometer over a stroke of 200 mm was observed with good repeatability. In terms of positioning resolution, an optical sensor from Heidenhain was used. The following measurement spectrum (Figures 7.27 and 7.28) shows a very well-resolved motion for

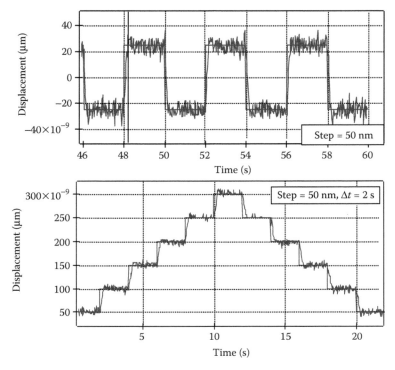

FIGURE 7.27 Reverse and continuous 50 nm step motion response.

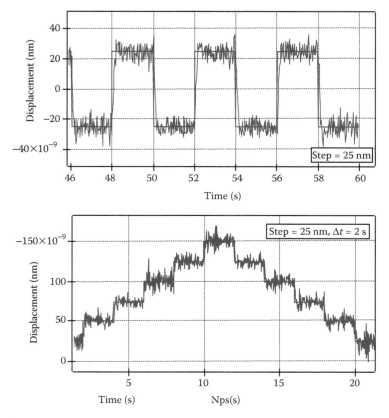

FIGURE 7.28 Reverse and continuous 25 nm step motion response.

continuous and reverse steps of 25 nm and 50 nm. A noisy response was observed for steps of 25 nm due to the 16 nm resolution of the optical probe. With a probe of a 4 µm resolution, the friction drive could achieve a well-resolved motion of 10 nm steps easily.

7.4.3 LEAD SCREW

Lead screw drives (Figure 7.29) are commonly used for precision applications. They are character-ized by mechanical advantage due to the lead screw pitch, which is useful for positioning heavy loads. However, they have many limitations, particularly for longer travels and high speeds. Any rotating shaft has a critical speed that may excite the first mode of whipping. This could be addressed

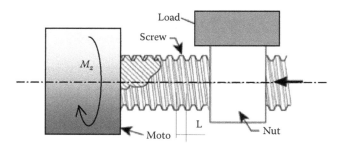

FIGURE 7.29 Lead screw with spring to minimize the backlash.

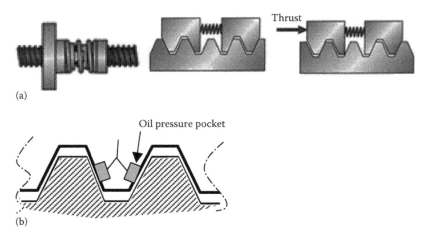

(a)

(b)

FIGURE 7.30 (a) Lead screw with spring to minimize the backlash and (b) hydrostatic lead screw.

by increasing the diameter of the screw, requiring more power and therefore more heat dissipation. Other disadvantages include mechanical backlash, pitch errors, and wind up. All of these can limit the accuracy and the repeatability.

Friction and torsion torque applied to the lead screw generate an angular torsion and then produce a backlash.

The hydrostatic lead screw (Figure 7.30b) has replaced this drive where oil pockets are added in the threads.

7.4.4 Ball Screws

The main disadvantage of this drive is the recirculating balls that lose their sphericity under the traction force. The repeatability achieved is of the order of $1\,\mu m$. For high-precision application, they can attain microinch resolution (Figure 7.31).

They could be used as bearings as well. The recirculating bearings result in a smaller footprint, and the bearing reaction forces are in a constant location with respect to the load. They can support high load capacities since there are typically several rows of bearings per puck. These bearings are not as smooth as linear bearings since the balls entering and exiting the raceway can create vibrations (Figures 7.32 and 7.33).

Obviously, a more precise recirculating actuator should be chosen for precision applications. The one shown in Figure 7.31 presents some errors due to lack of preload applied to the balls.

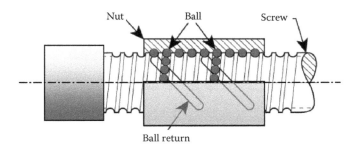

FIGURE 7.31 Lead screw with recirculating balls.

FIGURE 7.32   Leadscrew of the Takisawa milling machine.

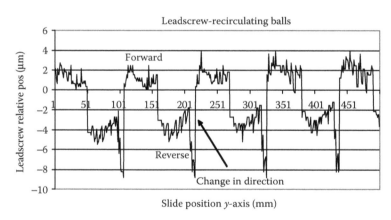

FIGURE 7.33   Positioning behavior of the leadscrew in Figure 7.32.

7.4.5 FLEXURES

The flexures (Figure 7.34) are flexible leaf springs used either single or in tandem and when bending they generate partial rotational motion. The "leaves" could be machined in a bulk of material to have monolithic flexures.

The moment of the cantilever is as follows:

$$M = EI \frac{\mathrm{d}^2 y}{\mathrm{d}x^2} \tag{7.16}$$

The generated motion could be either amplified or reduced depending on the linkage designed for the application. Subnanometer could be obtained in MEMS applications.

FIGURE 7.34   Fundamental element of a flexure plate.

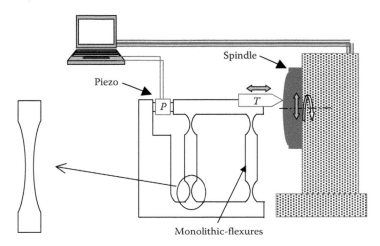

FIGURE 7.35   Linear motion with monolithic flexure.

The flexures could provide the following objectives:

1. Precise positioning within a short-range motion
2. Precise displacement in application of a specific applied force
3. Accurately known force (controlled) in applications of a specific applied displacement (Figure 7.35).

Advantages

1. Wear free as line of action will remain constant
2. Smooth and continuous displacement
3. Complete mechanisms can be produced from a single monolithic element
4. A linear relationship between applied force and displacement as predictable forces can be generated by controlled displacement
5. Failure mechanism due to overloading/fatigue; safety critical systems available

Disadvantages

1. Restricted to small displacements and small forces
2. Out of plane stiffness tends to be so low; drive axis should be collinear with desired motion
3. Specific force for a given deflection depends upon the elastic modulus necessary to calibrate after fabrication
4. Hysteresis depends upon the stress level and temperature

7.5   SENSORS

In motion control applications, the primary sensors used in electric drives consist of torque, position, speed/velocity, and acceleration sensors. A brief overview of the common sensors used for these variables is covered in this section.

7.5.1   POSITION MEASUREMENT

Common types of position sensors include the encoder, potentiometer, reflective optotransducer, linear variable differential transducer (LVDT), and the strain gage.

7.5.1.1   Encoder

An encoder, as presented in Figure 7.7, is a device that provides a digital output as a result of a displacement. The positioning information of an encoder is categorized into

1. Incremental encoder, detecting the displacement from some points
2. Absolute encoder, detecting the actual position of the encoder (Figure 7.36)

The rotating disk has a number of apertures through which light beam can pass from the light source to the light detector. When the disk rotates along with the shaft, pulsed output is produced by the sensor. The angular displacement of the shaft can then be determined by the number of the output pulses. The typical resolution of encoder employed in industry varies between 6° and 0.3°.

7.5.1.2   Potentiometer

A potentiometer consists of a resistance element with a sliding contact that is movable over the length of the element. Figure 7.37 shows the working diagram of a potentiometer.

The operating principle of a potentiometer is based on the relationship between resistance and length. The sliding contact of the potentiometer varies the effective length of this device. The longer

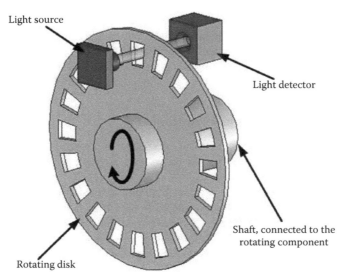

FIGURE 7.36   Encoder.

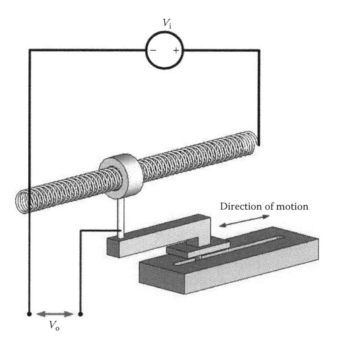

FIGURE 7.37 Potentiometer.

the effective length, the higher the resistance, and hence the output voltage V_o. The movable sliding contact can be attached to a mechanism, and therefore the position of that device can be determined from the output voltage.

7.5.1.3 Optotransducer

Figure 7.38 shows the construction of a reflective optotransducer, consisting of an infrared LED and phototransistor.   AQ7

The operating principle of an optotransducer is as follows. The beam from the three LEDs (in this case) is reflected back to the phototransistor when it hits the reflective surface of the rotating Gray-coded disk, which is configured as in Figure 7.39 and Table 7.4. On the other hand, the nonreflective surface breaks the beam. A reflection is indicated as 1, while a nonreflection is indicated as 0. The position of the rotating disk is then indicated by the combination of 1 s and 0 s.

7.5.1.4 Linear Variable Displacement Transformer

The construction and circuit arrangement of an LVDT are as shown in Figure 7.40. It consists of three coils mounted on a common coil former and having a magnetic core that is movable within the coils.

The operational principle of an LVDT can be explained as follows. The center coil is the primary and is supplied from an ac supply. The coils on either side are the secondary coils and labeled A and B in Figure 7.40. Coils A and B have equal number of turns and are connected in series in antiphase fashion so that the output voltage is the difference between the voltages induced in the coils. The output voltage changes in amplitude which increases with the movement from the neutral position to a maximum value. The phase changes depending on the direction of movement.

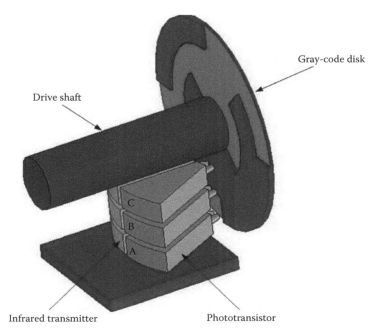

FIGURE 7.38 (See color insert following page 174.)   Optotransducer.

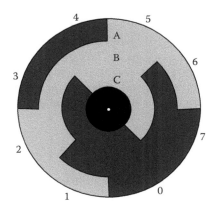

FIGURE 7.39   Gray-coded disk.

TABLE 7.4
Gray-Coded Disk Signal

Position	C	B	A
0	0	0	0
1	0	0	1
2	0	1	1
3	0	1	0
4	1	1	0
5	1	1	1
6	1	0	1
7	1	0	0

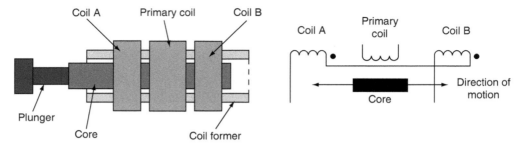

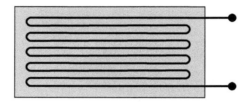

FIGURE 7.40   LVDT.

FIGURE 7.41   Strain gage.

7.5.1.5   Strain Gage

Strain gage is a sensor that measures the strain of a material. It can thus be used as a displacement sensor of a position sensing device indirectly. Figure 7.41 shows the construction of a strain gage, consisting of a grid of fine wire or semiconductor material bonded to a backing material.

The operating principle of the strain gage can be explained as follows. The strain gage unit is attached to the beam to be measured and is arranged so that the variation in length under loaded conditions is along the gage-sensitive axis. The loading on the beam increases the length of the gage wire and also reduces its cross-sectional area. Both of these effects will increase the resistance of the wire.

7.5.2   Velocity Measurement

The typical speed sensor used in electric drives is tachogenerator whose structure is presented in Figure 7.42. Tachogenerator is an electromechanical device that produces electrical signal from

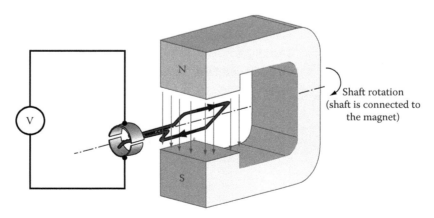

FIGURE 7.42 (See color insert following page 174.)   Tachogenerator structure.

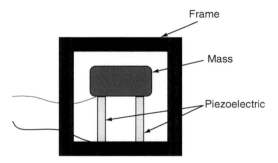

FIGURE 7.43   Accelerometer structure.

mechanical energy, which is the rotation of a shaft. It is essentially the reversal of an electric motor, and therefore the underlying working principles of those two devices are the same.

Tachogenerator consists of a coil mounted in a magnetic field. When the coil rotates along with the shaft, electromagnetic induction results in an alternating-induced emf across the coil. This emf is proportional to the rotational speed of the shaft, thus can be used as a measure of angular speed. Tachogenerator can typically measure up to 10,000 rpm.

7.5.3   Acceleration Measurement

For applications where the acceleration measurement is critical, accelerometer can be installed in the drive systems. Beside acceleration, an accelerometer can also be used to measure tilt motion, inclination, vibration, and collision. There are several types of accelerometer based on its working principle, for example hot gas, magnetic field, spring, and piezoelectric-based accelerometers. Figure 7.43 presents the structure of a typical piezoelectric-based accelerometer.

The main components of the accelerometer in Figure 7.43 are a piezoelectric crystal and a seismic mass, loaded along the polarization direction of the piezoelectric crystal. The piezoelectric crystal is capable of converting mechanical displacement or force into electric charge. When the accelerometer is subject to motion, the seismic mass will compress and stretch the piezoelectric crystal due to the mass inertia. This force causes an electric charge to be generated (because of piezoelectric property). Based on Newton's second law, the force is proportional to the acceleration and therefore the electric charge is also proportional to the acceleration. By closing the measurement circuit and adding necessary amplifiers, an electric voltage will be output.

As mentioned above, while the above-mentioned accelerometer is common, there are different types of accelerometers available as well. For example, the piezoelectric crystal may be replaced by a piezoresistive crystal or strain gage, which has a better sensitivity for low-acceleration measurements.

7.5.4   Torque Measurement

A common method of torque measurement is by sensing the deflection of the shaft caused by a twisting force. This can be accomplished with strain gages mounted on a rotating shaft, as presented in Figure 7.44. As the shaft rotates, deformation will take place on the shaft and this will be measured by the strain gages. A Wheatstone bridge that incorporates the strain gages will convert the measurement into a calibrated signal. Because the deformation is proportional to the output torque, the calibrated signal is also proportional to the torque. This method of torque measurement is a direct measurement, since the torque is indeed measured via a circuitry of sensors. Slip ring is often used to assist signal transmission.

Torque measurement can also be done with noncontacting sensors. Noncontacting sensors work based on radio telemetry communication between a stationary antenna in the frame and a loop

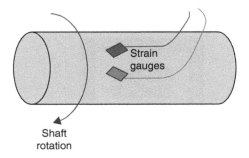

FIGURE 7.44   Torque sensor.

antenna on the rotating shaft. The power from the loop antenna activates the strain gages, allowing torque measurement to be conducted. The output signal is then retransmitted to the stationary antenna.

Torque can also be measured indirectly (inferred) by measuring the induced current of the motor. The direct relationship between current and torque allows this method to be used with satisfactory result. While this method is cost-effective, it is not as accurate as direct method with strain gages. Furthermore, as implies above, it is only possible in the region where the current is proportional to the torque. Current measurement, however, can be used to infer a torque measurement when an accurate torque observation is not so critical.

REFERENCES

1. D.F. Waechter, D. Liufu, M. Camirand, R. Blacow, and S.E. Prasad, Development of high-strain low-hysteresis actuators using electrostrictive lead magnesium niobate (PNM), *Proceedings of the 3rd CanSmart Workshop on Smart Materials and Structures*, pp. 31–36, 2000.
2. *IEEE Standard on Piezoelectricity, An American National Standard, Std.* 176, 1987.
3. R.D. Mindlin, Compliance of elastic bodies in contact, *J. Appl. Mech. Trans ASME*, 71, 259–268, 1949.
4. H. Mizumoto, M. Yabuya, T. Shimizu, and Y. Kami, An angstrom-positioning system using a twist-roller friction drive, *Prec. Eng.*, 17(1), 57–62, 1995.
5. S. Mekid, A non-linear model for pre-rolling friction contact in ultra precision positioning, *J. Eng. Tribol. Proc. IMechE*, 218 Part J, 305–311, 2004.
6. S. Mekid and O. Olejniczak, High precision linear slide. Part_2: Control and measurements, *Int. J. Mach. Tools Manuf.*, 40(7), 1051–1064, 2000.
7. S. Mekid and M. Bonis, *Numerical Resolution of the Contact Vibration under Harmonic Loads, Contact Mechanics—Computational Techniques*, Vol. 1, Brebbia, C.A., and Aliabadi, M.H. (eds), Computational Mechanics Publications, Wessex Institute of Technology, Southampton, U.K., 1993, pp. 61–67.

8 Current Issues in Error Modeling—3D Volumetric Positioning Errors

Charles Wang

CONTENTS

Machine tool accuracy is not linear. It is volumetric.

8.1   POSITIONING ERROR MODELING (SERIAL KINEMATICS MACHINES)

World competition requires good quality or accurate parts. Hence, the computer numerical controlled (CNC) machine tool positioning accuracy becomes very important. Twenty years ago, the largest machine tool positioning errors are leadscrew pitch error and thermal expansion error. Now, most of the above errors have been reduced by better leadscrew, linear encoder, and pitch error compensation. The largest machine tool positioning errors become squareness errors and straightness errors. Hence, to achieve higher 3D volumetric positioning accuracy, all three displacement errors, six straightness errors, and three squareness errors have to be measured. Using a conventional laser interferometer to measure these errors is rather difficult and costly. It usually takes days of machine downtime and experienced operator to perform these measurements.

It has been proposed to use the body diagonal displacement errors to define the volumetric positioning error [1]. However, the relations between the measured body diagonal displacement errors and the 21 rigid-body errors are not clear, and a more practical definition of a volumetric position error has been discussed but not defined yet. Hence, the current issues in machine errors modeling are to define and to determine the 3D volumetric positioning error of CNC machine tools. The definition should be directly linked to the 3D positioning errors and also practical to measure or determine such that it will be accepted by machine tool builders and used in the specification.

8.1.1   RIGID-BODY ERRORS

In general, the errors should be a function of all three coordinates, x, y, and z. For serial kinematics machines, the x-, y-, and z-axes are orthogonal and stacking on each other. To simplify the theory, it is reasonable to assume the motions are rigid-body motions. Hence, the errors become functions of a single coordinate instead of three coordinates. In the following, the rigid-body errors are derived based on the rigid-body assumption; for each axis, there are three linear errors and six angular errors as shown in Figure 8.1. In a three-axis machine, there are six errors per axis or a total of 18 errors plus three squareness errors as shown in Figure 8.2. These 21 rigid-body errors can be expressed as follows [2]:

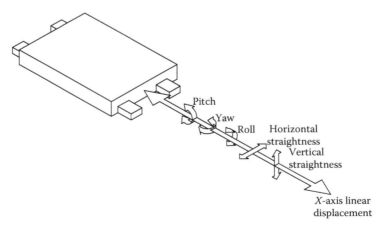

FIGURE 8.1   Linear and angular errors of a single axis.

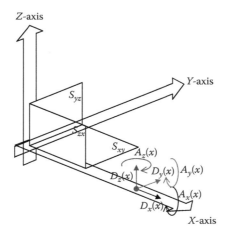

FIGURE 8.2 (See color insert following page 174.)   Squareness errors between axes.

Linear displacement errors: $D_x(x)$, $D_y(y)$, and $D_z(z)$
Vertical straightness errors: $D_y(x)$, $D_x(y)$, and $D_x(z)$
Horizontal straightness errors: $D_z(x)$, $D_z(y)$, and $D_y(z)$
Roll angular errors: $A_x(x)$, $A_y(y)$, and $A_z(z)$
Pitch angular errors: $A_y(x)$, $A_x(y)$, and $A_x(z)$
Yaw angular errors: $A_z(x)$, $A_z(y)$, and $A_y(z)$
Squareness errors: S_{xy}, S_{yz}, and S_{zx}

where D is the linear error, subscript is the error direction, and the position coordinate is inside the parenthesis; A is the angular error, subscript is the axis of rotation, and the position coordinate is inside the parenthesis.

Please note the positioning error caused by an angular error can be expressed as the Abbé offset times the angular errors. For example, positioning error in x-direction can be expressed as $zA_y(x) - yA_z(x)$, where z and y are the Abbé offset in the z- and y-directions, respectively.

8.1.2   Nonrigid-Body Errors

For nonrigid-body errors, they are also a function of the two other coordinate. To simplify the theory, assuming the variations is small and can be approximated by Taylor's expansion with the first-order term as the slope. The nonrigid-body errors become [3]

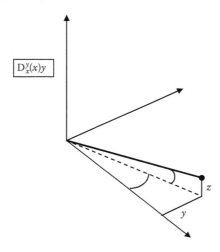

$$D_x(x)(y,\ z) = D_x(x) + D_x^y(x)y + D_x^z(x)z \tag{8.1}$$

$$D_y(x)(y,\ z) = D_y(x) + D_y^y(x)y + D_y^z(x)z \tag{8.2}$$

$$D_z(x)(y,\ z) = D_z(x) + D_z^y(x)y + D_z^z(x)z \tag{8.3}$$

$$D_x(y)(x,\ z) = D_x(y) + D_x^x(y)x + D_x^z(y)z \tag{8.4}$$

$$D_y(y)(x,\ z) = D_y(y) + D_y^x(y)x + D_y^z(y)z \tag{8.5}$$

$$D_z(y)(x,\ z) = D_z(y) + D_z^x(y)x + D_z^z(y)z \tag{8.6}$$

$$D_x(z)(x,\ y) = D_x(z) + D_x^x(z)x + D_x^y(z)y \tag{8.7}$$

$$D_y(z)(x,\ y) = D_y(z) + D_y^x(z)x + D_y^y(z)y \tag{8.8}$$

$$D_z(z)(x,\ y) = D_z(z) + D_z^x(z)x + D_z^y(z)y \tag{8.9}$$

Where D with a superscript is the slope and the superscript is the direction of the slope. There are a total of 27 parameters, 9 are the linear errors and 18 are the slopes of the nonrigid-body, that cause linear errors.

The angular errors are

$$A_x(x)(y,\ z) = A_x(x) + A_x^y(x)y + A_x^z(x)z \tag{8.10}$$

$$A_y(x)(y,\ z) = A_y(x) + A_y^y(x)y + A_y^z(x)z \tag{8.11}$$

$$A_z(x)(y,\ z) = A_z(x) + A_z^y(x)y + A_z^z(x)z \tag{8.12}$$

$$A_x(y)(x,\ z) = A_x(y) + A_x^x(y)x + A_x^z(y)z \tag{8.13}$$

$$A_y(y)(x,\ z) = A_y(y) + A_y^x(y)x + A_y^z(y)z \tag{8.14}$$

$$A_z(y)(x,\ z) = A_z(y) + A_z^x(y)x + A_z^z(y)z \tag{8.15}$$

$$A_x(z)(x,\ y) = A_x(z) + A_x^x(z)x + A_x^y(z)y \tag{8.16}$$

$$A_y(z)(x,\ y) = A_y(z) + A_y^x(z)x + A_y^y(z)y \tag{8.17}$$

$$A_z(z)(x,\ y) = A_z(z) + A_z^x(z)x + A_z^y(z)y \tag{8.18}$$

Where A with a superscript is the slope and the superscript is the direction of the slope. There are a total of 27 parameters: 9 are the angular errors and 18 are the slopes of the nonrigid-body caused angular errors.

For most machine tools, the structures are rather rigid. Hence, the nonrigid-body errors usually are small and negligible. However, for some large gantry type machines, because of the gravity and structure deformation, some nonrigid-body errors may not be negligible. The followings are two special cases as examples:

1. Horizontal milling machine of configuration *XFYZ* (see Section 8.1.3 for definition) with large counter weight along *y*-axis. All the slopes are negligible except the slopes in the *y*-direction.

$$D_x(x)(y, z) = D_x(x) + D_x^y(x)y \tag{8.19}$$

$$D_y(x)(y, z) = D_y(x) + D_y^y(x)y \tag{8.20}$$

$$D_z(x)(y, z) = D_z(x) + D_z^y(x)y \tag{8.21}$$

$$D_x(y)(x, z) = D_x(y) \tag{8.22}$$

$$D_y(y)(x, z) = D_y(y) \tag{8.23}$$

$$D_z(y)(x, z) = D_z(y) \tag{8.24}$$

$$D_x(z)(x, y) = D_x(z) \tag{8.25}$$

$$D_y(z)(x, y) = D_y(z) \tag{8.26}$$

$$D_z(z)(x, y) = D_z(z) \tag{8.27}$$

$$A_x(x)(y, z) = A_x(x) + A_x^y(x)y \tag{8.28}$$

$$A_y(x)(y, z) = A_y(x) + A_y^y(x)y \tag{8.29}$$

$$A_z(x)(y, z) = A_z(x) \tag{8.30}$$

$$A_x(y)(x, z) = A_x(y) \tag{8.31}$$

$$A_y(y)(x, z) = A_y(y) \tag{8.32}$$

$$A_z(y)(x, z) = A_z(y) \tag{8.33}$$

$$A_x(z)(x, y) = A_x(z) \tag{8.34}$$

$$A_y(z)(x, y) = A_y(z) \tag{8.35}$$

$$A_z(z)(x, y) = A_z(z) \tag{8.36}$$

There are five additional nonrigid-body errors. Here, for *XFYZ* configuration, the higher order nonrigid-body errors, $D_x^y(z)y$, $D_y^y(z)y$, $D_z^y(z)y$, $A_z^y(x)y$, $A_x^y(z)y$, $A_y^y(z)y$, and $A_y^y(z)y$, are negligible.

2. Large gantry vertical milling machine of configuration *XYFZ* and *X* » *Y*, *Z*. Here, the unbalanced weight shifting is along the *x*-direction. Hence, all the slopes are negligible except the slopes along the *x*-direction.

$$D_x(x)(y, z) = D_x(x) \tag{8.37}$$

$$D_y(x)(y, z) = D_y(x) \tag{8.38}$$

$$D_z(x)(y, z) = D_z(x) \tag{8.39}$$

$$D_x(y)(x, z) = D_x(y) + D_x^x(y)x \tag{8.40}$$

$$D_y(y)(x,\ z) = D_y(y) + D_y^{\ x}(y)x \tag{8.41}$$

$$D_z(y)(x,\ z) = D_z(y) + D_z^{\ x}(y)x \tag{8.42}$$

$$D_x(z)(x,\ y) = D_x(z) + D_x^{\ x}(z)x \tag{8.43}$$

$$D_y(z)(x,\ y) = D_y(z) + D_y^{\ x}(z)x \tag{8.44}$$

$$D_z(z)(x,\ y) = D_z(z) + D_z^{\ x}(z)x \tag{8.45}$$

$$A_x(x)(y,\ z) = A_x(x) \tag{8.46}$$

$$A_y(x)(y,\ z) = A_y(x) \tag{8.47}$$

$$A_z(x)(y,\ z) = A_z(x) \tag{8.48}$$

$$A_x(y)(x,\ z) = A_x(y) + A_x^{\ x}(y)x \tag{8.49}$$

$$A_y(y)(x,\ z) = A_y(y) + A_y^{\ x}(y)x \tag{8.50}$$

$$A_z(y)(x,\ z) = A_z(y) \tag{8.51}$$

$$A_x(z)(x,\ y) = A_x(z) + A_x^{\ x}(z)x \tag{8.52}$$

$$A_y(z)(x,\ y) = A_y(z) + A_y^{\ x}(z)x \tag{8.53}$$

$$A_z(z)(x,\ y) = A_z(z) \tag{8.54}$$

There are 10 additional nonrigid-body errors. Here, the higher order nonrigid-body errors, $A_z^{\ x}(y)x$ and $A_z^{\ x}(z)x$, are negligible.

8.1.3 MACHINE CONFIGURATIONS AND POSITIONING ERRORS

In most cases, coordinate measuring machines and machine tools can be classified into four configu-rations [4]. They are the *FXYZ*, *XFYZ*, *XYFZ*, and *XYZF* as shown in Figure 8.3a–d, respectively. Here, the axes before *F* show available motion directions of the workpiece with respect to the base, and the letters after *F* show the available motion directions of the tool (or probe) with respect to the base. For example, in *FXYZ*, the workpiece is fixed, and in *XYZF*, the tool is fixed.

8.1.3.1 Position Vector and Rotation Matrix

The vector positions of each stage, *X*, *Y*, and *Z* can be expressed as column vectors [2,4]:

$$X = \begin{bmatrix} x + D_x(x) \\ D_y(x) \\ D_z(x) \end{bmatrix} \tag{8.55}$$

$$Y = \begin{bmatrix} D_x(y) \\ y + D_y(y) \\ D_z(y) \end{bmatrix} \tag{8.56}$$

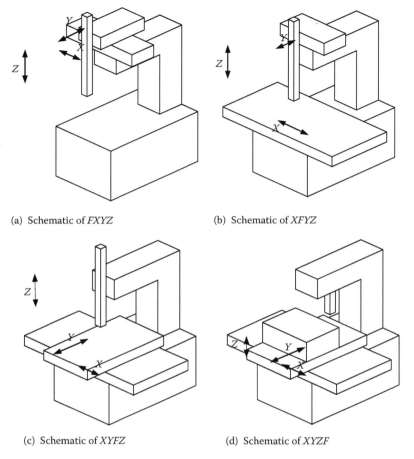

(a) Schematic of *FXYZ*   (b) Schematic of *XFYZ*

(c) Schematic of *XYFZ*   (d) Schematic of *XYZF*

FIGURE 8.3   Schematics of (a) *FXYZ*, (b) *XFYZ*, (c) *XYFZ*, and (d) *XYZF*.

$$Z = \begin{bmatrix} D_x(z) \\ D_y(z) \\ z + D_z(z) \end{bmatrix} \tag{8.57}$$

To simplify the calculation, the squareness errors can be included in the straightness errors by defining the new straightness error as the sum of the old straightness errors and the squareness errors as shown:

$$D_x(y) = D_x(y)(\text{old}) + S_{xy} * y \tag{8.58}$$

$$D_x(z) = D_x(z)(\text{old}) + S_{zx} * z \tag{8.59}$$

$$D_y(z) = D_y(z)(\text{old}) + S_{yz} * z \tag{8.60}$$

The tool offset can be expressed as a column vector:

$$T = \begin{bmatrix} X_t \\ Y_t \\ Z_t \end{bmatrix} \tag{8.61}$$

where X_t, Y_t, and Z_t are the tool offset.

The rotation matrix can be expressed as

$$R(u) = \begin{bmatrix} 1 & A_z(u) & -A_y(u) \\ -A_z(u) & 1 & A_x(u) \\ A_y(u) & -A_x(u) & 1 \end{bmatrix} \tag{8.62}$$

where $u = x$, y, or z.

Please note $A_u(u)$ is much smaller than 1 and also an odd function of u; hence, $\mathbf{R}(u)\mathbf{U} = \mathbf{U}\mathbf{R}(u)$, and $\mathbf{R}(-u) = \mathbf{R}\mathbf{I}(u)$, where $\mathbf{U}$ is a unit matrix and $\mathbf{R}\mathbf{I}$ is the inverse matrix of $\mathbf{R}$.

8.1.3.2 Vectors and Rotation Matrices Calculation

If the positions of the X, Y, and Z stages are represented by the vectors $\mathbf{X}$, $\mathbf{Y}$, and $\mathbf{Z}$, respectively. The angular errors of the X, Y, and Z stages are represented by the rotation matrices $\mathbf{R}(x)$, $\mathbf{R}(y)$, and $\mathbf{R}(z)$. The offset of the tool tip (or probe) are represented by the vector $\mathbf{T}(X_t, Y_t, Z_t)$. The actual positions with respect to the workpiece or machine coordinate can be represented by the vector $\mathbf{P}$. As shown in Ref. [5], the actual position vector $\mathbf{P}$ for the four configurations can be expressed in a machine coordinate as the followings [4]:

$$\text{For } FXYZ, \ \mathbf{P} = X + \mathbf{R}\mathbf{I}(x)\mathbf{Y} + \mathbf{R}\mathbf{I}(x)\mathbf{R}\mathbf{I}(y)\mathbf{Z} + \mathbf{R}\mathbf{I}(x)\mathbf{R}\mathbf{I}(y)\mathbf{R}\mathbf{I}(z)\mathbf{T} \tag{8.63}$$

$$\text{For } XFYZ, \ \mathbf{P} = \mathbf{R}\mathbf{I}(x)\mathbf{X} + \mathbf{R}\mathbf{I}(x)\mathbf{Y} + \mathbf{R}\mathbf{I}(x)\mathbf{R}\mathbf{I}(y)\mathbf{Z} + \mathbf{R}\mathbf{I}(x)\mathbf{R}\mathbf{I}(y)\mathbf{R}\mathbf{I}(z)\mathbf{T} \tag{8.64}$$

$$\text{For } XYFZ, \ \mathbf{P} = \mathbf{R}\mathbf{I}(y)\mathbf{R}\mathbf{I}(x)\mathbf{X} + \mathbf{R}\mathbf{I}(y)\mathbf{Y} + \mathbf{R}\mathbf{I}(y)\mathbf{R}\mathbf{I}(x)\mathbf{Z} + \mathbf{R}\mathbf{I}(x)\mathbf{R}\mathbf{I}(y)\mathbf{R}\mathbf{I}(z)\mathbf{T} \tag{8.65}$$

$$\text{For } XYZF, \ \mathbf{P} = \mathbf{R}\mathbf{I}(z)\mathbf{R}\mathbf{I}(y)\mathbf{R}\mathbf{I}(x)\mathbf{X} + \mathbf{R}\mathbf{I}(z)\mathbf{R}\mathbf{I}(y)\mathbf{Y} + \mathbf{R}\mathbf{I}(z)\mathbf{Z} + \mathbf{R}\mathbf{I}(x)\mathbf{R}\mathbf{I}(y)\mathbf{R}\mathbf{I}(z)\mathbf{T} \tag{8.66}$$

The actual tool tip position can be expressed as a column vector:

$$P = \begin{bmatrix} P_x \\ P_y \\ P_z \end{bmatrix} \tag{8.67}$$

8.1.3.3 Positioning Errors in Four Configurations

Substitute the position vectors, Equations 8.55 through 8.57, and Equations 8.61 and 8.62 into the above equations; we obtain the actual tool tip position, Equation 8.67 in the following [5,6]:

For $FXYZ$ configuration, Equation 8.63 becomes

$$P_x - x = [D_x(x) - y * A_z(x) + z * A_y(x)] + [D_x(y) + z * A_y(y)] + [D_x(z)] \tag{8.68}$$

$$P_y - y = [D_y(x) - z * A_x(x)] + [D_y(y) - z * A_x(y)] + [D_y(z)] \tag{8.69}$$

$$P_z - z = [D_z(x) + y * A_x(x)] + [D_z(y)] + [D_z(z)] \tag{8.70}$$

where $P_x - x$, $P_y - y$, and $P_z - z$ are the positioning errors in the x-, y-, and z-directions, respectively. Additional errors caused by a tool offset of X_t, Y_t, and Z_t are

$$P_{tx} = X_t + [-Y_t * A_z(x) + Z_t * A_y(x)] + [-Y_t * A_z(y) + Z_t * A_y(y)] + [-Y_t * A_z(z) + Z_t * A_y(z)] \tag{8.71}$$

$$P_{ty} = Y_t + [X_t * A_z(x) - Z_t * A_x(x)] + [X_t * A_z(y) - Z_t * A_x(y)] + [X_t * A_z(z) - Z_t * A_x(z)] \tag{8.72}$$

$$P_{tz} = Z_t + [-X_t * A_y(x) + Y_t * A_x(x)] + [-X_t * A_y(y) + Y_t * A_x(y)] + [-X_t * A_y(z) + Y_t * A_x(z)] \quad (8.73)$$

Similarly, for *XFYZ* configuration, Equation 8.64 becomes

$$P_x - x = [D_x(x) - y * A_z(x) + z * A_y(x)] + [D_x(y) + z * A_y(y)] + [D_x(z)] \quad (8.74)$$

$$P_y - y = [D_y(x) + x * A_z(x) - z * A_x(x)] + [D_y(y) - z * A_x(y)] + [D_y(z)] \quad (8.75)$$

$$P_z - z = [D_z(x) - x * A_y(x) + y * A_x(x)] + [D_z(y)] + [D_z(z)] \quad (8.76)$$

The errors caused by a tool offset are the same as in *FXYZ*, Equations 8.71 through 8.73.
Similarly, for *XYFZ* configuration, Equation 8.65 becomes

$$P_x - x = [D_x(x) + z * A_y(x)] + [D_x(y) - y * A_z(y) + z * A_y(y)] + [D_x(z)] \quad (8.77)$$

$$P_y - y = [D_y(x) + x * A_z(x) - z * A_x(x)] + [D_y(y) + x * A_z(y) - z * A_x(y)] + [D_y(z)] \quad (8.78)$$

$$P_z - z = [D_z(x) - x * A_y(x)] + [D_z(y) - x * A_y(y) + y * A_x(y)] + [D_z(z)] \quad (8.79)$$

The errors caused by a tool offset are the same as in *FXYZ*, Equations 8.71 through 8.73.
Finally for *XYZF* configuration, Equation 8.66 becomes,

$$P_x - x = [D_x(x)] + [D_x(y) - y * A_z(y)] + [D_x(z) - y * A_z(z) + z * A_y(z)] \quad (8.80)$$

$$P_y - y = [D_y(x) + x * A_z(x)] + [D_y(y) + x * A_z(y)] + [D_y(z) + x * A_z(z) - zA_x(z)] \quad (8.81)$$

$$P_z - z = [D_z(x) - x * A_y(x)] + [D_z(y) - x * A_y(y) + y * A_x(y)] + [D_z(z) - x * A_y(z) + y * A_x(z)] \quad (8.82)$$

The above results can also be derived by the Stacking model and Abbé offset. The displacement errors caused by the pitch, yaw, and roll angular errors are the Abbé offset times the angular errors. The sign is determined by the right-hand rule.

For the configuration *FXYZ*, x-axis is mounted on a fixed base, y-axis is mounted on the x-axis, and z-axis is mounted on the y-axis. Hence, for x-axis movement, there is no Abbé offset on x and the angular error terms are $y^*A_x(x)$, $y^*A_z(x)$, $-z^*A_x(x)$, and $z^*A_y(x)$; for y-axis movement, there are no Abbé offset on x and y and the angular error terms are $-z^*A_x(y)$ and $z^*A_y(y)$; for z-axis movement, there are no Abbé offsets on x, y, and z and there is no angular error term. The results are the same as Equations 8.68 through 8.70.

Similarly, for the configuration *XFYZ*, x-axis is mounted on a fixed base, y-axis is mounted on the x-axis, and z-axis is mounted on the y-axis. Hence, for x-axis movement, there are all three Abbé offsets and the angular error terms are $-x^*A_y(x)$, $x^*A_z(x)$, $y^*A_x(x)$, $-y^*A_z(x)$, $-z^*A_x(x)$, and $z^*A_y(x)$; for y-axis movement, there are no Abbé offsets on x and y and the angular error terms are $-z^*A_x(y)$ and $z^*A_y(y)$; for z-axis movement, there are no Abbé offsets on x, y, and z and there is no angular error term. The results are the same as Equations 8.74 through 8.76.

Similarly, for the configuration *XYFZ*, x-axis is mounted on a fixed base, y-axis is mounted on the x-axis, and z-axis is mounted on a fixed base. Hence, for x-axis movement, there is no Abbé offset on y and the angular error terms are $-x^*A_y(x)$, $x^*A_z(x)$, $-z^*A_x(x)$, and $z^*A_y(x)$; for y-axis movement, there are all three Abbé offsets, and the angular terms are $x^*A_y(y)$, $x^*A_z(y)$, $y^*A_x(y)$, $-y^*A_z(y)$, $-z^*A_x(y)$, and $z^*A_y(y)$; for z-axis movement, there is no Abbé offset on x, y, and z and no angular term. The results are the same as Equations 8.77 through 8.79.

Finally for the configuration *XYZF*, x-axis is mounted on a fixed base, y-axis is mounted on the x-axis, and z-axis is mounted on the y-axis and the spindle is fixed. Hence, for x-axis movement, there are no Abbé offsets on x and y and the angular error terms are $-z^* A_x(x)$ and $z^* A_y(x)$; for y-axis movement, there is no Abbé offset on z, and the angular error terms are $-x^* A_y(y)$, $x^* A_z(y)$, $y^* A_x(y)$, $-y^* A_z(y)$; for z-axis movement, there are all three Abbé offsets and the angular error terms are $-x^* A_y(z)$, $x^* A_z(z)$, $y^* A_x(z)$, $-y^* A_z(z)$, $-z^* A_x(z)$ and $z^* A_y(z)$. The results are the same as Equations 8.80 through 8.82.

8.2 POSITIONING ERROR COMPENSATION MODELING

The sum of all errors in the x-, y-, and z-directions are E_x, E_y, and E_z, respectively.

$$E_x(x,y,z) = D_x(x) + D_x(y) + D_x(z) + D_x^x(y)x - (z + Z_t) * A_y(x) + Y_t * A$$
$$-(z + Z_t) * [A_y(y) + A_y^x(y)x] + Y_t * A_z(y) - Z_t * A_y(z) + Y_t * A_z(z) \tag{8.83}$$

$$E_y(x,y,z) = D_y(y) + D_y^x(y)x + D_y(z) + D_y(x) - (x + X_t) * A_z(y) + (z + Z_t)$$
$$* [A_x(y) + A_x^x(y)x] - X_t * A_z(z) + Z_t * A_x(z) - X_t * A_z(x) + (z + Z_t) * A_x(x) \tag{8.84}$$

$$E_z(x,y,z) = D_z(z) + D_z(x) + D_z(y) + D_z^x(y)x - Y_t * A_x(z) + X_t * A_y(z) - Y_t$$
$$* A_x(x) + X_t * A_y(x) - Y_t * [A_x(y) + A_x^x(y)x] + (x + X_t) * [A_y(y) + A_y^x(y)x] \tag{8.85}$$

For the case the reference point is the tool tip, then $X_t = Y_t = Z_t = 0$. Hence, the sums of errors, Equations 8.83 through 8.85, reduce to the followings:

$$E_x(x,y,z) = D_x(x) + D_x(y) + D_x(z) + D_x^x(y)x - z * A_y(x) - z * [A_y(y) + A_y^x(y)x] \tag{8.86}$$

$$E_y(x,y,z) = D_y(y) + D_y^x(y)x + D_y(z) + D_y(x) - x * A_z(y) + z * [A_x(y) + A_x^x(y)x] + z * A_x(x) \tag{8.87}$$

$$E_z(x,y,z) = D_z(z) + D_z(x) + D_z(y) + D_z^x(y)x + A_x^x(y)x + x * [A_y(y) + A_y^x(y)x] \tag{8.88}$$

8.2.1 DISPLACEMENT ERROR COMPENSATION

Most machine controllers can provide compensation for repeatable leadscrew or encoder errors on each axis of motion. Usually this is called pitch error compensation.

The errors in the x-, y-, and z-directions can be expressed as

$$E_x(x) = D_x(x) \tag{8.89}$$

$$E_y(y) = D_y(y) \tag{8.90}$$

$$E_z(z) = D_z(z) \tag{8.91}$$

8.2.2 SQUARENESS AND STRAIGHTNESS ERROR COMPENSATION

Many machines with advanced controllers can provide compensation for repeatable displacement errors (leadscrew or encoder errors), vertical and horizontal straightness errors (guide way flatness error), and squareness errors. The errors in the x-, y-, and z-directions can be expressed as

$$E_x(x,y,z) = D_x(x) + D_x(y) + D_x(z) \tag{8.92}$$

$$E_y(x,y,z) = D_y(x) + D_y(y) + D_y(z) \tag{8.93}$$

$$E_z(x,y,z) = D_z(x) + D_z(y) + D_z(z) \tag{8.94}$$

8.2.3 ANGULAR ERROR COMPENSATION

Most machine controllers do not have the capability of compensate angular error and usually the angular error times the Abbé offset is included in the measured straightness errors. However, many times the tool offsets and the effect of angular errors are different. The errors in the x-, y-, and z-directions can be expressed as

$$E_x(x,y,z) = D_x(x) + D_x(y) + D_x(z) + (z+Z_t)*A_y(x) + Y_t*A_z(x)$$
$$-(z+Z_t)*A_y(y) + Y_t*A_z(y) - Z_t*A_y(z) + Y_t*A_z(z)$$
(8.95)

$$E_y(x,y,z) = D_y(x) + D_y(y) + D_y(z) - (x+X_t)*A_z(y) + (z+Z_t)*A_x(y)$$
$$-X_t*A_z(z) + Z_t*A_x(z) - X_t*A_z(x) + (z+Z_t)*A_x(x)$$
(8.96)

$$E_z(x,y,z) = D_z(x) + D_z(y) + D_z(z) + D_z^x(y)x - Y_t*A_x(z) + X_t*A_y(z)$$
$$-Y_t*A_x(x) + X_t*A_y(x) - Y_t*A_x(y) + (x+X_t)*A_y(y)$$
(8.97)

8.2.4 NONRIGID-BODY ERROR COMPENSATION

Same as the angular error compensation, the nonrigid-body errors such as weight shift errors and counter weight errors are included in the measured straightness errors. The errors in the x-, y-, and z-directions can be expressed as

$$E_x(x,y,z) = D_x(x) + D_x(y) + D_x(z) + D_x^x(y)x - (z+Z_t)*A_y(x) + Y_t*A_z(x)$$
$$-(z+Z_t)*[A_y(y) + A_y^x(y)x] + Y_t*A_z(y) - Z_t*A_y(z) + Y_t*A_z(z)$$
(8.98)

$$E_y(x,y,z) = D_y(y) + D_y^x(y)x + D_y(z) + D_y(x) - (x+X_t)*A_z(y) + (z+Z_t)$$
$$*[A_x(y) + A_x^x(y)x] - X_t*A_z(z) + Z_t*A_x(z) - X_t*A_z(x) + (z+Z_t)*A_x(x)$$
(8.99)

$$E_z(x,y,z) = D_z(z) + D_z(x) + D_z(y) + D_z^x(y)x - Y_t*A_x(z) + X_t*A_y(z) - Y_t*A_x(x)$$
$$+X_t*A_y(x) - Y_t*[A_x(y) + A_x^x(y)x] + (x+X_t)*[A_y(y) + A_y^x(y)x]$$
(8.100)

8.2.5 3D GRID POINT ERROR COMPENSATION

In many advanced controllers, the nonrigid-body repeatable errors can be compensated by a 3D grid point error map. In such 3D error compensation, the error compensation for an arbitrary interior point P, shown in Figure 8.4, is interpolated by the surrounding eight error compensation grid points. The error values of these eight grid points are measured and input to the control. The following equation is used to calculate the error compensation [7]:

For error compensation in x-axis direction for point P

$$Cx = C1x \cdot (1-x)(1-y)(1-z) + C2x \cdot x(1-y)(1-z)$$
$$+ C3x \cdot xy(1-z) + C4x \cdot (1-x)y(1-z)$$
$$+ C5x \cdot (1-x)(1-y)z + C6x \cdot x(1-y)z$$
$$+ C7x \cdot xyz + C8x \cdot (1-x)yz$$
(8.101)

where, Cnx is the measured error value for X-axis at grid points ($n = 1, 2, …, 8$)

$$x = \frac{|Px - P1x|}{|P2x - P1x|}$$
(8.102)

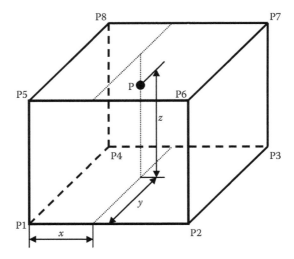

FIGURE 8.4   Schematic drawing showing the 3D grid point P.

$$y = \frac{|Py - P1y|}{|P2y - P1y|} \tag{8.103}$$

$$z = \frac{|Pz - P1z|}{|P2z - P1z|} \tag{8.104}$$

Error calculation (Equation 8.101) reflects the influence of the surrounding grid points. Similarly, for error compensation in y- and z-directions, the interpolation formulae are the same but errors are in y- and z-directions.

8.2.6   THERMAL EXPANSION AND DISTORTION COMPENSATION

The thermal behavior of a machine tool is one of the major factors influencing the final workpiece accuracy. In relation with the currently increasing power output of the machine spindles and the rising dynamics of all driven movements, the influence of the machine thermal state is continuously rising. The prediction of thermal deformations is hardly executable in the design phase of a new machine tool model. Theoretical work in the field of modeling of temperature distribution within the machine frame has not been satisfactorily concluded [8].

The causes of thermal deviations may be divided into two basic categories. The first part is from the thermal loadings resulting from the machine tool operation. The most significant sources of heat in a CNC machine tool are the spindle, ballscrew alternatively linear motor, and heat coming from the cutting process. The second part is represented by deviations raised from the thermal deformations of the machine frame caused by external influences—mainly the environmental temperature in the shop floor, temperature variations, airflow, direct sunshine, etc.

Currently machine tool builders are striving to increase machine tool accuracy and productivity by applying various methods covering machine frame design optimization, assembly work improvement, introduction of several cooling systems, etc. To further improve the machine positioning accuracy, an intelligent controller can be used to compensate these errors, provided that the 3D volumetric positioning errors and the machine temperature distributions can be measured. Furthermore, the measurement has to be performed in a short time such that the machine thermal state remains constant.

In a real machine shop environment, under various spindle loads and feed rates, the machine thermal expansion may cause large 3D volumetric positioning errors. Using the measured position

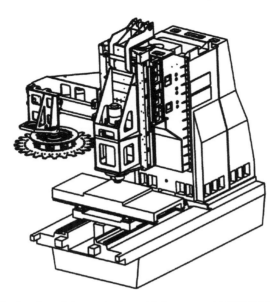

FIGURE 8.5   A schematic drawing of the vertical machining center MCFV5050LN.

errors, several error maps could be generated. Compensation tables at an actual thermal state can be interpolated to achieve higher accuracy at various thermal loadings.

Since the 3D volumetric position errors $E_x(x, y, z)$, $E_y(x, y, z)$, and $E_z(s, y, z)$ can be measured by the laser vector technique (Section 3.2.2) in less than 1 h, we can assume the machine temperature is constant during the measurement. The measured position errors at two different temperatures, T_m and T_n, can be expressed as $E_x(x, y, z, T_m)$, $E_y(x, y, z, T_m)$, and $E_z(x, y, z, T_m)$, and $E_x(x, y, z, T_n)$, $E_y(x, y, z, T_n)$, and $E_z(x, y, z, T_n)$. Assuming the machine errors are linear between T_m and T_n, for a temperature T_u, where $T_m > T_u > T_n$, the position errors at T_u can be interpolated as the follows:

$$E_x(x,y,z,T_u) = [E_x(x,y,z,T_m) - E_x(x,y,z,T_n)] * (T_u - T_n)/(T_m - T_n) + E_x(x,y,z,T_n) \quad (8.105)$$

$$E_y(x,y,z,T_u) = [E_y(x,y,z,T_m) - E_y(x,y,z,T_n)] * (T_u - T_n)/(T_m - T_n) + E_y(x,y,z,T_n) \quad (8.106)$$

$$E_z(x,y,z,T_u) = [E_z(x,y,z,T_m) - E_z(x,y,z,T_n)] * (T_u - T_n)/(T_m - T_n) + E_z(x,y,z,T_w) \quad (8.107)$$

To cover a large operational temperature range, we may need to measure the errors at several thermal states.

To demonstrate this, considerable work has been performed by Svoboda et al. [8,9] by measuring the 3D volumetric positioning errors and machine temperature distributions at various spindle rpm, feed rates, and ambient temperatures in a machining center of *XYFZ* configuration as shown in Figure 8.5. Some of the results are shown below.

8.2.7 Temperature History

The temperature data is displayed in Figure 8.6 for the sensor located on the spindle, z-column, x-middle, and y-front at six measurement runs. It is clear that the main heating occurs in parts close to the spindle. The temperatures were continuously increased due to the spindle heating and rapid xyz-axes motion. These temperature changes caused different thermal deformations of the z-column, and the xy-bed yielding into the measured variations of the 3D volumetric positioning accuracy.

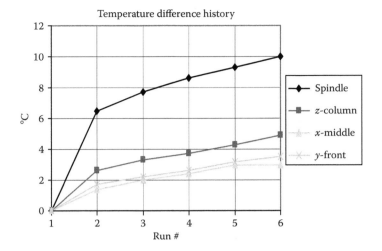

FIGURE 8.6 Measured temperature history at four locations.

8.2.8 STRAIGHTNESS ERRORS

Straightness errors were measured for all x-, y-, and z-axes and the changes were relatively small over the temperature range. Figure 8.7 shows the x-axis straightness errors $D_y(x)$ under various thermal states Run #1 to Run #6. The maximums varied between −0.010 and −0.007 mm. Figure 8.8 shows the y-axis straightness errors $D_x(y)$ under the same various thermal states. The maximums varied between −0.012 and −0.003 mm. Figure 8.9 shows the z-axis straightness errors $D_x(z)$ under the same various thermal states. The maximums varied between 0.009 and 0.003 mm. Figure 8.10 shown the z-axis straightness errors $D_y(z)$ under the same various thermal states. The maximums varied between 0.017 and 0.013 mm.

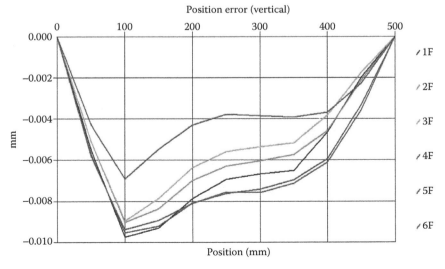

FIGURE 8.7 (See color insert following page 174.) Measured x-axis straightness errors $D_y(x)$ at six different thermal conditions, Run #1 to Run #6.

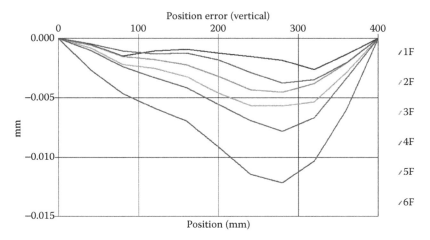

FIGURE 8.8 (See color insert following page 174.)   Measured y-axis straightness errors $D_x(y)$ at six different thermal conditions, Run #1 to Run #6.

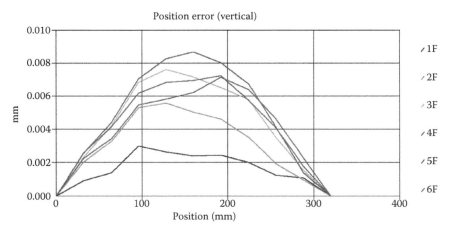

FIGURE 8.9 (See color insert following page 174.)   Measured z-axis straightness errors $D_x(z)$ at six different thermal conditions, Run #1 to Run #6.

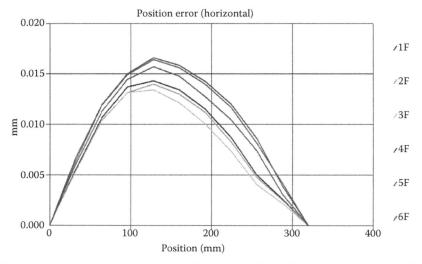

FIGURE 8.10   Measured z-axis straightness errors $D_y(z)$ at six different thermal conditions, Run #1 to Run #6.

8.2.9 Temperature Correlation and Linear Interpolation

It is noted that the maximum errors between Run #6 and Run #1 were not linear. Hence, at least three error maps, such as at Run #6, Run #3, and Run #1, are needed to get a better interpolation over the temperature range. A new error map can be generated at any thermal state by linear interpolation between two maps. Based on the correlation calculation, for linear displacement errors, use the temperature measured at x-middle (same as the y-front) for $D_x(x)$, and $D_y(y)$; use the temperature at z-column for $D_z(z)$. For squareness errors, use the temperature measured at x-middle for xy-plane; use the temperature at z-column for yz-plane and zx-plane. For straightness errors, use the temperature measured at x-middle for $D_y(x)$, $D_z(x)$, $D_x(y)$, and $D_z(y)$; and the temperature at z-column for $D_x(z)$ and $D_y(z)$. Using three error maps and the correlated temperatures for linear interpolation, the position errors can be reduced considerably.

It is concluded that large machine temperature changes caused somewhat small straightness error changes but large squareness error changes. Using the measured position errors, several error maps could be generated. Compensation tables at an actual thermal state can be interpolated to achieve higher accuracy at various thermal loadings.

8.3 POSITIONING ERROR MEASUREMENT USING LASER INTERFEROMETERS

Laser interferometers are becoming more popular and widely used in the machine tool industries for the calibration, compensation, and certifying machine tool positioning accuracy. Most of the laser interferometers are based on Michelson interferometer. Briefly, a laser beam is split into two beams by a beam splitter. One beam is reflected back from a fixed reference retroreflector and the other is reflected back from a target retroreflector attached to the machine's moving element. The two reflected beams are recombined by another beam splitter in front of a detector. The interference of these two beams generated a fringe pattern. When the target moves, the corresponding fringe pattern also moves. For each passing fringe, the detector will measure a cycle of high and low intensities. This one cycle corresponds to one count which is equal to a displacement of one half-wavelength. Hence, the total distance moved is equal to the half-wavelength times the number of counts.

A typical laser interferometer consists of a laser beam, remote interferometer, target retroreflector, photodetector, and electronics. For a conventional laser interferometer, the exit laser beam and the return laser beam are parallel but displaced by about 1 in., as shown in Figure 8.11. The laser Doppler displacement meter (LDDM) is a two-frequency AC interferometer [10]. It uses a single-aperture optical arrangement, the output laser beam, and the return laser beam passing through the same aperture as shown in Figure 8.11. Hence, a small retroreflector or a flat mirror can be used as target. Therefore, the laser system becomes very compact and versatile.

8.3.1 Direct Measurement of Positioning Errors

Using a conventional laser interferometer, the linear displacement errors and angular errors can easily be measured. However, the straightness error and the squareness errors are very difficult to measure. This is because very complex and expensive optics, such as Wollaston prism, are used. With the complex optics, it is very difficult to set up and align. It usually takes days of machine downtime and experienced operator to perform these measurements. Hence, the measurement of 21 rigid-body errors is very difficult and time consuming.

8.3.2 Indirect Measurement of Positioning Errors

Direct measurement means each measurement is independent and the difference between measured position and the targeted position is the positioning error. Indirect measurement means several

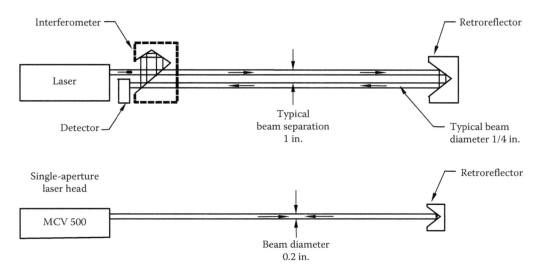

FIGURE 8.11 A comparison of a laser interferometer and a single-aperture laser Doppler system.

measurements are required to determine the final positioning errors, and the measurement error is limited by the repeatability of the machine. Based on this concept and also the concept that the measurement direction is not parallel to the movement direction, Optodyne has developed a laser vector technique for the measurement of 3D volumetric positioning errors, including three linear displacement errors, six straightness errors, and three squareness errors in a very short time [13,14].

8.3.2.1 Body Diagonal Displacement Measurement

Using a conventional laser interferometer to measure the straightness and squareness errors is rather difficult and costly. It usually takes days of machine downtime and experienced operator to perform these measurements. For these reasons, the body diagonal displacement error defined in the ASME B5.54 or ISO 230-6 standard is a good quick check of the volumetric error [11,12]. Furthermore, it has been used by Boeing Aircraft Company and many others for many years with very good results and success.

Briefly, similar to a laser linear displacement measurement, instead of pointing the laser beam in the axis direction, it points the laser beam in the body diagonal direction as shown in Figure 8.12. Mount a retroreflector on the spindle and move the spindle in the body diagonal direction. Starting from the zero position and at each increment of the three axes, which are moved together to reach the new position along the diagonal, the displacement error is measured. There are four body diagonal directions as shown in Figure 8.13. The accuracy of each position along the diagonal depends on the positioning accuracy of the three axes, including the straightness errors, angular errors, and squareness errors. Hence, the four body diagonal displacement measurements are a good method for the machine verification.

The relations between the measured four body diagonal displacement errors and the 21 rigid-body errors can be derived by the formulae in Section 1.3.3. For the *FXYZ*, the measured error DR at each increment can be expressed as [5,6]

$$
\begin{aligned}
DR_{ppp} &= a/r * D_x(x) + b/r * D_y(x) + c/r * D_z(x) + a/r * [D_x(y) + y S_{xy}] \\
&+ b/r * D_y(y) + c/r * D_z(y) + a/r * [D_x(z) + z S_{zx}] \\
&+ b/r * [D_y(z) + z S_{yz}] + c/r * D_z(z) + A_y(x) * ac/r \\
&- A_z(x) * ab/r + A_y(y) * ac/r - A_x(y) * bc/r
\end{aligned}
\tag{8.108}
$$

Body diagonal measurement

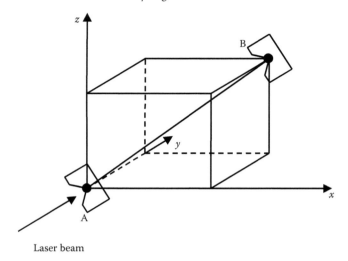

FIGURE 8.12 The body diagonal displacement measurement.

$$
\begin{aligned}
\text{DR}_{npp} = {}&-a/r * D_x(x) + b/r * D_y(x) + c/r * D_z(x) - a/r * [D_x(y) + y\, S_{xy}]\\
&+ b/r * D_y(y) + c/r * D_z(y) - a/r * [D_x(z) + z\, S_{zx}]\\
&+ b/r * [D_y(z) + z\, S_{yz}] + c/r * D_z(z) - A_y(x) * ac/r\\
&+ A_z(x) * ab/r - A_y(y) * ac/r - A_x(y) * bc/r
\end{aligned}
\tag{8.109}
$$

$$
\begin{aligned}
\text{DR}_{pnp} = {}&a/r * D_x(x) - b/r * D_y(x) + c/r * D_z(x) + a/r * [D_x(y) + y\, S_{xy}]\\
&- b/r * D_y(y) + c/r * D_z(y) + a/r * [D_x(z) + z\, S_{zx}]\\
&- b/r * [D_y(z) + z\, S_{yz}] + c/r * D_z(z) + A_y(x) * ac/r\\
&+ A_z(x) * ab/r + A_y(y) * ac/r + A_x(y) * bc/r
\end{aligned}
\tag{8.110}
$$

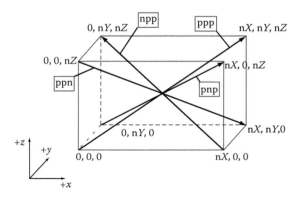

FIGURE 8.13 (See color insert following page 174.) Four body diagonal directions.

$$\begin{aligned}
\mathrm{DR}_{\mathrm{ppn}} = {}& a/r * D_x(x) + b/r * D_y(x) - c/r * D_z(x) + a/r * [D_x(y) + y\, S_{xy}] \\
& + b/r * D_y(y) - c/r * D_z(y) + a/r * [D_x(z) + z\, S_{zx}] \\
& + b/r * [D_y(z) + z\, S_{yz}] - c/r * D_z(z) - A_y(x) * ac/r \\
& - A_z(x) * ab/r - A_y(y) * ac/r + A_x(y) * bc/r
\end{aligned}$$

(8.111)

where the subscript ppp means body diagonal with all x, y, and z positive; npp means body diagonal with x negative, y and z positive; pnp means body diagonal with y negative, x and z positive; and ppn means body diagonal with z negative, x and y positive. Also a, b, c, and r are increments in x, y, z, and body diagonal directions, respectively. The body diagonal distance can be expressed as $r^2 = a^2 + b^2 + c^2$.

In the *FXYZ* configuration shown in Equations 8.108 through 8.111, there are four angular error terms, $A_y(x)^*ac/r$, $-A_z(x)^*ab/r$, $A_y(y)^*ac/r$, and $-A_x(y)^*bc/r$. In the *XFYZ* configuration, most of the angular error terms are cancelled and only two angular error terms, $A_y(y)^*ac/r$ and $-A_x(y)^*bc/r$, are left. Similarly, in the *XYFZ* configuration, only two angular error terms, $A_z(x)^*ab/r$ and $-A_x(x)^*bc/r$, are left. Finally, in the *XYZF* configuration, there are four angular error terms, $A_y(x)^*ac/r$, $-A_z(x)^*ab/r$, $A_y(y)^*ac/r$, and $-A_x(y)^*bc/r$ exactly the same as in the *FXYZ* configuration. Since the configurations for most common horizontal machining centers and vertical machining centers are *XFYZ* and *XYFZ*, respectively, we can conclude that the body diagonal displacement measurement is not sensitive to angular errors.

It is noted that if the four body diagonal displacement errors are small, then the machine errors are most likely very small. If the four body diagonal displacement errors are large, then the machine errors are large. However, because there are only four sets of data and there are nine sets of errors, we do not have enough information to determine which errors are large. In order to determine where the large errors are, the sequential step diagonal measurement or laser vector technique [13,14] has been developed by Optodyne to collect 12 sets of data with the same four diagonal setups. Based on these data, all three displacement errors, six straightness errors, and three squareness errors can be determined. Furthermore, the measured positioning errors can also be used to generate a 3D volumetric compensation table to correct the positioning errors and achieve higher positioning accuracy. Hence, 3D volumetric positioning errors can be measured without incurring high costs and long machine tool downtime.

The four body diagonal displacement errors shown in Equations 8.108 through 8.111 are sensitive to all of the nine linear errors and some angular errors. Hence, it is a good measurement of the 3D volumetric positioning errors. The errors in the above equations may be positive or negative and they may cancel each other. However, the errors are statistical in nature, the probability that all of the errors will be cancelled in all of the positions and in all of the four body diagonals are theoretically possible but very unlikely. Hence, it is indeed a quick measurement of volumetric positioning accuracy.

8.3.2.2 Vector or Sequential Step Diagonal Displacement Measurement

To overcome the limitations in the four body diagonal displacement measurement, a sequential step diagonal or vector technique [13–16] has been developed by Optodyne. The basic concept of the vector method is that the laser beam direction (or the measurement direction) is not parallel to the motion of the linear axis. Hence, the measured displacement errors are sensitive to the errors both parallel and perpendicular to the direction of the linear axis. More precisely, the measured linear errors are the vector sum of errors, namely, the displacement errors (parallel to the linear axis), the vertical straightness errors (perpendicular to the linear axis), and horizontal straightness errors (perpendicular to the linear axis and the vertical straightness error direction), projected to the direction of the laser beam. Furthermore, collect data with the laser beam pointing in three different diagonal directions; all nine error components can be determined. Since the errors of each axis of motion are the vector sum of the three perpendicular error components, we call this measurement a "vector" method.

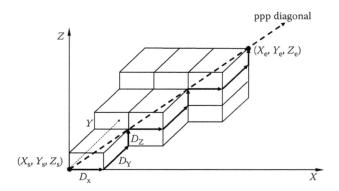

FIGURE 8.14 Vector measurement trajectory, the laser is pointing in the ppp diagonal direction. Move D_x, stop, collect data; move D_y, stop; and move D_z, stop, collect data, and so on.

For conventional body diagonal displacement measurement, all three axes move simultaneously, the displacement is a straight line along the body diagonal; hence, a laser interferometer can be used to do the measurement. However, for the vector measurement described here, the displacements are along the x-axis, y-axis, and z-axis. The trajectory of the target or the retroreflector is not parallel to the diagonal direction as shown in Figure 8.14. The deviations from the body diagonal are proportional to the size of the increment X, Y, or Z. A conventional laser interferometer will be a way-out of alignment even with an increment of a few millimeters.

To tolerate such large lateral deviations, an LDDM using a single-aperture laser head and a flat mirror as the target can be used [10]. This is because any lateral movement or movement perpendicular to the normal direction of the flat mirror will not displace the laser beam. Hence, the alignment is maintained. After three movements, the flat-mirror target will move back to the center of the diagonal again; hence, the size of the flat mirror has only to be larger than the largest increment. A schematic showing the flat-mirror positions during the measurement steps is shown in Figure 8.15. Here, the flat-mirror target is mounted on the machine spindle and it is perpendicular to the laser beam direction.

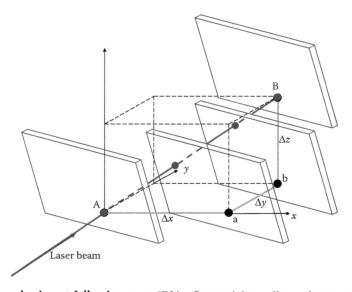

FIGURE 8.15 (See color insert following page 174.)   Sequential step diagonal or vector technique.

For this reason, three times more data is collected and also the positioning error due to each single axis movement can be separated. The collected data can be processed as the projection of the displacement of each single axis along the diagonal. In summary, in a conventional body diagonal measurement all three axes move simultaneously along a body diagonal and collect data at each preset increment. In the vector measurement all three axes move in sequence along a body diagonal and collect data after each axis is moved. Hence, not only three times more data are collected, the error due to the movement of each axis can also be separated.

In practice, first, point the laser beam in one of the body diagonal directions, similar to the body diagonal displacement measurement in the ASME B5.54 standard. However, instead of programming the machine to move, x, y, and z continuously to the next increment, stop and take a measurement, the machine is now programmed to, move the x-axis, stop and take a measurement, then move the y-axis, stop and take a measurement, then move the z-axis, stop and take a measurement. A typical setup on a CNC machining center is shown in Figure 8.16.

As compared to the conventional body diagonal measurement where only one data point is collected at each increment in the diagonal direction, the vector measurement collects three data points, one after x-axis movement, one after y-axis movement, and one after z-axis movement. Hence, three times more data is collected.

Second, point the laser beam in another body diagonal direction and repeat the same until all four body diagonals are measured. Since each body diagonal measurement collected three sets of data, there are 12 sets of data. Hence, there are enough data to solve the three displacement errors, six straightness errors, and the three squareness errors. The setup is simple and easy and the measurement can be performed in a few hours instead of a few days using a conventional laser interferometer.

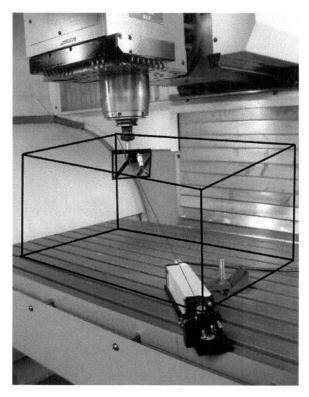

FIGURE 8.16 (See color insert following page 174.) A photo of actual laser setup for the vector measurement.

8.4 APPLICATIONS

Many machine tool controllers can provide volumetric positioning error compensations for repeatable linear position and straightness errors on each linear axis of motion. For most controllers, there are compensations for linear errors (or pitch errors) and straightness error (or cross errors, droop errors, sag errors, nonlinear errors).

8.4.1 SIEMENS CONTROLLER—SAG COMPENSATION

The Siemens 840 controllers has 18 tables for linear and sag compensations. For each axis, there are displacement error forward, displacement error backward, horizontal straightness forward, horizontal straightness backward, vertical straightness forward, and horizontal straightness backward, a total of six tables for each axis and a total of 18 tables for three axes. A sample compensation file is shown in Box 8.1.

8.4.2 FANUC CONTROLLER—PITCH ERRORS AND STRAIGHTNESS ERRORS COMPENSATION

The Fanuc 30i/31i/32i controller with 128-point option has nine compensation tables. The first three tables are for displacement errors (or pitch errors), for x-axis (DXX), y-axis (DYY), and z-axis (DZZ). The next six tables are for straightness errors, moving axis = X, compensation axis = $Y(DXY)$; moving axis = Y, compensation axis = $X(DYX)$; moving axis = Z, compensation axis = $X(DZX)$; and moving axis = Z, compensation axis = $Y(DZY)$, moving axis = X, compensation axis = $Z(DXZ)$; and moving axis = Y, compensation axis = $Z(DYZ)$. For all nine tables, the unit, the comp unit, the comp algorithm, comp digits, and travel direction should all be the same. The increment and the reference should be the same for x-, y-, and z-axes. A typical compensation file is shown in Box 8.2.

8.4.3 HEIDENHAIN CONTROLLER—NONLINEAR COMPENSATION

The Heidenhain controller can compensate linear pitch error and volumetric positioning error (called by Heidenhain as nonlinear error compensation). The volumetric compensation has three tables for linear displacement error (pitch error) compensation and six tables for the straightness error (nonlinear error) compensation. These are $Dx(X)$, $Dy(X)$, $Dz(X)$, $Dx(Y)$, $Dy(Y)$, $Dz(Y)$, $Dx(Z)$, $Dy(Z)$, and $Dz(Z)$. A configuration file with the same name but with an extension. CMA will be generated. A typical configuration file and comp file are shown in Box 8.3.

8.4.4 MDSI CONTROLLER—POSITION COMPENSATION

MDSI controller is a software-based open system CNC machine tool controller that meets the CNC machine control needs of small and large manufacturers in all industries. It does not use proprietary hardware and possesses the standards required for true open-architecture controls as established by OMAC (open modular architecture controls) and OSACA (open system architecture for controls within automation systems). It allows the integration of commercial off-the-shelf hardware and software components. It was built with components found in the open market, clearly defined and published in form of specification. An open application programming interface (API) is available in the MDSI open controller for customer to integrate third party application. The controller is user installable, configurable, and maintainable.

MDSI has been designed to be programmable at low level to achieve tool and fixtures offset compensation and leadscrew error compensation (LSEC). LSEC feature allows precise measurements along each axis. The deviation from the expected value due to irregularities in the leadscrew can then be compensated for. This could be done by entering variables that correspond to the measured deviations into the tune file that tell OpenCNC how to compensate the errors.

BOX 8.1   SIEMENS 840D
COMPENSATION FILE FORMAT

```
%_N_NC_CECfINI
CHANDATA(1)
$AN_CEC[0,0]=0.0000
$AN_CEC[0,1]=+0.0001
$AN_CEC[0,2]=+0.0000
$AN_CEC[0,3]=+0.0005
$AN_CEC[0,4]=+0.0024
$AN_CEC[0,5]=+0.0036
.............................................. . .
.............................................. . .
............................................. .
$AN_CEC[0,37]=-0.0019
$AN_CEC[0,38]=-0.0016
$AN_CEC[0,39]=-0.0025
$AN_CEC[0,40]=-0.0037
$AN_CEC_INPUT_AXIS[0]=(AX1)
$AN_CEC_OUTPUT_AXIS[0]=(AX1)
$AN_CEC_STEP[0]=+50.0000
$AN_CEC_MIN[0]=-2100.0000
$AN_CEC_MAX[0]=-100.0000
$AN_CEC_DIRECTION[0]=1
$AN_CEC_MULT_BY_TABLE[0]=0
$AN_CEC_IS_MODULO[0]=0
$AN_CEC[17,0]=+0.0001
$AN_CEC[17,1]=+0.0005
$AN_CEC[17,2]=+0.0007
$AN_CEC[17,3]=-0.0015
$AN_CEC[17,4]=-0.0010
..............................................................
.......................................................... . .
.......................................................... . .
$AN_CEC[23,40]=-0.0085
$AN_CEC_INPUT_AXIS[23]=(AX3)
$AN_CEC_OUTPUT_AXIS[23]=(AX10)
$AN_CEC_STEP[23]=+25.0000
$AN_CEC_MIN[23]=-1125.0000
$AN_CEC_MAX[23]=-125.0000
$AN_CEC_DIRECTION[23]=-1
$AN_CEC_MULT_BY_TABLE[23]=0
$AN_CEC_IS_MODULO[23]=0
M23
```

8.4.4.1   Off-line Error Compensation in MDSI

An understanding of the LSEC system can be applied in the MDSI at higher level to achieve geometric error compensation. A sample implementation of the geometric error compensation system using the LSEC approach is discussed here. Compensation data required for the LSEC system is saved in the parameter file called "Tune file" which is read by OpenCNC when the controller starts up. This file has specific format comprising variables names with values as shown in Box 8.4.

BOX 8.2   FANUC 15
COMPENSATION FILE FORMAT

```
%
N3620Q1A1P6A2P100A3P200
N3621Q1A1P0A2P100A3P200
N3622Q1A1P6A2P106A3P206
N3623Q1A1P1A2P1A3P1
N3624Q1A1P6000A2P3000A3P3000
N5711Q1P1
N5712Q1P2
N5713Q1P2
N5714Q1P3
N5715Q1P3
N5716Q1P1
N5721Q1P2
N5722Q1P1
N5723Q1P3
N5724Q1P1
N5725Q1P2
N5726Q1P3
N13381Q1P300
N13382Q1P400
N13383Q1P500
N13384Q1P600
N13385Q1P700
N13386Q1P800
N13391Q1P1
N13392Q1P1
N13393Q1P1
N13394Q1P1
N13395Q1P1
N13396Q1P1
%
Comp Value
%
N10000P1
N10001P0
N10002P0
N10003P0
N10004P1
N10005P0
N10006P0
N10100P0
N10101P0
N10102P0
N10103P1
..........
..........
..........
N10805P0
N10806P0
%
```

BOX 8.3   HEIDENHAIN COMPENSATION FILE FORMAT

```
BEGIN       CompHeidenhainConfig.CMA     ACT:0
NR          1                            2                   3
0           CompHeidenhainX              CompHeidenhainY     CompHeidenhainZ
[END]
BEGIN       CompHeidenhainX.com          DATUM: +0.0         DIST: 13
NR          1=f()
0           +0.0                         +0
1           +0.8192                      -0.00027
2           +1.6384                      -0.00036
3           +2.4576                      -0.00036
4           +3.2768                      -0.00034
5           +4.096                       -0.00036
6           +4.9152                      -0.0004
```

BOX 8.4   TAKISAWA COMPENSATION FILE FORMAT

```
# Compensation tune file for X
stage
#
axLSCompCount[0] 11
axLSCompSpacing[0] 5000000
axLSCompPosMin[0] 0
#
axLSCompDirNeg [0] [0]            544
axLSCompDirNeg[1][0]             230
axLSCompDirNeg[2][0]             107
axLSCompDirNeg[3][0]             -63
axLSCompDirNeg[4][0]             -309
axLSCompDirNeg[5][0]             -578
axLSCompDirNeg[6][0]             -812
axLSCompDirNeg[7][0]             -997
axLSCompDirNeg[8][0]             -1161
axLSCompDirNeg[9][0]             -1332
axLSCompDirNeg[10][0]            -1457
#
axLSCompDirPos[0] [0]            28
axLSCompDirPos[1] [0]            -28
axLSCompDirPos[2][0]             -125
axLSCompDirPos[3][0]             -304
axLSCompDirPos[4][0]             -558
axLSCompDirPos[5][0]             -858
axLSCompDirPos[6][0]             -1174
axLSCompDirPos[7][0]             -1487
axLSCompDirPos[8][0]             -1808
axLSCompDirPos[9][0]             -2188
axLSCompDirPos[10][0]            -2720
#
#
```

These are applied to existing axes of the machine tool and referred to from zero to n with an increment of one for each new axis. For Takisawa CNC machine* x, y, and z are assigned, respectively, 0,1, and 2. The arrays are described as follows:

- axLSCompDirPos and axLSCompDirNeg: Two arrays for the direction of the movement representing LSEC data; the first dimension subscript references the compensation data points and the second references the axis number.
- axLSCompCount: The number of measured axis positions for each axis.
- axLSCompSpacing: The spacing increment between measured axis positions; axLS CompSpacing[0] comprises the distance between measurement points for axis 0 and axLSCompSpacing [1] for axis 1.
- axLSCompPosMin: The starting locations of the measured axis positions.

By convention, the starting location of the measurements is always at the most negative axis location. Each of these arrays uses the subscript to designate the machine axis the data applies to. A MATLAB program has been written to convert measurement into OpenCNC format. The Tune file is then generated and ready to be stored for the controller start-up. It is worth noting that this file is read once and kept independent from the system files to avoid any confusion.

As the axes are moved under program control, the current axis positions and directions of motion are obtained and used to determine the corresponding locations within the appropriate error arrays. The values from the arrays are used in the error equations to calculate the amounts to move each of the axes, and the modified position command is sent to the motor amplifiers. The flow diagram for obtaining the modified position command is in Scheme 8.7.

These steps must occur during each system interrupt. The volumetric error components are complex equations which require a corresponding CPU time to estimate the interrupt range that could be supported.

8.4.4.2 Real-Time Error Compensation

The robust and fully documented API available in MDSI OpenCNC could be used to facilitate and implement real-time error compensation and in-process gauging. This is because it provides the

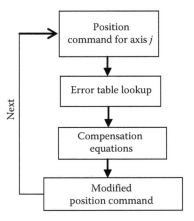

SCHEME 8.1 Compensation flow diagram.

---

* Programme developed at the University of Manchester.

capability to integrate other software products or technologies as programers can write hard real-time programs using Microsoft Visual Basic.

8.4.5 OTHER CONTROLLERS—CROSS COMPENSATION

The Fagor 8055 controller has three tables for linear displacement error (pitch error) compensation and three tables for the straightness error compensation. These are $D_x(X)$, $D_y(Y)$, $D_z(Z)$, $D_x(Y)$, $D_x(Z)$, and $D_y(Z)$. The file format is as follows:

Filename: Filename.FGR
P15 LSCRWCOM = ON, P16 NPOINTS = _____, P8 INCHES = 0, MM (G71)
1, INCH (G70)
P31 NPCROSS = _____, P54 NPCROSS2 = _____, P57 NPCROSS3 =_____,
P32 MOVAXIS = 2, P55 MOVAXIS2 = 3, P58 MOVAXIS3 = 3,
P33 COMPAXIS = 1, P56 COMPAXIS2 = 1, P59 COMPAXIS3 = 2.
Where NPOINTS=number of linear comp points, NPCROSS=number of cross comp points, MOVAXIS=moving axis, COMPAXIS=comp axis, $X = 1$, $Y = 2$ and $Z = 3$.

8.5 CURRENT ISSUES IN MACHINE ERRORS MODELING

The volumetric error more accurately reflects the accuracy to be expected from a machine tool than any other measurement that can be made. Hence, the volumetric error should be determined and listed on the specification sheet of every machine tool offered to industry. On the other hand, the measuring of the 21 rigid-body errors is challenging and time consuming. Hence, a definition or a method of approximating true volumetric error that correlates well to true 3D positioning error, but is less difficult to measure, is very important [1].

Traditionally, manufacturers have ensured part accuracy by linear calibration of each machine tool axis. The conventional definition of the 3D volumetric positioning error is the root mean square (RMS) of the three-axis displacement error. Twenty years ago, the dominate error is the leadscrew pitch error of three axes. This definition is adequate. However, now with better leadscrew, linear encoder, and compensation, the pitch error has been reduced considerably. The dominate errors are the squareness errors and straightness errors. Hence, the above definition is inadequate. Furthermore, using a conventional laser interferometer to measure straightness and squareness errors can be relatively difficult and time consuming.

During the past 3 years, the industry has seen demand emerge for the "volumetric accuracy" specification on machine tools. One hurdle remains: A standard definition so that everyone measures volumetric accuracy with the same yardstick. The issue has been discussed in many standards committees, machine tool builders, and the metrology community. In general, they fall into two camps: One for a definition that would define precise volumetric accuracy for all machine tools, the other for a method, being used by Boeing and others, called "body-diagonal" measurement, which gives accurate volumetric measurements for most equipment.

Beyond the 21 rigid-body errors are additional nonrigid-body errors. These consider that a machine tool is not a stationary mass. Weight shifts, which mean tool position, change ever so slightly from one end of the axis to the other. For most machines, the displacement is too small for a worthwhile measurement. But for a massive gantry machine, the error can make a small, detectable difference. To determine all rigid- and nonrigid-body errors takes 45 measurements, which can take days away from production, so it is not practical for most environments.

On the other hand, the body diagonal displacement method, while not measuring all the rigid-body errors, is very representative of 3D volumetric positioning accuracy, and certainly has more potential for production. Measuring for such errors follows the natural evolution of machine-tool technology. Like wanting a car to get better gas mileage and drive faster, industry is demanding machines cut faster and more accurately at the same time. With volumetric-error compensation, manufacturers may well get just that.

8.5.1   Definitions of 3D Volumetric Error

Volumetric accuracy for movements in X, Y, Z, $V(XYZ)$, is the maximum range of relative deviations between actual and ideal position in X, Y, Z and orientations in A, B, C for X, Y, Z movement in the volume concerned, where the deviations are relative deviations between the tool side and the work-piece side of the machine tool.

Assuming rigid-body motion, the formulae for the six errors in the directions of X, Y, and Z, and rotary axes A, B, and C as follows [12]:

$$V(XYZ, X) = \text{deviations in } X \text{ at } XYZ \tag{8.112}$$
$$= EXX + EXY + EXZ$$

$$V(XYZ, Y) = \text{deviations in } Y \text{ at } XYZ \tag{8.113}$$
$$= EYX + EYY + EYZ$$

$$V(XYZ, Z) = \text{deviations in } Z \text{ at } XYZ \tag{8.114}$$
$$= EZX + EZY + EZZ$$

$$V(XYZ, A) = \text{angular deviations around } X \text{ at } XYZ \tag{8.115}$$
$$= EAX + EAY + EAZ$$

$$V(XYZ, B) = \text{angular deviations around } Y \text{ at } XYZ \tag{8.116}$$
$$= EBX + EBY + EBZ$$

$$V(XYZ, C) = \text{angular deviations around } Z \text{ at } XYZ \tag{8.117}$$
$$= ECX + ECY + ECZ$$

Here, the squareness errors are included in the straightness errors. The angular errors are small and can be treated as scalar.

Definition 8.1

The amplitude of the volumetric error can be defined as the RMS of the three linear deviations and the amplitude of the volumetric angular error can be defined as the RMS of the three angular deviations.

$$V(XYZ, R) = \text{SQRT}\{V(XYZ, X) * V(XYZ, X) + V(XYZ, Y) * V(XYZ, Y) \\ + V(XYZ, Z) * V(XYZ, Z)\} \tag{8.118}$$

$$V(XYZ, W) = \text{SQRT}\{V(XYZ, A) * V(XYZ, A) + V(XYZ, B) * V(XYZ, B) \\ + V(XYZ, C) * V(XYZ, C)\} \tag{8.119}$$

The volumetric accuracy and volumetric angular accuracy can be defined as the maximum range over the working space.

$$R_{max} = \text{Max}\{V(XYZ, R)\} \tag{8.120}$$

$$W_{max} = \text{Max}\{V(XYZ, W)\} \tag{8.121}$$

Definition 8.2

The maximum range of error in each direction can be expressed as

$$X_{max} = \text{Max}\{V(XYZ,X)\} - \min\{V(XYZ,X)\} \tag{8.122}$$

$$Y_{max} = \text{Max}\{V(XYZ,Y)\} - \min\{V(XYZ,Y)\} \tag{8.123}$$

$$Z_{max} = \text{Max}\{V(XYZ,Z)\} - \min\{V(XYZ,Z)\} \tag{8.124}$$

$$A_{max} = \text{Max}\{V(XYZ,A)\} - \min\{V(XYZ,A)\} \tag{8.125}$$

$$B_{max} = \text{Max}\{V(XYZ,B)\} - \min\{V(XYZ,B)\} \tag{8.126}$$

$$C_{max} = \text{Max}\{V(XYZ,C)\} - \min\{V(XYZ,C)\} \tag{8.127}$$

The volumetric accuracy and volumetric angular accuracy can be defined as the RMS of the maximum range of error in each direction.

$$R_{max} = \text{SQRT}\{X_{max} * X_{max} + Y_{max} * Y_{max} + Z_{max} * Z_{max}\} \tag{8.128}$$

$$W_{max} = \text{SQRT}\{A_{max} * A_{max} + B_{max} * B_{max} + C_{max} * C_{max}\} \tag{8.129}$$

The above two equations are valid definitions. However, to determine the volumetric error, it requires extensive and time-consuming measurement. A third definition is to use the four body diagonal displacement errors to define the volumetric accuracy.

8.5.2 New Definition of 3D Volumetric Error Based on the Body Diagonal Errors

The performance or accuracy of a machine tool is determined by 3D volumetric positioning error, which includes linear displacement error, straightness error, angular error, and thermally induced error. The body diagonal displacement error defined in ASME B5.54 or ISO 230-6 is a good quick check of volumetric error. All the errors will contribute to the four body diagonal displacement errors. The B5.54 tests have been used by Boeing Aircraft Co. and others for years.

When using body diagonal displacement error measurement, body diagonal error (Ed) does not include squareness errors. But Ed is currently defined in ISO 230-6 and ASME B5.54 as a measure of volumetric error. Squareness errors can be included, and our new proposed measure volumetric error, ESd, includes squareness errors.

Some definitions: ppp/nnn indicates body diagonal direction with the increments in X, Y, and Z all positive/negative, and npp/pnn indicates the increments in X, Y, and Z are negative/positive, positive/negative, and positive/negative, etc. Body diagonal errors in each direction are Dr(r) ppp/nnn, Dr(r) npp/pnn, Dr(r) pnp/npn, and Dr(r) ppn/nnp. Based on the definition in ISO 230-6, E is defined as

$$E_{ppp/nnn} = \max[Dr(r)ppp / nnn] - \min[Dr(r)ppp / nnn] \tag{8.130}$$

$$E_{npp/pnn} = \max[Dr(r)npp / pnn] - \min[Dr(r)npp / pnn] \tag{8.131}$$

$$E_{pnp/npn} = \max[Dr(r)pnp / npn] - \min[Dr(r)pnp / npn] \tag{8.132}$$

$$E_{ppn/nnp} = \max[Dr(r)ppn / nnp] - \min[Dr(r)ppn / nnp] \tag{8.133}$$

And volumetric error is defined as

$$Ed = max[E_{ppp/nnn}, E_{npp/pnn}, E_{pnp/npn}, E_{ppn/nnp}]$$

This definition does not include squareness errors. To include squareness errors, define the volumetric error as

$$ESd = Max[Dr(r)ppp/nnn, Dr(r)npp/pnn, Dr(r)pnp/npn, Dr(r)ppn/nnp]$$
$$-min[Dr(r)pp/nnn, Dr(r)npp/pnn, Dr(r)pnp/npn, Dr(r)ppn/nnp]$$

The definition ELv is still commonly used as the definition of a 3D volumetric error, and ELSv including straightness and squareness errors is a true volumetric error. The Ed is defined in ISO 230-6 and ASME B5.54 as a measure of volumetric error. The ESd, including squareness errors, should be a good measure of volumetric error.

To demonstrate this new definition, measurements were performed on 10 selected CNC machine tools, representing the modern mid-size CNC machining centers [1]. Eight were made by the German manufacturer Deckel Maho Gildemeister (DMG), 1 by the U.K. Bridgeport, and 1 by the Czech company Kovosvit MAS. The DMG machines are for better illustration inscribed with a number behind each type description (e.g., DMU80T-2). A brief description of the 10 machines is in Table 8.1.

The measurement results are shown in Table 8.2. Measurements according to ISO 230-2 were performed along the three edges of the machine working volume. These are identified by the marks I, II, and III. The angular errors are derived from the linear positioning by respecting the Abbé offsets. The diagonal positioning accuracy is described by the parameter Ed (diagonal systematic deviation of positioning) according to ISO 230-6. The remaining geometric errors were evaluated from the laser vector method.

The 3D volumetric errors, such as ELv, ELSv, Ed, and ESd, are calculated and tabulated in Table 8.2. As compared with the true 3D volumetric error ELSv, the ELv and the Ed underestimate the 3D volumetric error, but the Ed varies with the squareness errors. The ESd also underestimates the 3D volumetric position error but relatively stable and not effected by the squareness errors.

TABLE 8.1
Machine Parameters of the 10 Selected Modern Machining Centers

Machine no.	1	2	3	4	5
Machine id.	DMC60H-1	DMC60H-2	DMC65V-1	DMC65V-2	DMU80T-1
Manufacturer	DMG	DMG	DMG	DMG	DMG
Type	Horizontal	Horizontal	Vertical	Vertical	Vertical
Axis stroke (X/Y/Z) mm	600/560/560	600/560/560	650/500/500	650/500/500	880/630/630
Control sys.	Sinumerik 840D	Sinumerik 840D	Sinumerik 840D	Sinumerik 840D	Heidenhein iTNC530
Service hours	2589	1655	3550	3338	2847
Machine no.	6	7	8	9	10
Machine id.	DMU80T-2	DMU80T-3	DMU80T-4	VMC500	MCV1000
Manufacturer	DMG	DMG	DMG	Bridgeport	MAS
Type	Vertical	Vertical	Vertical	Vertical	Vertical
Axis stroke (X/Y/Z) mm	880/630/630	880/630/630	880/630/630	650/500/500	1016/610/720
Control sys.	Heidenhein TNC430	Heidenhein iTNC530	Heidenhein TNC430	Heidenhein TNC410	Heidenhein iTNC530
Service hours	4081	1672	3723	892	437

TABLE 8.2
Measurement Results

Measurement Technique	Error Type	Positions	Machine No.									
			1	2	3	4	5	6	7	8	9	10
			Maximal deviation (µm);resp. (µm/m)									
ISO 230-2	$D_x(x)$	I	9.5	5.3	16.5	24.0	35.8	23.5	10.7	20.5	7.6	12.3
		II	7.2	7.3	31.1	22.5	47.7	24.1	12.0	54.3	—	—
		III	—	—	19.2	19.0	51.6	28.4	—	29.4	—	—
ISO 230-2	$D_z(z)$	I	—	36.3	10.9	10.6	14.0	16.5	6.8	6.6	23.3	14.1
		II	14.3	17.8	14.9	7.1	15.5	19.2	8.4	8.7	—	—
		III	25.2	21.1	10.1	7.7	18.0	15.2	7.7	15.7	—	—
Calc	$A_x(z)$		—	−99.0	−2.0	5.0	−5.0	−8.0	−3.0	−15.0	—	—
	$A_y(z)$		—	−72.0	5.0	−14.0	7.0	15.0	−7.0	−6.0	—	—
ISO 230-6	Ed		15.9	33.4	34.4	38.3	45.4	31.8	15.8	41.5	33.2	26.9
Laser vector method	$D_x(x)$		2.7	8.4	20.2	7.8	18.6	11.8	1.7	16.8	12.8	6.9
	$D_y(x)$		2.9	2.9	7.5	2.9	3.9	5.6	6.2	2.8	7.1	15.6
	$D_z(x)$		2.4	3.4	9.2	4.1	2.5	3.1	2.4	1.9	8.5	6.6
	$D_y(y)$		2.2	8.2	15.2	8.3	14.0	8.9	1.5	12.6	8.2	9.4
	$D_z(y)$		2.3	2.8	2.3	1.2	2.0	4.0	7.3	3.3	2.3	3.5
	$D_x(y)$		2.4	8.8	6.7	11.9	5.2	4.5	10.3	4.0	18.4	7.9
	$D_z(z)$		2.6	9.7	10.8	4.2	10.8	6.8	2.7	9.7	15.3	7.8
	$D_y(z)$		6.1	13.1	5.3	23.3	25.1	23.9	5.2	9.3	27.5	21.3
	$D_x(z)$		15.9	28.2	7.2	5.2	5.2	2.1	8.5	15.8	25.6	6.4
	Bxy		−1	15	−18	−8	5	3	15	−8	56	11
	Bxz		41	−52	−31	−7	7	4	−18	−39	64	−37
	Byz		−18	−18	−8	−67	−53	−48	−16	−27	73	−7
3D Volumetric errors	ESd		30	33	33	33	46	34	27	45	54	44
	ELv		25.55	27.13	27.62	31.61	51.95	35.59	20.18	39.61	28.03	24.43
	ELSv		42.81	58.66	49.31	62.89	77.05	62.37	43.56	62.58	78.41	63.72
	Ed		15.9	33.4	34.4	38.3	45.4	31.8	15.8	41.5	33.2	26.9
	ELSv/ELv		1.67	2.16	1.79	1.99	1.48	1.75	2.16	1.58	2.80	2.61
	ELSv/Ed		2.69	1.76	1.43	1.64	1.70	1.96	2.76	1.51	2.36	2.37
	ELSv/ESd		1.43	1.78	1.49	1.91	1.67	1.83	1.61	1.39	1.45	1.45
Calc	$A_y(x)$		—	—	53.0	4.0	−12.0	−7.0	—	14.0	—	—
	$A_z(x)$		—	—	15.0	26.0	−36.0	5.0	—	84.0	—	—
ISO 230-2	$D_y(y)$	I	15.8	7.8	15.3	18.4	20.3	14.3	16.2	5.5	13.6	15.7
		II	12.0	8.7	4.9	20.4	18.3	19.2	17.1	6.3	—	—
		III	—	—	13.2	24.9	22.9	21.6	11.2	12.1	—	—
Calc	$A_x(y)$		—	—	60.0	−12.0	11.0	−3.0	7.0	−8.0	—	—
	$A_z(y)$		—	—	38.0	6.0	2.0	−3.0	−9.0	28.0	—	—

For more quantitative comparison, a multiple factor, M1, is defined as ELSv/ELv, M2 as ELSv/Ed, and M3 as ELSv/ESd. Hence, the true 3D volumetric error ELSv can be obtained by multiplying the ELv by M1, the Ed by M2, and the ESd by M3. The multiple factors M1, M2, and M3 for various

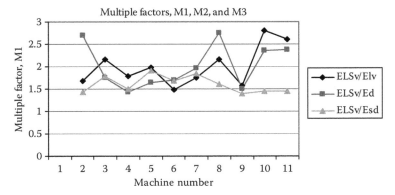

FIGURE 8.17 Multiple factors for various definitions of volumetric errors.

machine tools are plotted in Figure 8.17. The M1 varies from 1.4 to 2.8, the M2 from 1.43 to 2.76, and the M3 from 1.4 to 1.9. The range of variations for M1 and M2 are relatively large, while it is relatively small for M3. Hence, the ESd is a good estimate of 3D volumetric position error.

8.6 SUMMARY AND CONCLUSION

Four definitions of the 3D volumetric positioning error have been provided. The positioning errors of 10 CNC machine tools have been measured. Based on these measurement results, the 3D volumetric errors using various definitions can be calculated. It is concluded that the laser body diagonal displacement measurement in the ASME B5.54 or ISO 230-6 machine tool performance measurement standards is a quick check of the volumetric positioning error and the value ESd is a good measure of the volumetric error.

Measurements performed on 10 mid-size machining centers reveal that when compared to true 3D volumetric error ELSv, ELv underestimates volumetric error. The Ed underestimates true volumetric error and varies with squareness errors. Finally, ESd underestimates 3D volumetric position error but is relatively stable and not influenced by squareness errors. Thus ESd is a good measure of volumetric error.

REFERENCES

1. C. Wang, O. Svoboda, P. Bach, and G. Liotto, Definitions and correlations of 3D volumetric positioning errors of CNC machining centers, *Proceedings of the IMTS 2004 Manufacturing Conference*, Chicago, IL, September 8–10, 2004.
2. R. Schultschik, The components of the volumetric accuracy, *Annals of the CIRP*, 25(1), 223–228, 1977.
3. W. Beckwith Jr. and R.I. Warwick, Method for calibration of coordinate measuring machine, U.S. Patent 4,939,678, July 3, 1990.
4. G. Zhang, R. Quang, B. Lu, R. Hocken, R. Veale, and A. Donmez, A displacement method for machine geometry calibration, *CIRP*, 37(1), 515–518, 1988.
5. G. Ren, J. Yang, G. Liotto, and C. Wang, Theoretical derivations of 4 body diagonal displacement errors in 4 machine configurations, *Proceedings of the LAMDAMAP Conference,* Cransfield, UK, June 27–30, 2005.
6. C. Wang and G. Liotto, A theoretical analysis of 4 body diagonal displacement measurement and sequential step diagonal measurement, *Proceedings of the LAMDAMAP 2003*, Huddersfield, England, July 1–4, 2003.
7. *Fanuc 3D Grid Point Compensation Manual*, GE Fanuc Automation North America, Charlottesville, VA, 1997.
8. O. Svoboda, P. Bach, G. Liotto, and C. Wang, Machine tool 3D volumetric positioning error measurement under various thermal conditions, *Proceedings of the ISPMM 2006 Conference*, Urumqi, Xinjiang, China, August 2–6, 2006.

9. O. Svoboda, P. Bach, J. Yang, and C. Wang, Correlation of 3D volumetric positioning errors and temperature distributions—Theory and measurement, *Proceedings of the ISICT 2006 Conference*, Beijing, China, October 13–15, 2006.

10. C. Wang, Laser Doppler displacement measurement, *Lasers and Optronics*, 6, 69–71, 1987.

11. *Methods for Performance Evaluation of Computer Numerically Controlled Machining Centers, An American National Standard*, ASME B5.54-1992 by the American Society of Mechanical Engineers, New York, p. 69, 1992.

12. ISO 230-6: 2002 Test code for machine tools—Part 6: Determination of positioning accuracy on body and face diagonals (diagonal displacement tests), *An International Standard*, by International Standards Organization, 2002.

13. C. Wang, Laser vector measurement technique for the determination and compensation of volumetric positioning errors. Part I: Basic theory, *Review of Scientific Instruments*, 71(10), 3933–3937, 2000.

14. J. Janeczko, R. Griffin, and C. Wang, Laser vector measurement technique for the determination and compensation of volumetric position errors, Part II: Experimental verification, *Review of Scientific Instruments*, 71, 3938–3941, 2000.

15. O. Svoboda, P. Bach, G. Liotto, and C. Wang, Volumetric positioning accuracy: Measurement, compensation and verification, *Proceedings of the JUSFA 2004 Conference*, Denver, CO, July 19–21, 2004.

16. O. Svoboda, Milling machine tools' accuracy in thermally unbalanced conditions, Ph.D. Thesis, Czech Technical University in Prague, July 2007.

Index

Milton Keynes UK
Ingram Content Group UK Ltd.
UKHW050309111024
449327UK00049B/370